CONTEMPORARY MATHEMATICS

730

Model Theory of Modules, Algebras and Categories

International Conference
Model Theory of Modules, Algebras and Categories
July 28–August 2, 2017
Ettore Majorana Foundation and Centre for Scientific Culture, Erice, Sicily, Italy

Alberto Facchini
Lorna Gregory
Sonia L'Innocente
Marcus Tressl
Editors

AMS AMERICAN MATHEMATICAL SOCIETY

2010 *Mathematics Subject Classification.* Primary 03C60, 13D09, 13F30, 16D40, 16D70, 16G20; Secondary 18D10, 18E05, 37A35.

Library of Congress Cataloging-in-Publication Data

Names: International Conference on Model Theory of Modules, Algebras and Categories (2017 : Erice, Italy) | Facchini, Alberto, editor.

Title: Model theory of modules, algebras and categories : International Conference on Model Theory of Modules, Algebras and Categories, July 28–August 2, 2017, Ettore Majorana Foundation and Centre for Scientific Culture, Erice, Sicily, Italy / Alberto Facchini [and three others], editors.

Description: Providence, Rhode Island : American Mathematical Society, [2019] | Series: Contemporary mathematics ; volume 730 | Includes bibliographical references.

Identifiers: LCCN 2018047324 | ISBN 9781470443672 (alk. paper)

Subjects: LCSH: Modules (Algebra)–Congresses. | Algebra–Congresses. | AMS: Mathematical logic and foundations – Model theory – Model-theoretic algebra. msc | Commutative algebra – Homological methods – Derived categories. msc | Commutative algebra – Arithmetic rings and other special rings – Valuation rings. msc | Associative rings and algebras – Modules, bimodules and ideals – Free, projective, and flat modules and ideals. msc | Associative rings and algebras – Modules, bimodules and ideals – Structure and classification (except as in 16Gxx), direct sum decomposition, cancellation. msc | Associative rings and algebras – Representation theory of rings and algebras – Representations of quivers and partially ordered sets. msc | Category theory; homological algebra – Categories with structure – Monoidal categories (= multiplicative categories), symmetric monoidal categories, braided categories. msc | Category theory; homological algebra – Abelian categories – Preadditive, additive categories. msc | Dynamical systems and ergodic theory – Ergodic theory – Entropy and other invariants, isomorphism, classification. msc

Classification: LCC QA247.3 .I5855 2017 | DDC 512/.42–dc23

LC record available at https://lccn.loc.gov/2018047324

Contemporary Mathematics ISSN: 0271-4132 (print); ISSN: 1098-3627 (online)

DOI: https://doi.org/10.1090/conm/730

∞ The paper used in this book is acid-free and falls within the guidelines
established to ensure permanence and durability.
Visit the AMS home page at `https://www.ams.org/`

10 9 8 7 6 5 4 3 2 1 24 23 22 21 20 19

Contents

Contemporary Mathematics
Volume **730**, 2019
https://doi.org/10.1090/conm/730/14707

On isonoetherian and isoartinian modules

Alberto Facchini and Zahra Nazemian

ABSTRACT. We present the basic results concerning isoartinian and isonoetherian rings and modules. These rings and modules, which were essentially defined in two papers of ours recently published in the *Journal of Algebra*, generalize artinian and noetherian rings and modules. In particular, here we focus on the isoradical of a ring, which is a generalization of the Jacobson radical, and modules generated by isosimple modules, which generalize semisimple modules. Open problems are given at the end of each section.

1. Introduction

Let R be a ring and M a right R-module. We say that M is *isoartinian* if, for every descending chain $M \geq M_1 \geq M_2 \geq \dots$ of submodules of M, there exists an index $n \geq 1$ such that M_n is isomorphic to M_i for every $i \geq n$. Dually, M is called *isonoetherian* if, for every ascending chain $M_1 \leq M_2 \leq \dots$ of submodules of M, there exists an index $n \geq 1$ such that $M_n \cong M_i$ for every $i \geq n$. A module M is *isosimple* if it is non-zero and every non-zero submodule of M is isomorphic to M. The motivation to define these three classes of modules, which often have a surprising analogy with the "classical" case of the classes of artinian, noetherian and simple modules, originates from the article [**13**], a paper devoted to dualizing the concept of uniserial dimension [**29**]. Isosimple modules that are not simple are called anti-coHopfian in [**33**]. For a study of Hopfian modules, coHopfian modules and their generalizations, see [**2**, **8**, **12**, **17**, **20**] and [**32**].

In [**13**], some properties of isosimple (= anti-coHopfian) modules were studied. For example, in the proof of [**13**, Proposition 5.4], it was shown that an isosimple module has the cancellation property. That is, if S is an isosimple module and M, N are two modules such that $S \oplus M \cong S \oplus N$, then $M \cong N$. Cancellation property holds for modules whose endomorphism ring is semilocal, in particular local, in particular for simple modules.

In Section 2 of the paper, we consider isosimple modules and their endomorphism rings. We show that the endomorphism ring of an isosimple module is an integral domain with the a. c. c. on principal right ideals. At the end of the section we present some new examples of isosimple modules, based on derivations and

2010 *Mathematics Subject Classification.* Primary 16D70.

The first author was partially supported by Dipartimento di Matematica "Tullio Levi-Civita" of Università di Padova (Project BIRD163492/16 "Categorical homological methods in the study of algebraic structures" and Research program DOR1690814 "Anelli e categorie di moduli".

The second author was supported by a grant from IPM.

partially due to Chirstian Lomp. He proved that if A is an algebra over a field k of characteristic zero and δ is a locally nilpotent derivation of A such that $\delta(x) = 1$ for some x in center of A, then A is an isosimple $A[z;\delta]$-module if and only if A^δ is a principal left ideal domain.

In Section 3, we consider isonoetherian and isoartinian modules and rings. Isonoetherian modules have finite uniform dimension (= Goldie dimension). A ring R is called *right isonoetherian* if R_R is an isonoetherian module. We present examples of isonoetherian valuation rings that are not noetherian. Recall that a commutative valuation domain is noetherian if and only if it is a discrete valuation domain of Krull dimension ≤ 1. A commutative valuation domain is isonoetherian if and only if it is a discrete valuation domain of Krull dimension ≤ 2 (Theorem 3.7). We say that a ring R is *right isoartinian* if the right module R_R is isoartinian. A (not necessarily commutative) integral domain D is right isoartinian if and only if the module D_D is isosimple, if and only if D is a principal right ideal domain (PRID, for short). Semiprime right isoartinian rings have a particularly good description. They are right non-singular rings with the a. c. c. on annihilator ideals (Lemmas 3.16 and 3.17). Moreover, they are right noetherian if and only if they are right Goldie, if and only if they have finite uniform dimension, if and only if they are a (direct) sum of isosimple right ideals, if and only if R is a finite direct product of rings of matrices over PRIDs (Theorems 3.18 and 3.19).

The weak radical of a ring was defined and studied by K. Koh and A. C. Mewborn in [**22–24**]. The *right weak radical* of a ring R, denoted by $\mathrm{W}(R)$, is the intersection of the annihilators of all rationally uniform homogeneous right R-modules [**19**]. Here, by *rationally uniform homogeneous module*, we mean a uniform module that can be embedded in all its non-zero submodules. In Section 5, we consider the *right isoradical* $\mathrm{I\text{-}rad}(R_R)$ of a ring R, which is the intersection of the annihilators of all isosimple right R-modules. Since isosimple modules are rationally uniform homogeneous, we have that

$$\mathrm{P}(R) \leq \mathrm{W}(R) \leq \mathrm{I\text{-}rad}(R_R) \leq \mathrm{rad}(R),$$

where $\mathrm{P}(R)$ denotes the *prime radical* of the ring R and $\mathrm{rad}(R)$ is the Jacobson radical of R. Unlike the Jacobson radical, which is the intersection of the annihilators of all simple right modules, the isoradical of a ring is not left/right symmetric, in the sense that the right isoradical and the left isoradical of a ring R can be two different two-sided ideals of R. In Theorem 4.6, we show that if all cyclic right R-modules are isoartinian, then $\mathrm{P}(R) = \mathrm{I\text{-}rad}(R)$. We also investigate modules generated by isosimple modules. Special cases are when a module is a sum or a direct sum of isosimple modules. A module that is a sum of isosimple modules is not necessarily a direct sum of isosimple modules, but if a module M is a sum of pairwise non-isomorphic isosimple modules, then the sum is direct (Lemma 4.11). A ring that is generated by isosimple right modules must be a semiprime right noetherian ring. We show in Theorem 4.13 that a ring R is generated by isosimple right modules if and only if it is a sum of isosimple right ideals, if and only if R is a finite direct product of prime right noetherian rings that are sums of isosimple right ideals. A right semihereditary ring R is generated by isosimple R-modules if and only if it is a finite direct product of rings of matrices over PRIDs (Theorem 4.14).

It is well known that a right module M is noetherian and artinian if and only if it is of finite length, that is, there is a chain $0 = M_0 \leq M_1 \leq \cdots \leq M_n = M$

of submodules of M with M_i/M_{i-1} a simple module for each $1 \leq i \leq n$. As a generalization of this, modules of finite I-length are considered in Section 5. Any right module that is isoartinian and isonoetherian is of finite I-length (Lemma 5.2). A module of finite I-length must be of finite uniform dimension, and if its I-length is equal to its uniform dimension, then it is direct sum of cyclic uniform modules (Proposition 5.4). In Theorem 5.5, we show that a module M that is a finite direct sum of isosimple modules is isoartinian if and only if, for every submodule K of M, I-length $(K) = \mathrm{u.dim}(K)$, where u. dim denotes the uniform dimension.

Some generalizations of isoartinian and isonoetherian modules have been recently considered. See for instance [**7**]. Also see [**11**], [**21**] and [**30**].

We consider right (or left) unitary modules over an associative ring R with identity.

2. Isosimple modules

A simple right module over a ring R is a non-zero right R-module without any non-zero proper submodule. So it is natural to generalize this notion, defining isosimple modules as follows.

DEFINITION 2.1. A right R-module M_R is *isosimple* if it is non-zero and every non-zero submodule of M_R is isomorphic to M_R.

Clearly, every isosimple module M_R is cyclic, and all the submodules of M_R are cyclic. In particular, the module M_R turns out to be noetherian. Thus M_R contains a uniform submodule, so that M_R itself is uniform. Hence every isosimple module is a cyclic noetherian uniform module. For example, the $\mathbb{Z}$-module $\mathbb{Z}$ is isosimple. More generally, the module R_R is isosimple whenever R is a PRID (principal right ideal domain). Clearly, simple modules are isosimple and, over some rings, a module is simple if and only if it is isosimple. Semiartinian rings and commutative Von Neumann regular rings [**5**, Theorem 2.13] are examples of such rings.

Recall that a *uniserial module* is a module whose lattice of submodules is linearly ordered by inclusion. Let ω denote the first infinite ordinal number. For uniserial isosimple modules, we have that:

LEMMA 2.2. *If a right module M is uniserial and isosimple, then its lattice of submodules is order anti-isomorphic to the ordinal number ω^γ, where γ is the Krull dimension* K.dim(M) *of M.*

PROOF. Let M be uniserial and isosimple. Then M is uniserial and noetherian, so that its lattice of submodules is order anti-isomorphic to an ordinal number α. Since M is a critical module, if the Krull dimension K.dim(M) of M is γ, then $\alpha = \omega^\gamma$ [**28**, 6.2.22, p. 199] □

A non-zero left module ${}_SM$ over a ring S is said to be *completely prime* if the left annihilator $\mathrm{l.ann}(m)$ of m in S is equal to the left annihilator $\mathrm{l.ann}(M)$ of M in S for every non-zero element $m \in M$. A proper submodule N of a non-zero left S-module M is said to be *completely prime* if M/N is a completely prime S-module.

Let R be a ring and I a proper right ideal of R. Recall that the endomorphism ring of a non-zero cyclic right R-module R/I is canonically isomorphic to E_I/I, where $E_I := \{ r \in R \mid rI \subseteq I \}$ is the *idealizer* of I.

PROPOSITION 2.3. [**10**, Proposition 3.3] *Let R be a ring. A right R-module M_R is isosimple if and only if M_R is isomorphic to R/I, where:* (a) *I is a proper right* *ideal of* R, (b) *for every right ideal* J *of* R *containing* I, *there exists an element* r *in the idealizer* E_I *of* I *with* $J = rR+I$, *and* (c) I *is a completely prime submodule of* ${}_{E_I}R$ *(the module* R *as a left module over the ring* E_I*).*

It is well known that the endomorphism ring of a simple module is a division ring. Now we will investigate the endomorphism ring of an isosimple module.

THEOREM 2.4. (1) *Every non-zero endomorphism of an isosimple module is an injective mapping.*

(2) *The endomorphism ring of an isosimple module is a right Ore domain.*

(3) *The endomorphism ring of an isosimple module satisfies the ascending chain condition on principal right ideals.*

PROOF. (1) Let f be a non-zero endomorphism of an isosimple module M. Then $M/\ker f \cong f(M) \cong M$. Since M is noetherian, $\ker(f) = 0$ by [**9**, Lemma 2.17].

(2) An isosimple module M is compressible, i.e, can be embedded in its non-zero submodules. Since M is noetherian, M must have Krull dimension, so that its endomorphism ring is a right Ore domain by [**16**, Corollary 2.8].

(3) See [**10**, Corollary 3.5]. □

Recall that a module M is *intrinsically projective* if, for every submodule N of M, every integer $n \geq 0$, every homomorphism $f\colon M \to N$ and every epimorphism $g\colon M^n \to N$, there exists a homomorphism $h\colon M \to M^n$ such that $f = gh$ [**35**, p. 30]. The module M is *semiprojective* if this condition holds only for $n = 1$.

COROLLARY 2.5. *The following conditions are equivalent for the endomorphism ring* E *of an isosimple module* M*:*

(1) *The lattice* $\mathcal{L}_p(E_E)$ *of principal right ideals of* E *is a sublattice of the lattice* $\mathcal{L}(E_E)$ *of right ideals of* E.

(2) E *is a right Bézout domain.*

(3) E *is a* PRID.

(4) M *is an intrinsically projective module.*

PROOF. (1) $\Rightarrow$ (2) is clear.

(2) $\Rightarrow$ (3). By Theorem 2.4(3), the ring E satisfies the a. c. c. on principal right ideals. So if E is right Bézout, E is a PRID.

(3) $\Rightarrow$ (1) is trivial, because in this case all right ideals are principal.

(3) $\Rightarrow$ (4) From Theorem 2.4(1), it is easy to see that an isosimple module is semiprojective, and so $\mathrm{Hom}(M, IM) = I$ for every (principal) right ideal I of E.

(4) $\Rightarrow$ (3) Let I be a right ideal of E. There exists an $f \in E$ such that $IM = fM$, because either $IM = 0$ or $IM \cong M$. Since M is intrinsically projective, $I = \mathrm{Hom}(M, IM) = fE$ by [**35**, Theorem 5.7]. □

Recall that a right R-module M is a *self-generator* if every submodule of M can be generated by M, that is, is a homomorphic image of a direct sum of copies of M.

LEMMA 2.6. *A right* R*-module* M *with endomorphism ring* S *is isosimple if and only if*

(1) M *is a self-generator, and*

(2) *for every non-zero submodule* N *of* M, *the right ideal* $\mathrm{Hom}(M, N)$ *of* S *is of the form* fS *for some injective mapping* $f \in S$.

In particular, a self-generator with endomorphism ring a PRID *is isosimple.*

PROOF. Assume M isosimple. Clearly, we have (1). To prove (2), fix a non-zero submodule N of M and consider the right ideal $I := \mathrm{Hom}(M, N)$ of S. Since M is isosimple, there exists an isomorphism $f\colon M \to N$. For every $g\colon M \to N$, we have that $g = f(f^{-1}g) \in fS$. This proves (2).

Conversely, suppose that (1) and (2) hold. Let N be a non-zero submodule of M. Thus there exists a set I and an epimorphism $g\colon M^{(I)} \longrightarrow N$, and $\mathrm{Hom}(M, N) = fS$ for some injective $f \in S$. We want to show that $f(M) = N$. If $y \in N$, then there exists $x = (m_i)_{i\in I} \in M^{(I)}$ such that $y = g(x)$. Let J be the finite subset of I of all $i \in I$ with $m_i \neq 0$. Let $\varepsilon_i\colon M \to M^{(I)}$ be the i-th embedding of M in $M^{(I)}$. Then $y = g(x) = \sum_{i\in J}(g \circ \varepsilon_i)(m_i)$. Now $g \circ \varepsilon_i \in \mathrm{Hom}(M, N) = fS$, so that there exists $s_i \in S$ with $g \circ \varepsilon_i = fs_i$. Hence $y = \sum_{i\in J}(fs_i)(m_i) \in f(M)$. Thus f induces an isomorphism $M \cong N$, so that M is isosimple. □

If every cyclic submodule of a module M is M*-cyclic*, that is, is a homomorphic image of M, then M is a self-generator. So, as a corollary of Lemma 2.6, we have that:

COROLLARY 2.7. *Let M_R be a right R-module. Suppose that* $\mathrm{End}(M)$ *is a* PRID. *Then M_R is isosimple if and only if every cyclic submodule of M is M-cyclic.*

We do not know whether the endomorphism ring of any isosimple right module is a PRID. In order to construct further examples of isosimple modules, in the hope of finding isosimple modules whose endomorphism is not a PRID, if they exist, we now present some results, mostly due to Christian Lomp. We are grateful to him for the permission of including these results here.

Recall that if δ is a derivation on ring A, then a left ideal I of A is δ*-stable* if $\delta(I) \subseteq I$. From here to the end of this section, we consider left modules. If δ is a derivation on ring A, let $B := A[z; \delta]$ denote the ring of differential operators. Then A is a left B-module defining $z^n a = \delta^n(a)$ for every $a \in A$. We will denote by A^δ the subring of A consisting of all elements $a \in A$ with $\delta(a) = 0$.

Recall that an element a of a ring A is a *right zerodivisor* if it is non-zero and there exists $b \in R$, $b \neq 0$ such that $ba = 0$.

LEMMA 2.8. *Let A be a ring with a derivation δ and let $A[z; \delta]$ be the ring of differential operators. Then A is an isosimple left $A[z; \delta]$-module if and only if*

(i) *every δ-stable left ideal I of A is of the form $I = Ac$ for some $c \in A^\delta$; and*
(ii) *the elements of A^δ are not right zerodivisors in A.*

PROOF. Suppose that A is an isosimple $A[z; \delta]$-module. Let I be a non-zero $A[z; \delta]$-submodule of A and $\Theta\colon A \longrightarrow I$ be an $A[z; \delta]$-module isomorphism. Set $c := \Theta(1)$. Then

$$\delta(c) = zc = z\Theta(1) = \Theta(z \cdot 1) = \Theta(0) = 0.$$

So $c \in A^\delta$. Since Θ is surjective, $I = Ac$. As $\mathrm{End}_{A[z;\delta]}(A)$ is isomorphic to the opposite of the ring A^δ, we can identify elements of A^δ with endomorphisms of A. But any non-zero endomorphism of an isosimple module is injective, so that every (non-zero) element of A^δ is not a right zerodivisor in A.

Conversely, suppose conditions (i) and (ii) hold. Let I be a left $A[z; \delta]$-submodule of A. Then I is a δ-stable left ideal of A. By (i), there exists an element $c \in A^\delta$

(2) By (1), I-rad$(R_R) = \mathrm{P}(R)$. If $\mathrm{P}(R)$ is not left T-nilpotent, choose $a_1, a_2, \dots$ in $\mathrm{P}(R)$ such that $a_1 \dots a_n \neq 0$ for every n. Let L be a right ideal of R maximal among the right ideals I with $a_1 \dots a_n \notin I$ for every n. As in (1), let K/L be an isosimple right ideal of R/L. There exists an index m such that $x = a_1 \dots a_m \in K$. Since $a_{m+1} \in \mathrm{P}(R) =$ I-rad(R_R), we have that $a_{m+1} \in \mathrm{ann}(K/L)$. Thus $xa_{m+1} \in L$, a contradiction.

(3) Set $I := \mathrm{P}(R)$. Consider the ascending chain of ideals

$$\mathrm{r.ann}(I) \leq \mathrm{r.ann}(I^2) \leq \dots$$

Then we have the descending chain $\mathrm{l.ann}(\mathrm{r.ann}(I)) \geq \mathrm{l.ann}(\mathrm{r.ann}(I^2)) \geq \dots$ of ideals of R. Since R is right isoartinian, there exists an index n such that $\mathrm{l.ann}(\mathrm{r.ann}(I^n))$ is isomorphic to $\mathrm{l.ann}(\mathrm{r.ann}(I^i))$ as a right R-module for every $i \geq n$. So $\mathrm{r.ann}(I^n) = \mathrm{r.ann}(I^i)$ for all $i \geq n$. We claim that $I^{2n} = 0$. If not, then, for some $a_1 \in I^n$, we would have $I^n a_1 \neq 0$. Thus $I^{2n} a_1 \neq 0$ and so, for some $a_2 \in I^n$, we have $I^n a_2 a_1 \neq 0$. By continuing in this manner, we get a contradiction to the fact that I is right T-nilpotent. □

Now we will study the class of right modules over a ring R generated by isosimple right R-modules. Let $\mathcal{U}$ denote the class of all isosimple right R-modules and $\mathrm{Gen}(\mathcal{U})$ the class of all right modules M for which there exist an indexed set $(U_\alpha)_{\alpha \in A}$ in $\mathcal{U}$ and an epimorphism $\oplus_{\alpha \in A} U_\alpha \to M$. It is clear that the class of semisimple modules, i.e., the class of modules generated by all simple right modules, is a subclass of $\mathrm{Gen}(\mathcal{U})$, and that these two classes coincide if and only if isosimple right modules are simple. Using the notation of [**1**], for any right module M, set $\mathrm{Tr}_M(\mathcal{U}) := \sum\{\, h(U) \mid h \colon U \to M$ is a homomorphism for some $U \in \mathcal{U}\,\}$. Thus $M \in \mathrm{Gen}(\mathcal{U})$ if and only if $\mathrm{Tr}_M(\mathcal{U}) = M$. It is convenient to denote $\mathrm{Tr}_M(\mathcal{U})$ by I-soc(M).

REMARK 4.7. (1) Clearly, $\mathrm{soc}(M) \leq$ I-soc(M). The equality does not necessarily holds. For example, if R is PRID that is not a division ring, then I-soc$(R_R) = R$ and $\mathrm{soc}(R_R) = 0$.

(2) If $M = M_1 \oplus M_2$, then I-soc$(M) =$ I-soc$(M_1) \oplus$ I-soc(M_2).

(3) If $M \in \mathrm{Gen}(\mathcal{U})$, then M is a finitely generated module if and only if it is noetherian.

(4) $\mathrm{Gen}(\mathcal{U})$ is closed under homomorphic images and direct sums.

(5) If I-soc$(M) = M$ (that is, $M \in \mathrm{Gen}(\mathcal{U})$) and M is projective, then M is a direct summand of a direct sum of isosimple modules.

The following lemmas will be repeatedly used in the sequel.

LEMMA 4.8. *If M is a nonsingular module, then* I-soc$(M) = M$ *if and only if M is the sum of its isosimple submodules.*

PROOF. ($\Leftarrow$) is clear.

($\Rightarrow$). Suppose I-soc$(M) = M$. Then $M = \sum\{\, h(U) \mid h \colon U \to M$ is a homomorphism for some $U \in \mathcal{U}\,\}$. Since M is nonsingular, $h(U)$ must be isomorphic to U for every $h \colon U \to M$. This proves that M is a sum of isosimple modules. □

LEMMA 4.9. *Let R and S be two Morita equivalent rings and F be a category equivalence from the category R-*Mod *to the category S-*Mod. *If M is an R-module such that* I-soc$(M) = M$, *then* I-soc$(F(M)) = F(M)$.

PROOF. Being isosimple is a Morita invariant property of modules (Lemma 3.10). Thus the assertion follows from the fact that F preserves epimorphisms and direct sums. □

For a ring R, we indicate I-soc(R_R) by I-soc(R). Clearly I-soc(R) $= R$ if and only if I-soc(M) $= M$ for every right R-module M. Thus, by Lemma 4.9, we have:

COROLLARY 4.10. *Let R and S be two Morita equivalent rings. Then* I-soc(R) $= R$ *if and only if* I-soc(S) $= S$.

The class of modules that are sums of isosimple modules is contained in the class of modules generated by isosimple modules. The next lemma and Example 4.12 were suggested to us by Professor G. Bergman.

LEMMA 4.11. *Suppose $M = S_1 + \cdots + S_n$, where the modules S_i are isosimple and pairwise non-isomorphic. Then the sum is a direct sum.*

PROOF. The proof is by induction on n. The case $n = 1$ is clear. Assume that $n > 1$ and the assertion holds for $n - 1$. If the sum $M = S_1 + \cdots + S_n$ is not direct, there exist elements $a_i \in S_i$, $i = 1, 2, \ldots, n$, not all zero, such that $a_1 + a_2 + \cdots + a_n = 0$. By the inductive hypothesis, $a_i \neq 0$ for every $i = 1, \ldots, n$. Let us show that there is a well defined isomorphism $a_1R \to a_2R$ that maps a_1r to a_2r for every $r \in R$. To this end, it suffices to prove that, for every $r \in R$, $a_1r = 0$ if and only if $a_2r = 0$. Now $a_1r = 0$ implies $(a_2 + \cdots + a_n)r = 0$. But $a_2 + \cdots + a_n \in S_2 + \cdots + S_n$, which is a direct sum by the inductive hypothesis. Thus $a_2r = 0$. Conversely, if $a_2r = 0$, then $(a_1 + a_3 + \cdots + a_n)r = 0$. Now $a_1 + a_3 + \cdots + a_n \in S_1 + S_3 + \cdots + S_n$, which is a direct sum by the inductive hypothesis. Thus $a_1r = 0$. This proves that the non-zero submodules a_1R of S_1 and a_2R of S_2 are isomorphic, so that the isosimple modules S_1 and S_2 are isomorphic, a contradiction. □

EXAMPLE 4.12. Consider the $\mathbb{Z}$-module $\mathbb{Z} \oplus (\mathbb{Z}/n\mathbb{Z})$. Then $M = S_1 + S_2$, where S_1 is generated by $(1, 0)$ and S_2 is generated by $(1, [1])$. Thus S_1 and S_2 are isomorphic to $\mathbb{Z}$. If $n = p^2$ for some prime number p, then M cannot be direct sum of isosimple modules. But in case n is square free, M is direct sum of possibly more than two non-isomorphic isosimple modules.

THEOREM 4.13. *The following conditions are equivalent for a ring R:*

(1) I-soc(R) $= R$.

(2) *R is a sum of isosimple right ideals.*

(3) *For any right R-module M,* I-soc(M) $= M$.

(4) *R is a finite direct product of prime right noetherian rings, each of which is a sum of isosimple right ideals.*

PROOF. (2) $\Rightarrow$ (1), (1) $\Leftrightarrow$ (3) and (4) $\Rightarrow$ (2) are trivial.

(1) $\Rightarrow$ (2). From Remark 4.7((3) and (5)), we see that R is right noetherian and I-rad(R_R) $= 0$. From the inclusions (4.1) after Lemma 4.3, we get that R is semiprime. By [**14**, Corollary 3.32], every right noetherian semiprime ring is right non-singular, and so, by Lemma 4.8, R is a sum of isosimple right ideals.

(1) $\Rightarrow$ (4). As we have just noticed, R is a semiprime right noetherian ring. Let n be the uniform dimension of R_R. Notice that R has at most n isosimple right ideals up to isomorphism, because if $S_1, \ldots, S_t$ are pairwise non-isomorphic isosimple right ideals, then, by Lemma 4.11, $t = \mathrm{u.dim}(S_1 \oplus \cdots \oplus S_t) \leq \mathrm{u.dim}(R_R) =$

n. Thus R contains only finitely many non-isomorphic isosimple right ideals, $S_1, \dots, S_m$ say. For each i, let I_{S_i} be the sum of all the isosimple right ideals isomorphic to S_i. By Lemma 3.14, I_{S_i} is a two-sided ideal of R. Suppose $i \neq j$. Fix an element $x \in S_i$. If $xS_j \neq 0$, then $S_i \cong xS_j$. On the other hand, $xS_j \cong S_j/(\mathrm{r.ann}(x) \cap S_j)$. Now R is right non-singular, and so we must have that $\mathrm{r.ann}(x) \cap S_j = 0$. This shows that $S_i \cong S_j$, which is a contradiction. Thus $I_{S_i} I_{S_j} = 0$. By Lemma 4.8, every non-zero right ideal of R contains an isosimple right ideal. So if $N := I_{S_i} \cap \sum_{j=1, j\neq i}^n I_{S_j}$ is non-zero for some index i, then we can take an isosimple right ideal T contained in N. Thus $T^2 \leq I_{S_i} \sum_{j=1,j\neq i}^n I_{S_j} = 0$, which is a contradiction. This shows that $R = I_{S_1} \oplus \cdots \oplus I_{S_m}$. Now it is enough to show that each I_{S_i} is a prime ring. Let $i = 1, \dots, m$ and let $I, J \leq I_{S_i}$ be two non-zero right ideals of R with $IJ = 0$. Without loss of generality, we can suppose I and J isosimple, hence isomorphic. Then $IJ = 0$ implies $I^2 = 0$, which is a contradiction. Thus I_{S_i} is a prime ring. □

THEOREM 4.14. *The following conditions are equivalent for a ring R:*
(1) *R is a finite direct product of rings of matrices over* PRID*s.*
(2) *R is a direct sum of isosimple right ideals.*
(3) *R is right semihereditary and* I-soc$(R_R) = R$.
(4) *R is a semiprime right isoartinian right noetherian ring.*
(5) *R is a semiprime right isoartinian ring and R_R is of finite uniform dimension.*

PROOF. (1) ⇒ (2). Let $S := \mathrm{Mat}_n(D)$ be the ring of matrices with entries in a principal right ideal domain D. Under the standard Morita equivalence between D and S, the free D-module D^n corresponds to S. Thus S is a direct sum of n pairwise isomorphic isosimple right ideals.

(2) ⇒ (1). Assume that R is a direct sum of isosimple right ideals. Then these isosimple right ideals are projective, and so R is right isoartinian by Proposition 3.9. On the other hand, since R is a semiprime right noetherian ring by Theorem 4.13, R is isomorphic to a product of rings of matrices over PRIDs by Theorem 3.19(1).

(3) ⇒ (1). We first show that R is right isoartinian. There exist finitely many isosimple right modules $U_1, \dots, U_n$ and an epimorphism $h \colon U_1 \oplus \cdots \oplus U_n \to R$. For each i, $h(U_i)$ is a right ideal of R, and so it is a projective R-module. Since the modules U_i are uniform and R is right nonsingular, the modules U_i must be projective. Then $U_1 \oplus \cdots \oplus U_n$ is an isoartinian module by Proposition 3.9. On the other hand, R_R is a direct summand of $U_1 \oplus \cdots \oplus U_n$, and therefore R_R is an isoartinian right module. Thus R is a finite direct product of rings of matrices over PRIDs by Theorem 3.19 (1).

(1) ⇒ (3) is clear.

(1) ⇔ (4) ⇔ (5) follow from Theorem 3.18. □

As we have said above, we do not know whether the condition right noetherian in statement (4) is necessary. We also do not know whether the condition right semihereditary in statement (3) is necessary. In the rest of this section, we will consider this problem.

PROPOSITION 4.15. *Let R be a ring Morita equivalent to a domain. If* I-soc$(R_R) = R$*, then R is right hereditary.*

PROOF. Let R be Morita equivalent to a domain D. Then $\text{I-soc}(D_D) = D$ by Corollary 4.10. By Lemma 4.8, D contains an isosimple right ideal. Any isosimple right ideal of D must be isomorphic to D. Therefore D is a PRID. Thus R is right hereditary, because being right hereditary is a Morita invariant property [**14**, p. 13]. □

PROPOSITION 4.16. *Let R be a simple ring.*

(1) *If R has a nonsingular isosimple right module, then* $\text{I-soc}(R_R) = R$.

(2) *If R has a projective isosimple right module, then R is a ring of matrices over a simple* PRID.

PROOF. (1). Suppose $M = mR$ is a nonsingular isosimple right module. Let I be the right annihilator $\text{r.ann}(m)$ of m. Since M is non-singular, I is not an essential right ideal in R, and so $I \cap K = 0$ for some non-zero right ideal K of R. Thus K is an isosimple right ideal isomorphic to M. Since $Z(R_R)$ is a two-sided ideal of the simple ring R, it follows that $Z(R_R) = 0$. Therefore R is right nonsingular, and so $R = \text{I-soc}(R_R)$ by Lemma 3.14.

(2). By [**18**], any simple noetherian ring with a projective uniform right ideal is Morita equivalent to a domain. The statement follows from Theorem 4.14 and Proposition 4.15. □

Open problems. Let R be a ring with $\text{I-soc}(R_R) = R$ (or, equivalently, R be a sum of isosimple right ideals). Is R right semihereditary (or, equivalently, a direct sum of isosimple right ideals)?

In [**31**], Stafford gave a negative answer to an important question posed in [**6**, Question 1, p. 113]. The question was whether simple right noetherian rings are Morita equivalent to domains. For further details about Stafford's example, see [**15**]. Later, another example of such a simple right noetherian ring was given in [**27**]. Is there a simple right noetherian ring R containing an isosimple right ideal such that R is not Morita equivalent to any domain? Note that for such a ring R, if it exists, we would have that $\text{I-soc}(R_R) = R$ by Proposition 4.16(1), and this would give a negative answer to the problem above. By Theorem 4.13, we can suppose R prime.

5. Modules of finite I-length

In this section, we will introduce and study modules of finite I-length.

DEFINITION 5.1. We say that a chain $0 = P_0 < P_1 < \cdots < P_n = M$ of length n of submodules of M is an I-*series* for M if $P_i \not\cong P_{i-1}$ for every $i = 1, 2, \dots, n$ and, for every submodule K of M with $P_{i-1} \leq K \leq P_i$, we have that either $K \cong P_{i-1}$ or $K \cong P_i$. A module M is said to be *of finite* I-*length* if it has an I-series of length n for some non-negative integer n. In this case, the least such n is called the I-*length* of M, $0 = P_0 < P_1 < \cdots < P_n = M$ is called an I-*length series for* M, and we write $\text{I-length}(M) = n$.

LEMMA 5.2. *Let M be a module that is both isonoetherian and isoartinian. Then M is of finite* I-*length.*

PROOF. Since M is isoartinian, it has an isosimple submodule, P_1 say. Let $\mathcal{A}_1$ be the set of all submodules N of M such that $N > P_1$ and $N \not\cong P_1$. If $\mathcal{A}_1$ is empty, then $\text{I-length}(M) = 1$. If $\mathcal{A}_1$ is non-empty, then, by Lemma 3.8, there exists

$P_2 \in \mathcal{A}_1$ that is minimal in $\mathcal{A}_1$ up to isomorphism. That is, if K is a submodule with $P_1 \leq K \leq P_2$, then $K \cong P_1$ or $K \cong P_2$. By continuing in this manner, we find an ascending chain $P_1 < P_2 < \dots$ of submodules of M such that $P_n \ncong P_{n+1}$ for each n, and for every submodule K of M with $P_n \leq K \leq P_{n+1}$, either $K \cong P_n$ or $K \cong P_{n+1}$. Since M is isonoetherian, this chain is stationary. So $P_n = M$ for some n. □

REMARK 5.3. (1) I-length$(M) = 1$ if and only if M is isosimple.

(2) *Terse* modules are the modules whose distinct submodules are not isomorphic [**34**]. Cyclic modules over a commutative regular ring and artinian uniserial modules are terse modules. One can easily see that, for a terse module, being of finite length and being of finite I-length are equivalent.

(3) A factor module of a module of finite I-length is not necessarily of finite I-length. For example, if D is a DVR that is not field and Q is its field of fractions, then Q is of I-length 2 as a D-module, but Q/D is a terse module, and therefore Q/D is not of finite I-length.

PROPOSITION 5.4. *Let M be a module of finite* I-*length. Then:*

(1) *M is of finite uniform dimension and* $\mathrm{u.dim}(M) \leq$ I-length(M).

(2) *If* $\mathrm{u.dim}(M) =$ I-length$(M) = n$*, then M is noetherian and is direct sum of n uniform cyclic submodules.*

PROOF. (1) The proof is by induction on $n :=$ I-length(M). The case $n = 1$ is clear. Suppose $n > 1$. Let $0 < P_1 < \cdots < P_n = M$ be an I-length series for M. Since P_{n-1} is a module of I-length $n-1$, by the inductive hypothesis $\mathrm{u.dim}(P_{n-1}) \leq n-1$. If P_{n-1} is essential in M, then $\mathrm{u.dim}(M) = \mathrm{u.dim}(P_{n-1}) \leq n-1$ and we are done. If P_{n-1} is not essential in M, then there exists a cyclic submodule T of M such that $P_{n-1} \cap T = 0$, and so $P_{n-1} \oplus T \cong M$. If T is not of finite uniform dimension, then T contains an infinite direct sum of non-zero cyclic submodules, $\oplus_{i=1}^{\infty} T_i$ say. Since $P_{n-1} \cap (\oplus_{i=1}^{\infty} T_i) = 0$, we have that $M \cong P_{n-1} \oplus \oplus_{i=1}^{\infty} T_i$. From $P_{n-1} \oplus T \cong M$, we conclude that M has a finitely generated essential submodule. Therefore $P_{n-1} \oplus \oplus_{i=1}^{\infty} T_i$ has a finitely generated essential submodule, which is a contradiction. Thus T is of finite uniform dimension. By taking a uniform submodule U of T, we have $P_{n-1} \oplus U \cong M$. It follows that $\mathrm{u.dim}(M) \leq n$.

(2) Let $0 < P_1 < \cdots < P_n = M$ be an I-length series for M. As we have seen, P_i is not essential in P_{i+1} for every $i = 1, \dots, n-1$. Since M has finite uniform dimension, there exists a cyclic uniform submodule U_i of P_{i+1} such that $P_{i+1} \cong P_i \oplus U_i$. Note that P_1 is uniform and noetherian. To complete the proof, it is enough to show that each U_i, where $i = 1, \dots, n-1$, is noetherian. If K is a non-zero submodule of U_1, then $P_2 \cong P_1 \oplus K$. On the other hand, $P_2 \cong P_1 \oplus U_1$, and so P_2 is finitely generated. Thus K is finitely generated and U_1 is noetherian. With a similar proof, one can see by induction that U_i is noetherian for every $i = 1, \dots, n-1$. □

THEOREM 5.5. *Suppose $M = U_1 \oplus \cdots \oplus U_n$, where the modules U_i are isosimple. Then:*

(1) I-length$(M) = n$.

(2) *A submodule K of M is essential in M if and only if K contains a submodule isomorphic to M.*

(3) *M is isoartinian if and only if* I-length$(K) = \mathrm{u.dim}(K)$ *for every submodule K of M.*

PROOF. (1) Consider the chain $0 < U_1 < U_1 \oplus U_2 < \cdots < M$. One can check that this is an I-series for M. So M is of finite I-length $\leq n$. Now Proposition 5.4(1) shows that I-length$(M) = \mathrm{u.dim}(M) = n$.

(2) The implication ($\Leftarrow$) is easy, because isosimple modules are uniform. For the implication ($\Rightarrow$), let K be an essential submodule of M. Then $K_i := U_i \cap K$ is a non-zero submodule of U_i and $K_1 \oplus \cdots \oplus K_n$ is a submodule of K isomorphic to M.

(3) See [**11**, Theorem 5.8].

□

Open problem. Is a module of finite I-length an isoartinian and isonoetherian module?

Acknowledgment

We are grateful to Christian Lomp for the examples of isosimple modules presented at the end of Section 2.

References

[1] F. W. Anderson and K. R. Fuller, *Rings and categories of modules*, 2nd ed., Graduate Texts in Mathematics, vol. 13, Springer-Verlag, New York, 1992. MR1245487

[2] E. P. Armendariz, J. W. Fisher, and R. L. Snider, *On injective and surjective endomorphisms of finitely generated modules*, Comm. Algebra **6** (1978), no. 7, 659–672, DOI 10.1080/00927877808822263. MR0469974

[3] I. Borges and C. Lomp, *Irreducible actions and compressible modules*, J. Algebra Appl. **10** (2011), no. 1, 101–117, DOI 10.1142/S0219498811004446. MR2784756

[4] J. Clark and D. v. Huynh, *When is a self-injective semiperfect ring quasi-Frobenius?*, J. Algebra **165** (1994), no. 3, 531–542, DOI 10.1006/jabr.1994.1128. MR1275918

[5] F. Couchot, *Modules with RD-composition series over a commutative ring*, Comm. Algebra **31** (2003), no. 7, 3171–3194, DOI 10.1081/AGB-120022218. MR1990267

[6] J. Cozzens and C. Faith, *Simple Noetherian rings*, Cambridge Tracts in Mathematics, vol. 69, Cambridge University Press, Cambridge-New York-Melbourne, 1975. MR0396660

[7] R. Dastanpour and A. Ghorbani, *Divisibility on chains of submodules*, Comm. Algebra **46** (2018), no. 6, 2305–2318, DOI 10.1080/00927872.2017.1376217. MR3778392

[8] S. Deo and K. Varadarajan, *Hopfian and co-Hopfian groups*, Bull. Austral. Math. Soc. **56** (1997), no. 1, 17–24, DOI 10.1017/S0004972700030690. MR1464045

[9] A. Facchini, *Module theory: Endomorphism rings and direct sum decompositions in some classes of modules*, Progress in Mathematics, vol. 167, Birkhäuser Verlag, Basel, 1998. MR1634015

[10] A. Facchini and Z. Nazemian, *Modules with chain conditions up to isomorphism*, J. Algebra **453** (2016), 578–601, DOI 10.1016/j.jalgebra.2016.01.025. MR3465365

[11] A. Facchini and Z. Nazemian, *Artinian dimension and isoradical of modules*, J. Algebra **484** (2017), 66–87, DOI 10.1016/j.jalgebra.2017.03.039. MR3656713

[12] A. Ghorbani and A. Haghany, *Generalized Hopfian modules*, J. Algebra **255** (2002), no. 2, 324–341, DOI 10.1016/S0021-8693(02)00124-2. MR1935502

[13] A. Ghorbani, S. K. Jain, and Z. Nazemian, *Indecomposable decomposition and couniserial dimension*, Bull. Math. Sci. **5** (2015), no. 1, 121–136, DOI 10.1007/s13373-014-0062-6. MR3319982

[14] K. R. Goodearl, *Ring theory: Nonsingular rings and modules*, Pure and Applied Mathematics, vol. 33, Marcel Dekker, Inc., New York-Basel, 1976. MR0429962

[15] K. R. Goodearl, *Simple Noetherian rings—the Zalesskiĭ-Neroslavskiĭ examples*, Ring theory (Proc. Conf., Univ. Waterloo, Waterloo, 1978), Lecture Notes in Math., vol. 734, Springer, Berlin, 1979, pp. 118–130. MR548126

[16] R. Gordon and J. C. Robson, *Krull dimension*, Memoirs of the American Mathematical Society, No. 133, American Mathematical Society, Providence, R.I., 1973. MR0352177

[17] A. Haghany, *Hopficity and co-Hopficity for Morita contexts*, Comm. Algebra **27** (1999), no. 1, 477–492, DOI 10.1080/00927879908826443. MR1668301

[18] R. Hart, *Simple rings with uniform right ideals*, J. London Math. Soc. **42** (1967), 614–617, DOI 10.1112/jlms/s1-42.1.614. MR0218390

[19] A. G. Heinicke, *Some results in the theory of radicals of associative rings*, Ph. D. Thesis, University of British Columbia, Vancouver, 1968.

[20] Y. Hirano and I. Mogami, *Modules whose proper submodules are non-Hopf kernels*, Comm. Algebra **15** (1987), no. 8, 1549–1567, DOI 10.1080/00927878708823487. MR884760

[21] M. Khalifa, *Isonoetherian power series rings*, Comm. Algebra **46** (2018), no. 6, 2451–2458, DOI 10.1080/00927872.2017.1384004. MR3778404

[22] K. Koh, *On almost maximal right ideals*, Proc. Amer. Math. Soc. **25** (1970), 266–272, DOI 10.2307/2037203. MR0265393

[23] K. Koh and A. C. Mewborn, *A class of prime rings*, Canad. Math. Bull. **9** (1966), 63–72, DOI 10.4153/CMB-1966-008-0. MR0197491

[24] K. Koh and A. C. Mewborn, *The weak radical of a ring*, Proc. Amer. Math. Soc. **18** (1967), 554–559, DOI 10.2307/2035498. MR0213388

[25] T. Y. Lam, *A first course in noncommutative rings*, Graduate Texts in Mathematics, vol. 131, Springer-Verlag, New York, 1991. MR1125071

[26] T. Y. Lam, *Lectures on modules and rings*, Graduate Texts in Mathematics, vol. 189, Springer-Verlag, New York, 1999. MR1653294

[27] M. Lorenz, *K_0 of skew group rings and simple Noetherian rings without idempotents*, J. London Math. Soc. (2) **32** (1985), no. 1, 41–50, DOI 10.1112/jlms/s2-32.1.41. MR813383

[28] J. C. McConnell and J. C. Robson, *Noncommutative Noetherian rings*, Revised edition, Graduate Studies in Mathematics, vol. 30, American Mathematical Society, Providence, RI, 2001. MR1811901

[29] Z. Nazemian, A. Ghorbani, and M. Behboodi, *Uniserial dimension of modules*, J. Algebra **399** (2014), 894–903, DOI 10.1016/j.jalgebra.2013.09.054. MR3144618

[30] S. Prakash and A. K. Chaturvedi, *Iso-noetherian rings and modules*, Comm. Algebra (2019), Published online: 11 Jan 2019, DOI `https://doi.org/10.1080/00927872.2018.1492591`.

[31] J. T. Stafford, *A simple Noetherian ring not Morita equivalent to a domain*, Proc. Amer. Math. Soc. **68** (1978), no. 2, 159–160, DOI 10.2307/2041762. MR0466210

[32] K. Varadarajan, *Hopfian and co-Hopfian objects*, Publ. Mat. **36** (1992), no. 1, 293–317, DOI 10.5565/PUBLMAT_36192_21. MR1179618

[33] K. Varadarajan, *Anti Hopfian and anti co-Hopfian modules*, Noncommutative rings, group rings, diagram algebras and their applications, Contemp. Math., vol. 456, Amer. Math. Soc., Providence, RI, 2008, pp. 205–218, DOI 10.1090/conm/456/08892. MR2416152

[34] W. D. Weakley, *Modules whose distinct submodules are not isomorphic*, Comm. Algebra **15** (1987), no. 8, 1569–1586, DOI 10.1080/00927878708823488. MR884761

[35] R. Wisbauer, *Modules and algebras: bimodule structure and group actions on algebras*, Pitman Monographs and Surveys in Pure and Applied Mathematics, vol. 81, Longman, Harlow, 1996. MR1396313

Dipartimento di Matematica, Università di Padova, 35121 Padova, Italy
Email address: `facchini@math.unipd.it`

School of Mathematics, Institute for Research in Fundamental Sciences (IPM), P. O. Box: 19395-5746, Tehran, Iran
Email address: `z_nazemian@yahoo.com`

Contemporary Mathematics
Volume **730**, 2019
https://doi.org/10.1090/conm/730/14708

Derived categories for Grothendieck categories of enriched functors

Grigory Garkusha and Darren Jones

Dedicated to Mike Prest on the occasion of his 65th birthday

ABSTRACT. The derived category $\mathbf{D}[\mathcal{C},\mathcal{V}]$ of the Grothendieck category of enriched functors $[\mathcal{C},\mathcal{V}]$, where $\mathcal{V}$ is a closed symmetric monoidal Grothendieck category and $\mathcal{C}$ is a small $\mathcal{V}$-category, is studied. We prove that if the derived category $\mathbf{D}(\mathcal{V})$ of $\mathcal{V}$ is a compactly generated triangulated category with certain reasonable assumptions on compact generators or $\mathbf{K}$-injective resolutions, then the derived category $\mathbf{D}[\mathcal{C},\mathcal{V}]$ is also compactly generated triangulated. Moreover, an explicit description of these generators is given.

1. Introduction

Enriched categories generalize the idea of a category by replacing Hom-sets with objects from a monoidal category. In practice the Hom-sets often have additional structure that should be respected, e.g., that of being a topological space of morphisms, or a chain complex of morphisms. They have plenty of uses and applications. For example, Bondal–Kapranov [**BK**] construct enrichments of some triangulated categories over chain complexes ("DG-categories") to study exceptional collections of coherent sheaves on projective varieties. Today, DG-categories have become an important tool in many branches of algebraic geometry, non-commutative algebraic geometry, representation theory, and mathematical physics (see a survey by Keller [**Kel**]). There are also applications in motivic homotopy theory. For example, Dundas–Röndigs–Østvær [**DRO1**, **DRO2**] use enriched category theory to give a model for the Morel–Voevodsky category $SH(k)$. In [**GP1**, **GP2**, **GP3**] enrichments of smooth algebraic varieties over symmetric spectra have been used in order to develop the theory of "K-motives" and solve a problem for the motivic spectral sequence.

In [**AG**] the category of enriched functors $[\mathcal{C},\mathcal{V}]$ was studied, where $\mathcal{V}$ is a closed symmetric monoidal Grothendieck category and $\mathcal{C}$ is a small category enriched over $\mathcal{V}$. It was shown that $[\mathcal{C},\mathcal{V}]$ is a Grothendieck $\mathcal{V}$-category with a set of generators $\{\mathcal{V}(c,-)\oslash g_i \mid c\in \operatorname{Ob}\mathcal{C}, i\in I\}$, where $\{g_i\}_I$ is a set of generators of $\mathcal{V}$. The category $[\mathcal{C},\mathcal{V}]$ is called in [**AG**] the *Grothendieck category of enriched functors*. Basic examples are given by categories of additive functors $(\mathcal{B},\mathrm{Ab})$ or DG-modules

2010 *Mathematics Subject Classification.* Primary 13D09; Secondary 18D10, 18D20.
Key words and phrases. Grothendieck categories of enriched functors, derived categories, compactly generated triangulated categories.

$\operatorname{Mod}\mathcal{A}$ over a DG-category $\mathcal{A}$. An advantage of this result is that we can recover some well-known theorems for Grothendieck categories in the case where $\mathcal{V} = \mathrm{Ab}$. Another advantage is that $\mathcal{V}$ can also contain some rich homological or homotopical information, which is extended to the category of enriched functors $[\mathcal{C}, \mathcal{V}]$. This homotopical information is of great utility to study the derived category $\mathbf{D}(\mathcal{C}_R)$ of the category of generalized modules $\mathcal{C}_R = (\operatorname{mod} R, \mathrm{Ab})$ over a commutative ring R. It was proven in [**AG**] that $\mathbf{D}(\mathcal{C}_R)$ is essentially the same as a unital algebraic stable homotopy category in the sense of Hovey–Palmieri–Strickland [**HPS**] except that the compact objects do not have to be strongly dualizable, but must have a duality. Moreover, this duality is nothing but the classical Auslander–Gruson–Jensen Duality extended to compact objects of $\mathbf{D}(\mathcal{C}_R)$ (see [**AG**] for details).

In this paper we investigate the problem of when the derived category $\mathbf{D}[\mathcal{C}, \mathcal{V}]$ of the Grothendieck category $[\mathcal{C}, \mathcal{V}]$ is compactly generated triangulated and give an explicit description of the compact generators. The importance of this problem is that the general localization theory of compactly generated triangulated categories becomes available for $\mathbf{D}[\mathcal{C}, \mathcal{V}]$ in that case. Namely, we prove the following result (see Theorem 6.2):

THEOREM. *Let $(\mathcal{V}, \otimes, e)$ be a closed symmetric monoidal Grothendieck category such that the derived category of $\mathcal{V}$ is a compactly generated triangulated category with compact generators $\{P_j\}_{j\in J}$. Further, suppose we have a small $\mathcal{V}$-category $\mathcal{C}$ and that any one of the following conditions is satisfied:*

1. *each P_j is **K**-projective, in the sense of Spaltenstein [**Sp**];*
2. *for every **K**-injective $Y \in \mathbf{Ch}[\mathcal{C}, \mathcal{V}]$ and every $c \in \mathcal{C}$, the complex $Y(c) \in \mathbf{Ch}(\mathcal{V})$ is **K**-injective;*
3. *$\mathbf{Ch}(\mathcal{V})$ has a model structure, with quasi-isomorphisms being weak equivalences, such that for every injective fibrant complex $Y \in \mathbf{Ch}[\mathcal{C}, \mathcal{V}]$ the complex $Y(c)$ is fibrant in $\mathbf{Ch}(\mathcal{V})$.*

Then $\mathbf{D}[\mathcal{C}, \mathcal{V}]$ is a compactly generated triangulated category with compact generators $\{\mathcal{V}_{\mathcal{C}}(c, -) \oslash Q_j \mid c \in \mathcal{C}, j \in J\}$ where, if we assume either (1) or (2), $Q_j = P_j$ or if we assume (3) then $Q_j = P_j^c$ a cofibrant replacement of P_j.

The formulations of the first two statements of the theorem have nothing to do with model categories and use the terminology of the classical homological algebra only. However, in practice these statements are normally covered by the situation when $\mathbf{Ch}(\mathcal{V})$ is equipped with a "projective model structure with certain finiteness conditions" or when every evaluation functor $Ev_c : \mathbf{Ch}[\mathcal{C}, \mathcal{V}] \to \mathbf{Ch}(\mathcal{V})$, $c \in \mathcal{C}$, is right Quillen. In this case we should be able to extend homological/homotopical information from $\mathbf{Ch}(\mathcal{V})$ to $\mathbf{Ch}[\mathcal{C}, \mathcal{V}]$. To this end, we need the following result proved in Theorems 3.2 and 5.4.

THEOREM. *Let $\mathcal{V}$ be a closed symmetric monoidal Grothendieck category and $\mathcal{C}$ be a small $\mathcal{V}$-category. Then the category of chain complexes $\mathbf{Ch}(\mathcal{V})$ is closed symmetric monoidal Grothendieck and the category $\mathbf{Ch}[\mathcal{C}, \mathcal{V}]$ is naturally isomorphic to the category $[\mathcal{C}, \mathbf{Ch}(\mathcal{V})]$, where $\mathcal{C}$ is enriched over $\mathbf{Ch}(\mathcal{V})$ by the obvious complexes concentrated in degree zero.*

As an application of the theorems, we can generate numerous (closed symmetric monoidal) compactly generated triangulated categories which are of independent interest. Moreover, several important results of [**AG**] are extended from $\mathbf{D}(\mathcal{C}_R)$ to

$\mathbf{D}[\mathcal{C},\mathcal{V}]$. Other applications are expected in the study of pure-injectivity of compactly generated triangulated categories, in the telescope conjecture for compactly generated triangulated categories and in the study of Voevodsky's triangulated categories of motives. The flexibility of the theorems is that we can vary $\mathcal{V}$ in practice. Furthermore, $\mathcal{V}$ itself can contain rich homological/homotopical structures, in which case we can use the homological algebra and the Bousfield localization theory of $\mathbf{D}[\mathcal{C},\mathcal{V}]$ together with homological/homotopical structures of $\mathcal{V}$. In order to operate with such structures in practice, we need the above theorems.

2. Enriched Category Theory

In this section we collect basic facts about enriched categories we shall need later. We refer the reader to [**Bor2**, **R**] for details. Throughout this paper the quadruple $(\mathcal{V},\otimes,\underline{\mathrm{Hom}},e)$ is a closed symmetric monoidal category with monoidal product $\otimes$, internal Hom-object $\underline{\mathrm{Hom}}$ and monoidal unit e. We sometimes write $[a,b]$ to denote $\underline{\mathrm{Hom}}(a,b)$, where $a,b\in\operatorname{Ob}\mathcal{V}$. We have structure isomorphisms

$$a_{abc}:(a\otimes b)\otimes c\to a\otimes(b\otimes c),\quad l_a:e\otimes a\to a,\quad r_a:a\otimes e\to a$$

in $\mathcal{V}$ with $a,b,c\in\operatorname{Ob}\mathcal{V}$.

DEFINITION 2.1. A *$\mathcal{V}$-category* $\mathcal{C}$, or *a category enriched over* $\mathcal{V}$, consists of the following data:

(1) a class $\operatorname{Ob}(\mathcal{C})$ of objects;
(2) for every pair $a,b\in\operatorname{Ob}(\mathcal{C})$ of objects, an object $\mathcal{V}_{\mathcal{C}}(a,b)$ of $\mathcal{V}$;
(3) for every triple $a,b,c\in\operatorname{Ob}(\mathcal{C})$ of objects, a composition morphism in $\mathcal{V}$,

$$c_{abc}:\mathcal{V}_{\mathcal{C}}(a,b)\otimes\mathcal{V}_{\mathcal{C}}(b,c)\to\mathcal{V}_{\mathcal{C}}(a,c);$$

(4) for every object $a\in\mathcal{C}$, a unit morphism $u_a:e\to\mathcal{V}_{\mathcal{C}}(a,a)$ in $\mathcal{V}$.

These data must satisfy the natural associativity and unit axioms.

When $\operatorname{Ob}\mathcal{C}$ is a set, the $\mathcal{V}$-category $\mathcal{C}$ is called a *small $\mathcal{V}$-category*.

DEFINITION 2.2. Given $\mathcal{V}$-categories $\mathcal{A},\mathcal{B}$, a *$\mathcal{V}$-functor* or an *enriched functor* $F:\mathcal{A}\to\mathcal{B}$ consists in giving:

(1) for every object $a\in\mathcal{A}$, an object $F(a)\in\mathcal{B}$;
(2) for every pair $a,b\in\mathcal{A}$ of objects, a morphism in $\mathcal{V}$,

$$F_{ab}:\mathcal{V}_{\mathcal{A}}(a,b)\to\mathcal{V}_{\mathcal{B}}(F(a),F(b))$$

in such a way that the following axioms hold:
- ◇ for all objects $a,a',a''\in\mathcal{A}$, diagram (2.1) below commutes (composition axiom);
- ◇ for every object $a\in\mathcal{A}$, diagram (2.2) below commutes (unit axiom).

$$\begin{array}{ccc}\mathcal{V}_{\mathcal{A}}(a,a')\otimes\mathcal{V}_{\mathcal{A}}(a',a'') & \xrightarrow{c_{aa'a''}} & \mathcal{V}_{\mathcal{A}}(a,a'') \\ \downarrow{\scriptstyle F_{aa'}\otimes F_{a'a''}} & & \downarrow{\scriptstyle F_{aa''}} \\ \mathcal{V}_{\mathcal{B}}(Fa,Fa')\otimes\mathcal{V}_{\mathcal{B}}(Fa',Fa'') & \xrightarrow[c_{Fa,Fa',Fa''}]{} & \mathcal{V}_{\mathcal{B}}(Fa,Fa'')\end{array}\tag{2.1}$$

where $C'_i \otimes_R C''_j$ denotes the tensor product of R-modules C'_i and C''_j. Then the triple $(\mathbf{Ch}(\operatorname{Mod} R), \otimes_R, R)$ is a closed symmetric monoidal category. It is Grothendieck by [**AG**, 3.4]. Here R is regarded as a complex concentrated in the zeroth degree.

(4) $(\operatorname{Mod} kG, \otimes_k, k)$ is closed symmetric monoidal Grothendieck category, where k is a field and G is a finite group.

(5) Given a field F, the category $NSwT/F$ of Nisnevich sheaves with transfers [**SV**, Section 2] is a closed symmetric monoidal Grothendieck category with

$$\{\mathbb{Z}_{tr}(X) \mid X \text{ is an } F\text{-smooth algebraic variety}\}$$

a family of generators.

In this section we prove the following natural fact, as the authors were unable to find a complete account in the literature. We find it necessary to give such a complete account as it will be important to our analysis. The authors do not pretend to originality here.

THEOREM 3.2. *Let $\mathcal{V}$ be a closed symmetric monoidal Grothendieck category. Then the category of chain complexes over $\mathcal{V}$, denoted* $\mathbf{Ch}(\mathcal{V})$, *is closed symmetric monoidal Grothendieck.*

PROOF. Firstly, by [**AG**, 3.4] given $\mathcal{V}$ Grothendieck, we have that $\mathbf{Ch}(\mathcal{V})$ is also Grothendieck. It remains to define the closed symmetric monoidal structure on $\mathbf{Ch}(\mathcal{V})$. Denote the tensor product of $\mathcal{V}$ by $\otimes$ and its unit object by e. Further denote the associativity isomorphism a, the left unitor isomorphism by l and the right unitor by r respectively. We also assign sw to mean the symmetry isomorphism in $\mathcal{V}$.

Given $X, Y \in \mathbf{Ch}\,\mathcal{V}$ we define $X \odot Y$ as the chain complex with entries

$$(X \odot Y)_n := \bigoplus_{n=p+q} X_p \otimes Y_q.$$

Throughout this proof we tacitly assume that the category $\operatorname{Gr}\mathcal{V}$ of $\mathbb{Z}$-graded objects in $\mathcal{V}$ is closed symmetric monoidal. This follows from Day's theorem [**Day**] and literally repeats [**AG**, Example 4.5]. The differential $d_n^{X \odot Y} : (X \odot Y)_n \to (X \odot Y)_{n-1}$ determined by its action on each summand as

$$d_{(p,q)}^{X \odot Y} : X_p \otimes Y_q \to (X_{p-1} \otimes Y_q) \oplus (X_p \otimes Y_{q-1})$$

followed by inclusion into $(X \odot Y)_{n-1}$ such that

$$d_{(p,q)}^{X \odot Y} = d_p^X \otimes \mathrm{id}_q^Y + (-1)^p \,\mathrm{id}_p^X \otimes d_q^Y.$$

It does indeed define a chain complex as we see by

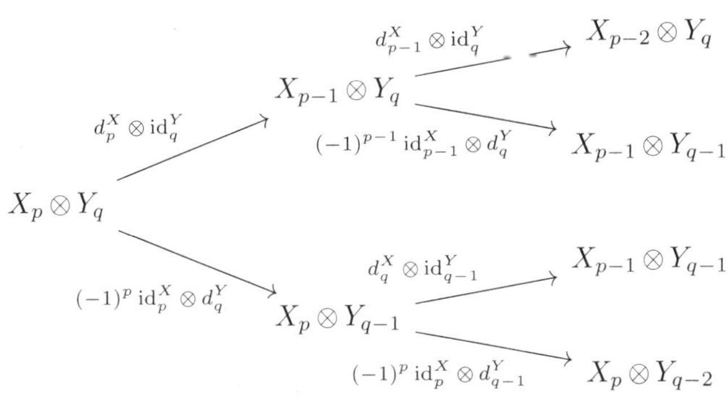

and thus, we are able to calculate

$$d^{X\odot Y}_{(p-1,q)}\circ d^{X\odot Y}_{n}+d^{X\odot Y}_{(p,q-1)}\circ d^{X\odot Y}_{n}=0+(-1)^{p-1}d^X_p\otimes d^Y_q+(-1)^p d^X_p\otimes d^Y_q+0=0$$

for every $p,q\in\mathbb{Z}$, hence $d\circ d=0$.

Next, given chain maps $f:X\to X'$ and $g:Y\to Y'$, we define

$$(f\odot g)_n:=\bigoplus_{n=p+q} f_p\otimes g_q$$

for each component $n\in\mathbb{Z}$ and then consider the following diagram

$$\begin{array}{ccc} X_p\otimes Y_q & \xrightarrow{d^X_p\otimes \mathrm{id}^Y_q+(-1)^p\,\mathrm{id}^X_p\otimes d^Y_q} & (X_{p-1}\otimes Y_q)\oplus(X_p\otimes Y_{q-1}) \\ \downarrow{\scriptstyle f_p\otimes g_q} & & \downarrow{\scriptstyle (f_{p-1}\otimes g_q)\oplus(f_p\otimes g_{q-1})} \\ X'_p\otimes Y'_q & \xrightarrow{d^{X'}_p\otimes \mathrm{id}^{Y'}_q+(-1)^p\,\mathrm{id}^{X'}_p\otimes d^{Y'}_q} & (X'_{p-1}\otimes Y'_q)\oplus(X'_p\otimes Y'_{q-1}) \end{array}$$

which commutes on each summand for all choices p,q, hence $f\odot g$ is consistent with the differential and as $\odot$ is clearly a functor on graded objects, we can thus conclude that $\odot$ is a bifunctor $\mathbf{Ch}(\mathcal{V})\times\mathbf{Ch}(\mathcal{V})\to\mathbf{Ch}(\mathcal{V})$.

Now we are in a position to define our structure isomorphisms. Given chain complexes $X,Y,Z\in\mathbf{Ch}(\mathcal{V})$ we define an associativity isomorphism

$$\alpha:(X\odot Y)\odot Z\to X\odot(Y\odot Z).$$

For $n\in\mathbb{Z}$ we define $\alpha_n=\bigoplus_{n=i+j+k}a^{X_i,Y_j,Z_k}$, where $a^{X_i,Y_j,Z_k}:(X_i\otimes Y_j)\otimes Z_k\to X_i\otimes(Y_j\otimes Z_k)$ is the component of the natural associativity isomorphism in $\mathcal{V}$. Since we know that a is a natural isomorphism, a direct sum of its components is also a natural isomorphism, and further we can say that this α will satisfy the relevant coherence conditions as it will hold at each degree. However, we need to check that these α_n give a chain map, i.e. these are consistent with the differential. We have:

$$\begin{aligned} d^{X\odot(Y\odot Z)}_{(i,j,k)} &= d^X_i\otimes \mathrm{id}^{Y\odot Z}_{j+k}+(-1)^i\,\mathrm{id}^X_i\otimes d^{Y\odot Z}_{j+k}\\ &= d^X_i\otimes \mathrm{id}^{Y\odot Z}_{j+k}+(-1)^i\,\mathrm{id}^X_i\otimes(d^Y_j\otimes\mathrm{id}^Z_k+(-1)^j\,\mathrm{id}^Y_j\otimes d^Z_k)\\ &= d^X_i\otimes \mathrm{id}^{Y\odot Z}_{j+k}+(-1)^i\,\mathrm{id}^X_i\otimes(d^Y_j\otimes\mathrm{id}^Z_k)+(-1)^{i+j}\,\mathrm{id}^X_i\otimes(\mathrm{id}^Y_j\otimes d^Z_k)\\ &= d^X_i\otimes(\mathrm{id}^Y_j\otimes\mathrm{id}^Z_k)+(-1)^i\,\mathrm{id}^X_i\otimes(d^Y_j\otimes\mathrm{id}^Z_k)+(-1)^{i+j}\,\mathrm{id}^X_i\otimes(\mathrm{id}^Y_j\otimes d^Z_k) \end{aligned}$$

and

$$\begin{aligned} d^{(X\odot Y)\odot Z}_{(i,j,k)} &= d^{X\odot Y}_{i+j}\otimes\mathrm{id}^Z_k+(-1)^{i+j}\,\mathrm{id}^{X\odot Y}_{i+j}\otimes d^Z_k\\ &= (d^X_i\otimes\mathrm{id}^Y_j+(-1)^i\,\mathrm{id}^X_i\otimes d^Y_j)\otimes\mathrm{id}^Z_k+(-1)^{i+j}\,\mathrm{id}^{X\odot Y}_{i+j}\otimes d^Z_k\\ &= (d^X_i\otimes\mathrm{id}^Y_j)\otimes\mathrm{id}^Z_k+(-1)^i(\mathrm{id}^X_i\otimes d^Y_j)\otimes\mathrm{id}^Z_k+(-1)^{i+j}\,\mathrm{id}^{X\odot Y}_{i+j}\otimes d^Z_k\\ &= (d^X_i\otimes\mathrm{id}^Y_j)\otimes\mathrm{id}^Z_k+(-1)^i(\mathrm{id}^X_i\otimes d^Y_j)\otimes\mathrm{id}^Z_k+(-1)^{i+j}(\mathrm{id}^X_i\otimes\mathrm{id}^Y_j)\otimes d^Z_k \end{aligned}$$

which agree up to a change of brackets (i.e. by applying a^{X_i,Y_j,Z_k}), hence α_n-s give a chain map.

We define a unit object for our new tensor product, which we denote by ε as being the chain complex with e in zeroth degree and 0 in every other degree and note that

$$(X\odot\varepsilon)_n=X_n\otimes e\quad\text{and}\quad d^{X\odot\varepsilon}_n=d^X_n\otimes\mathrm{id}_e$$

for all $n \in \mathbb{Z}$ as tensoring with zero is zero and a direct sum is unchanged by adding zeros. Thus we define ρ^X with $\rho^X_n = r^{X_n}$, the fact that this is a chain map follows directly from the naturality of r and moreover is itself a natural transformation in $\mathbf{Ch}(\mathcal{V})$. Coherence conditions for the right unitor are satisfied at each degree by properties of $\mathcal{V}$, hence hold in $\mathbf{Ch}(\mathcal{V})$.

Similarly, note that

$$(\varepsilon \odot Y)_n = e \otimes Y_n \quad \text{and} \quad d_n^{\varepsilon \odot Y} = \mathrm{id}_e \otimes d_n^Y$$

and hence define the left unitor λ^Y as $\lambda^Y_n = l^Y_n$ which satisfies the relevant conditions by a similar argument.

Next, consider chain complexes $X, Y \in \mathbf{Ch}(\mathcal{V})$. We want to define a map

$$\sigma^{X,Y} : X \odot Y \to Y \odot X.$$

We shall consider this map to consist of $\sigma_n^{X,Y} : \bigoplus_{n=p+q} X_p \otimes Y_q \to \bigoplus_{n=q+p} Y_q \otimes X_p$ for $n \in \mathbb{Z}$. They are completely determined by its action on each summand

$$\sigma^{X,Y}_{(p,q)} : X_p \otimes Y_q \to Y_q \otimes X_p,$$

and these we define to be

$$\sigma^{X,Y}_{(p,q)} = (-1)^{pq} \, \mathrm{sw}^{X_p, Y_q},$$

the components of the symmetry isomorphism in $\mathcal{V}$ multiplied by $(-1)^{pq}$. Such a map is a natural isomorphism as it is in $\mathcal{V}$, and satisfies the coherence conditions as it will do so on each component. However, we need to check if this map is indeed a chain map. This is demonstrated by the following commutative diagram:

$$\begin{array}{ccc}
X_p \otimes Y_q & \xrightarrow{d_p^X \otimes \mathrm{id}_q^Y + (-1)^p \mathrm{id}_p^X \otimes d_q^Y} & (X_{p-1} \otimes Y_q) \oplus (X_p \otimes Y_{q-1}) \\
\downarrow{\scriptstyle (-1)^{pq} \mathrm{sw}^{X_p, Y_q}} & & \downarrow{\scriptstyle ((-1)^{(p-1)q} \mathrm{sw}^{X_{p-1}, Y_q}) \oplus ((-1)^{p(q-1)} \mathrm{sw}^{X_p, Y_{q-1}})} \\
Y_q \otimes X_p & \xrightarrow{(-1)^q \mathrm{id}_p^Y \otimes d_q^X + d_q^Y \otimes \mathrm{id}_p^X} & (Y_q \otimes X_{p-1}) \oplus (Y_{q-1} \otimes X_p).
\end{array}$$

Thus $\mathbf{Ch}(\mathcal{V})$ is symmetric monoidal and Grothendieck. We next define an internal Hom-object $\underline{\mathrm{Hom}}(X, Y)$ for $X, Y \in \mathbf{Ch}(\mathcal{V})$, as having in each degree $n \in \mathbb{Z}$

$$\underline{\mathrm{Hom}}(X, Y)_n := \prod_p [X_p, Y_{p+n}],$$

where $[X_p, Y_{p+n}] := \mathcal{V}(X_p, Y_{p+n})$. To define its differential $d_n^{\underline{\mathrm{Hom}}(X,Y)} : \underline{\mathrm{Hom}}(X, Y)_n \to \underline{\mathrm{Hom}}(X, Y)_{n-1}$, it is enough to define this map to each factor by first projecting onto p and $p - 1$, and then one sets

$$d_{(p,n)}^{\underline{\mathrm{Hom}}(X,Y)} : [X_p, Y_{p+n}] \times [X_{p-1}, Y_{p+n-1}] \to [X_p, Y_{p+n-1}]$$

to be

$$d_{(p,n)}^{\underline{\mathrm{Hom}}(X,Y)} = [\mathrm{id}_p^X, d_{p+n}^Y] - (-1)^n [d_p^X, \mathrm{id}_{p+n-1}^Y].$$

Again, we need to verify that this defines a differential. We check this on each factor p and $p-1$, using the following diagram

$$
\begin{array}{ccccc}
[X_p, Y_{p+n}] & \xrightarrow{[\mathrm{id}_p^X, d_{p+n}^Y]} & & & \\
 & & [X_p, Y_{p+n-1}] & & \\
[X_{p-1}, Y_{p-1+n}] & \xrightarrow{-(-1)^n[d_p^X, \mathrm{id}_{p+n-1}^Y]} & & \xrightarrow{[\mathrm{id}_p^X, d_{p+n-1}^Y]} & \\
 & & & & [X_p, Y_{p+n-2}] \\
[X_{p-1}, Y_{p-1+n}] & \xrightarrow{[\mathrm{id}_{p-1}^X, d_{p-1+n}^Y]} & & \xrightarrow{-(-1)^{n-1}[d_p^X, \mathrm{id}_{p+n-2}^Y]} & \\
 & & [X_{p-1}, Y_{(p-1)+n-1}] & & \\
[X_{p-2}, Y_{p-2+n}] & \xrightarrow{-(-1)^n[d_{p-1}^X, \mathrm{id}_{p+n-2}^Y]} & & &
\end{array}
$$

We have

$$
\begin{aligned}
&[\mathrm{id}_p^X, d_{p+n-1}^Y] \circ ([\mathrm{id}_p^X, d_{p+n}^Y] - (-1)^n[d_p^X, \mathrm{id}_{p+n-1}^Y]) \\
&\quad = 0 - (-1)^n[d_p^X, d_{p+n-1}^Y] \\
&- (-1)^{n-1}[d_p^X, \mathrm{id}_{p+n-2}^Y] \circ ([\mathrm{id}_{p-1}^X, d_{p-1+n}^Y] - (-1)^n[d_{p-1}^X, \mathrm{id}_{p-1+n-1}^Y]) \\
&\quad = -(-1)^{n-1}[d_p^X, d_{p+n-1}^Y] + 0
\end{aligned}
$$

which sums to zero. Hence $d \circ d = 0$ and $\underline{\mathrm{Hom}}(X, Y)$ is a chain complex.

To define a closed structure on $\mathbf{Ch}(\mathcal{V})$, it is necessary that $\underline{\mathrm{Hom}}(X, Y)$ is functorial. It is apparent that the internal Hom-object of $\mathcal{V}$ and the product are functors on graded objects. We need only to check consistency with differentials. Given $f' : X' \to X$ and $g : Y \to Y'$, define $\underline{\mathrm{Hom}}(f', g)_n := \prod_p [f'_p, g_{p+n}]$ at each degree $n \in \mathbb{Z}$ and then consider the following commutative diagram

$$
\begin{array}{ccc}
[X_p, Y_{p+n}] \times [X_{p-1}, Y_{p-1+n}] & \xrightarrow{[\mathrm{id}_p^X, d_{p+n}^Y] - (-1)^n[d_p^X, \mathrm{id}_{p+n-1}^Y]} & [X_p, Y_{p-1+n}] \\
\downarrow{\scriptstyle [f'_p, g_{p+n}] \times [f'_{p-1}, g_{p-1+n}]} & & \downarrow{\scriptstyle [f'_p, g_{p-1+n}]} \\
[X'_p, Y'_{p+n}] \times [X'_{p-1}, Y'_{p-1+n}] & \xrightarrow[{[\mathrm{id}_p^{X'}, d_{p+n}^{Y'}] - (-1)^n[d_p^{X'}, \mathrm{id}_{p+n-1}^{Y'}]}]{} & [X'_p, Y'_{p-1+n}]
\end{array}
$$

Lastly, we need to make sure that our definition of the internal Hom-object of chain complexes satisfies the following isomorphism

$$\varphi : \mathrm{Hom}(X \odot Y, Z) \cong \mathrm{Hom}(X, \underline{\mathrm{Hom}}(Y, Z)),$$

natural in $X, Y, Z \in \mathbf{Ch}(\mathcal{V})$.

Given a chain map $k : X \odot Y \to Z$, we know that this is uniquely determined on each degree by maps on each summand. The collection of maps $k_{(p,q)} : X_p \otimes Y_q \to Z_{p+q}$ for all $p, q \in \mathbb{Z}$ determines k uniquely. Using the closed structure of $\mathcal{V}$, we can derive a collection $\phi(k_{(p,q)}) : X_p \to [Y_q, Z_{p+q}]$, where $\phi(k_{(p,q)})$ is the adjunction map in $\mathcal{V}$ corresponding to $k_{(p,q)}$, which is sufficient information to define maps

$$\varphi(k)_p : X_p \to \prod_q [Y_q, Z_{p+q}]$$

and construct $\varphi(k) : X \to \underline{\mathrm{Hom}}(Y, Z)$ as $(\varphi(k)_p)_{p \in \mathbb{Z}}$. Thus we have established a one-to-one correspondence between k and $\varphi(k)$. As usual, we need to check that this identification is compatible with the differentials. More precisely, let us check that $\varphi(k)$ is a morphism of complexes. We know that $k \in \operatorname{Hom}(X \odot Y, Z)$ if and only if for all integers p, q

$$\begin{array}{ccc} X_p \otimes Y_q & \xrightarrow{\quad k_{(p,q)} \quad} & Z_{p+q} \\ \Big\downarrow{\scriptstyle d_p^X \otimes \mathrm{id}_q^Y + (-1)^p \mathrm{id}_p^X \otimes d_q^Y} & & \Big\downarrow{\scriptstyle d_{p+q}^Z} \\ (X_{p-1} \otimes Y_q) \oplus (X_p \otimes Y_{q-1}) & \xrightarrow[\quad k_{(p-1,q)} + k_{(p,q-1)} \quad]{} & Z_{p+q-1} \end{array}$$

commutes. Our adjunction in $\mathcal{V}$ will lead to equalities

$$\begin{aligned} [\mathrm{id}_q^Y, d_{p+q}^Z] \circ \phi(k_{(p,q)}) &= \phi(d_{p+q}^Z \circ k_{(p,q)}) \\ &= \phi(k_{(p-1,q)} \circ (d_p^X \otimes \mathrm{id}_q^Y) + (-1)^p k_{(p,q-1)} \circ (\mathrm{id}_p^X \otimes d_q^Y)) \\ &= \phi(k_{(p-1,q)}) \circ d_p^X + (-1)^p \phi(k_{(p,q-1)} \circ (\mathrm{id}_p^X \otimes d_q^Y)) \\ &= \phi(k_{(p-1,q)}) \circ d_p^X + (-1)^p [d_q^Y, \mathrm{id}_{p+q-1}^Z] \circ \phi(k_{(p,q-1)}). \end{aligned}$$

So we must have

$$\phi(k_{(p-1,q)}) \circ d_p^X = [\mathrm{id}_q^Y, d_{p+q}^Z] \circ \phi(k_{(p,q)}) - (-1)^p [d_q^Y, \mathrm{id}_{p+q-1}^Z] \circ \phi(k_{(p,q-1)}).$$

If we write it as a commutative diagram, we get

$$\begin{array}{ccc} X_p & \xrightarrow{\quad (\phi(k_{(p,q)}), \phi(k_{(p,q-1)})) \quad} & [Y_q, Z_{p+q}] \times [Y_{q-1}, Z_{p+q-1}] \\ {\scriptstyle d_p^X}\Big\downarrow & & \Big\downarrow{\scriptstyle [\mathrm{id}_q^Y, d_{p+q}^Z] - (-1)^p [d_q^Y, \mathrm{id}_{p+q-1}^Z]} \\ X_{p-1} & \xrightarrow[\quad \phi(k_{(p-1,q)}) \quad]{} & [Y_q, Z_{p+q-1}] \end{array}$$

for all integers p, q. In other words, $\varphi(k)$ is a chain map if and only if so is k, as required.

Next, we have to determine whether our identification is natural. Consider maps $f : X \to X'$, $g : Y \to Y'$ and $h : Z' \to Z$, in $\mathbf{Ch}(\mathcal{V})$ and $k : X' \odot Y' \to Z'$. Given that two chain maps are equal if and only if they are equal on each degree, we fix a degree $n \in \mathbb{Z}$ and calculate

$$\begin{aligned} \varphi(h \circ k \circ (f \odot g))_n &= \prod_q \phi(h_{n+q} \circ k_{(n,q)} \circ (f_n \otimes g_q)) \\ &= \prod_q [g_q, h_{n+q}] \circ \phi(k_{(n,q)}) \circ f_n \\ &= \underline{\mathrm{Hom}}(g, h)_n \circ \varphi(k)_n \circ f_n \\ &= (\underline{\mathrm{Hom}}(g, h) \circ \varphi(k) \circ f)_n. \end{aligned}$$

Thus we have the desired naturality. We also have automatically that these isomorphisms are additive, and hence an adjunction in the Grothendieck category $\mathbf{Ch}(\mathcal{V})$.

We conclude that if $\mathcal{V}$ is a closed symmetric monoidal Grothendieck category, then the category of chain complexes $\mathbf{Ch}(\mathcal{V})$ is closed symmetric monoidal Grothendieck with the structure detailed above, as was to be proved. □

The preceding theorem leads to the following natural definition.

DEFINITION 3.3. A category $\mathcal{C}$ enriched over $\mathbf{Ch}(\mathcal{V})$ is said to be a *differential graded $\mathcal{V}$-category* or just a DG *$\mathcal{V}$-category*. $\mathcal{C}$ is *small* if its objects form a set. Ordinary DG-categories are recovered as DG $\mathcal{V}$-categories with $\mathcal{V} = \mathrm{Ab}$.

The *category of differential graded $\mathcal{V}$-modules* or just DG *$\mathcal{V}$-modules* is the category $[\mathcal{C}, \mathbf{Ch}(\mathcal{V})]$ of enriched functors from a DG $\mathcal{V}$-category $\mathcal{C}$ to $\mathbf{Ch}(\mathcal{V})$. Ordinary DG-modules over a DG-category are recovered as DG $\mathcal{V}$-modules with $\mathcal{V} = \mathrm{Ab}$.

Given any complete closed symmetric monoidal category $\mathcal{V}$ and any small $\mathcal{V}$-category $\mathcal{C}$, $[\mathcal{C}, \mathcal{V}]$ is a closed $\mathcal{V}$-module by Lemma 2.5. We write $\oslash$ for the corresponding functor $[\mathcal{C}, \mathcal{V}] \otimes \mathcal{V} \to [\mathcal{C}, \mathcal{V}]$.

COROLLARY 3.4. *Given a closed symmetric monoidal Grothendieck category $\mathcal{V}$ with a family of generators $\{g_i\}_I$ and small differential graded $\mathcal{V}$-category $\mathcal{C}$, the category of differential graded $\mathcal{V}$-modules $[\mathcal{C}, \mathbf{Ch}(\mathcal{V})]$ is Grothendieck with the set of generators $\{\mathbf{Ch}(\mathcal{V})_{\mathcal{C}}(c, -) \oslash D^n g_i \mid c \in \mathcal{C}, i \in I, n \in \mathbb{Z}\}$, where each $D^n g_i \in \mathbf{Ch}(\mathcal{V})$ is the complex which is g_i in degree n and $n-1$ and 0 elsewhere, with interesting differential being the identity map.*

PROOF. By the preceding theorem $\mathbf{Ch}(\mathcal{V})$ is a closed symmetric monoidal Grothendieck category. By the proof of [**AG**, 3.4] its set of generators is given by the family of complexes $\{D^n g_i \mid i \in I, n \in \mathbb{Z}\}$. Our statement now follows from [**AG**, 4.2]. $\square$

4. The enriched structure

Suppose $\mathcal{V}$ is a closed symmetric monoidal Grothendieck category and $\mathcal{C}$ is a small $\mathcal{V}$-category. In order to get some information about $\mathbf{Ch}[\mathcal{C}, \mathcal{V}]$, we shall identify this category with $[\mathcal{C}, \mathbf{Ch}(\mathcal{V})]$ (see Theorem 5.4) if we regard $\mathcal{C}$ as trivially a $\mathbf{Ch}(\mathcal{V})$-category, where for each $a, b \in \mathcal{C}$ we define the chain $\mathbf{Ch}(\mathcal{V})_{\mathcal{C}}(a, b)$ as having in zeroth degree the $\mathcal{V}$-object $\mathcal{V}_{\mathcal{C}}(a, b)$ and zero in every other degree. But first we need to collect some facts about $\mathbf{Ch}(\mathcal{V})$.

It is known (see [**Bor2**]) that a closed symmetric monoidal category canonically carries the structure of a category enriched over itself. It will be important for us to describe the unit and composition morphisms in the case of $\mathbf{Ch}(\mathcal{V})$ explicitly, using the unit and composition morphisms belonging to $\mathcal{V}$.

We begin by describing the unit. Given $a \in \mathcal{C}$ and any $F \in [\mathcal{C}, \mathbf{Ch}(\mathcal{V})]$, the unit morphism $u_{F(a)} : \varepsilon \to \underline{\mathrm{Hom}}(F(a), F(a))$, where ε is the unit object of the tensor product on $\mathbf{Ch}(\mathcal{V})$ defined in the proof of Theorem 3.2, reduces to a single morphism in degree zero, $u_{F(a)} : e \to \prod_p [F(a)_p, F(a)_p]$ with e the unit object of the tensor product on $\mathcal{V}$. Moreover, $u_{F(a)} = (u_{F(a)_p})_{p \in \mathbb{Z}} : e \to \prod_p [F(a)_p, F(a)_p]$ where $u_{F(A)_p}$ is nothing but the unit morphism in $\mathcal{V}$ associated to the $\mathcal{V}$-object $F(a)_p$ for $p \in \mathbb{Z}$.

Next, in order to describe the composition morphism, we need to first understand the evaluation in $\mathbf{Ch}(\mathcal{V})$ in terms of evaluation in $\mathcal{V}$. Thus we consider $A, B \in \mathbf{Ch}(\mathcal{V})$, and denote $\underline{\mathrm{ev}}_{A,B} : \underline{\mathrm{Hom}}(A, B) \odot A \to B$ the evaluation morphism in $\mathbf{Ch}(\mathcal{V})$. This evaluation morphism is defined to be adjunct to the identity morphism on $\underline{\mathrm{Hom}}(A, B)$. Hence, we are able to calculate this morphism explicitly by maps in $\mathcal{V}$ as follows. Consider the projection maps

$$\mathrm{pr}_{s,s+t} : \prod_p [A_p, B_{p+t}] \to [A_s, B_{s+t}], \quad s, t \in \mathbb{Z}.$$

We can calculate $\underline{\text{ev}}_{A,B}$ by applying the adjunction ϕ^{-1} in $\mathcal{V}$ (see Theorem 3.2). We have then that

$$(\underline{\text{ev}}_{A,B})_n = \bigoplus_{s+t=n} \phi^{-1}(\text{pr}_{s,s+t}) : \bigoplus_{s+t=n} \Big(\prod_p [A_p, B_{p+t}]\Big) \otimes A_s \to B_{s+t}$$

Next, consider the following commutative diagram

$$\begin{array}{ccc} \prod_p [A_p, B_{p+t}] & \xrightarrow{\text{pr}_{s,s+t}} & [A_s, B_{s+t}] \\ \downarrow{\scriptstyle \text{pr}_{s,s+t}} & & \downarrow{\scriptstyle [\text{id},\text{id}]} \\ [A_s, B_{s+t}] & \xrightarrow{\text{id}} & [A_s, B_{s+t}] \end{array}$$

and apply the adjunction in $\mathcal{V}$, and deduce the following commutative diagram

$$\begin{array}{ccc} \big(\prod_p [A_p, B_{p+t}]\big) \otimes A_s & \xrightarrow{\phi^{-1}(\text{pr}_{(s,s+t)})} & B_{s+t} \\ \downarrow{\scriptstyle \text{pr}_{s,s+t} \otimes \text{id}} & & \downarrow{\scriptstyle \text{id}} \\ [A_s, B_{s+t}] \otimes A_s & \xrightarrow{\text{ev}_{A_s, B_{s+t}}} & B_{s+t}, \end{array}$$

where $\text{ev}_{A_s, B_{s+n}}$ is the evaluation morphism in $\mathcal{V}$. Thus

$$(\underline{\text{ev}}_{A,B})_n = \bigoplus_{s+t=n} \phi^{-1}(\text{pr}_{(s,s+t)}) = \bigoplus_{s+t=n} \text{ev}_{A_s, B_{s+t}} \circ (\text{pr}_{s,s+t} \otimes \text{id}).$$

We are now in a position to describe the composition morphism

$$\underline{\text{Hom}}(A, B) \odot \underline{\text{Hom}}(B, C) \to \underline{\text{Hom}}(A, C)$$

explicitly. Following Borceux [**Bor2**, Diagram 6.6] this map is defined to be adjoint to the composite

$$\underline{\text{ev}}_{B,C} \circ \sigma \circ (\underline{\text{ev}}_{A,B} \odot 1) \circ (\text{id}_{\underline{\text{Hom}}(A,B)} \odot \sigma) : \underline{\text{Hom}}(A, B) \odot \underline{\text{Hom}}(B, C) \odot A \to C,$$

where σ is the swapping chain isomorphism described in the proof of Theorem 3.2. Furthermore, this composite at degree $n \in \mathbb{Z}$ is determined by a collection of morphisms with $r + p + q = n$,

$$(\underline{\text{ev}}_{B,C})_{r+p+q} \circ (-1)^{(r+p)q} \,\text{sw} \circ \big((\underline{\text{ev}}_{A,B})_{r+p} \otimes \text{id}\big) \circ \big((-1)^{qr} \,\text{sw} \otimes \text{id}\big).$$

Using our description of evaluation we may consider the following diagram, where the rightmost path is any morphism from the collection above.

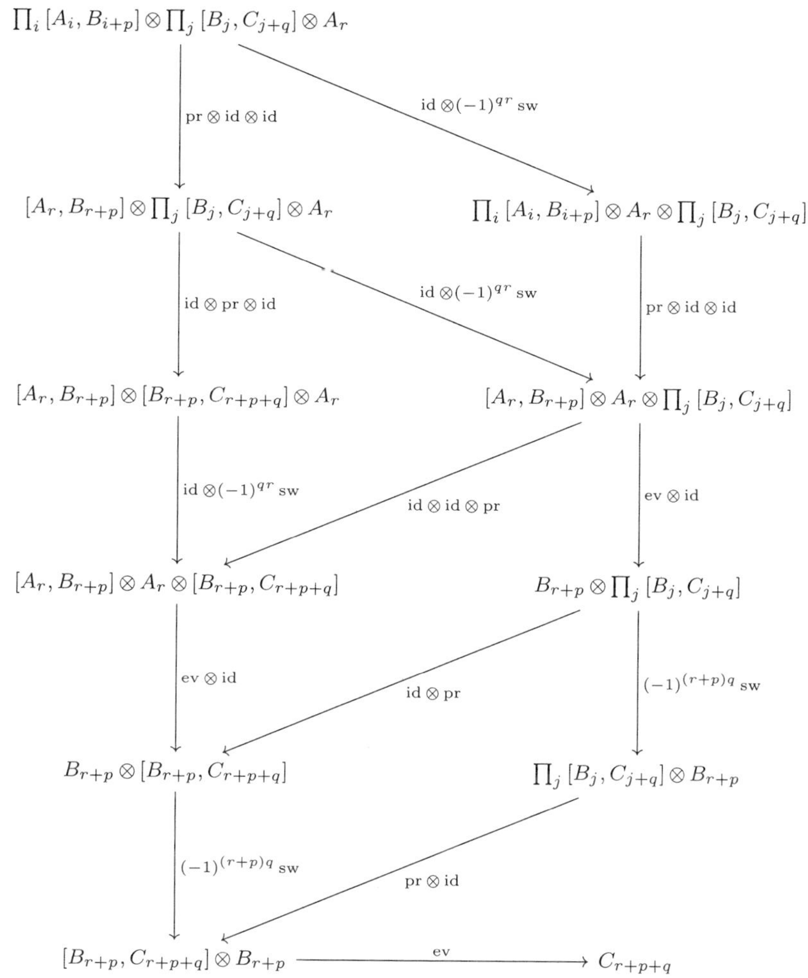

It is clear that this diagram is commutative, and we may take the leftmost path and apply the adjunction in $\mathcal{V}$. Thus we are able to conclude that the composition morphism in $\mathbf{Ch}(\mathcal{V})$ at each degree

$$(c_{\mathbf{Ch}\,\mathcal{V}})_n : \bigoplus_{p+q=n} \left(\prod_i [A_i, B_{i+p}] \otimes \prod_j [B_j, C_{j+q}] \right) \to \prod_r [A_r, C_{r+p+q}]$$

for $n \in \mathbb{Z}$ is determined by morphisms

$$(-1)^{pq} c_{A_r, B_{p+r}, C_{p+q+r}} \circ (\mathrm{pr}_{[A_r, B_{p+r}]} \otimes \mathrm{pr}_{[B_{p+r}, C_{p+q+r}]}). \tag{4.1}$$

Hence, composition in $\mathbf{Ch}(\mathcal{V})$ is the same as first taking projections and then composing in $\mathcal{V}$, up to a sign $(-1)^{pq} = (-1)^{qr}(-1)^{(r+p)q}$.

5. Identifying chain complexes with enriched functors

We shall work with a closed symmetric monoidal Grothendieck category $\mathcal{V}$, and consider a small $\mathcal{V}$-category $\mathcal{C}$. It is evident that $\mathcal{C}$ can be regarded as trivially a

$\mathbf{Ch}(\mathcal{V})$-category, where for each $a, b \in \mathcal{C}$ we define the chain $\mathbf{Ch}(\mathcal{V})_{\mathcal{C}}(a, b)$ as having in zeroth degree the $\mathcal{V}$-object $\mathcal{V}_{\mathcal{C}}(a, b)$ and zero in every other degree.

DEFINITION 5.1. Consider the trivial $\mathbf{Ch}(\mathcal{V})$-enrichment on $\mathcal{C}$ introduced above. We define the enriched functor category $[\mathcal{C}, \mathbf{Ch}(\mathcal{V})]$ as a category with objects $\mathbf{Ch}(\mathcal{V})$-functors $F : \mathcal{C} \to \mathbf{Ch}(\mathcal{V})$ and the morphisms in $[\mathcal{C}, \mathbf{Ch}(\mathcal{V})]$ are defined as $\mathbf{Ch}(\mathcal{V})$-natural transformations.

Note that for any $\mathbf{Ch}(\mathcal{V})$-functor $F : \mathcal{C} \to \mathbf{Ch}(\mathcal{V})$ and $a, b \in \mathcal{C}$, $F_{a,b} : \mathbf{Ch}(\mathcal{V})_{\mathcal{C}}(a, b) \to \underline{\mathrm{Hom}}(F(a), F(b))$ is, by definition, a morphism in $\mathbf{Ch}(\mathcal{V})$ of the form:

$$\begin{array}{ccccccccc}
\cdots & \longrightarrow & 0 & \xrightarrow{0} & \mathcal{V}_{\mathcal{C}}(a,b) & \xrightarrow{0} & 0 & \longrightarrow & \cdots \\
 & & \downarrow & & \downarrow & & \downarrow & & \\
\cdots & \longrightarrow & \underline{\mathrm{Hom}}(F(a),F(b))_1 & \xrightarrow{\partial_1} & \underline{\mathrm{Hom}}(F(a),F(b))_0 & \xrightarrow{\partial_0} & \underline{\mathrm{Hom}}(F(a),F(b))_{-1} & \longrightarrow & \cdots
\end{array}$$

Using the definition of the complex $\underline{\mathrm{Hom}}(F(a), F(b))$ (see the proof of Theorem 3.2), we see that $F_{a,b}$ reduces to the single non-trivial map

$$\mathcal{V}_{\mathcal{C}}(a, b) \to \prod_p [F(a)_p, F(b)_p]$$

in degree 0 with the property that

$$[\mathrm{id}_p^{F(a)}, d_p^{F(b)}] \circ (F_{a,b})_p - [d_p^{F(a)}, \mathrm{id}_{p-1}^{F(b)}] \circ (F_{a,b})_{p-1} = 0 \tag{5.1}$$

for every $p \in \mathbb{Z}$.

LEMMA 5.2. [**Bor2**, 6.2.8] *Given any closed symmetric monoidal category $\mathcal{V}$ and $\mathcal{V}$-functors $X, Y : \mathcal{C} \to \mathcal{V}$, a $\mathcal{V}$-natural transformation $\alpha : X \to Y$ can be defined as a collection of maps $\alpha(a) : X(a) \to Y(a)$ in $\mathcal{V}$ such that*

$$\begin{array}{ccc}
\mathcal{V}_{\mathcal{C}}(a,b) & \xrightarrow{X_{a,b}} & [X(a), X(b)] \\
\downarrow{\scriptstyle Y_{a,b}} & & \downarrow{\scriptstyle [\mathrm{id},\alpha(b)]} \\
[Y(a), Y(b)] & \xrightarrow{[\alpha(a),\mathrm{id}]} & [X(a), Y(b)]
\end{array}$$

commutes for all a, b in $\mathcal{C}$.

DEFINITION 5.3. The category of chain complexes $\mathbf{Ch}[\mathcal{C}, \mathcal{V}]$ over the category of enriched functors $[\mathcal{C}, \mathcal{V}]$ is defined as having objects G, consisting of collections of $\mathcal{V}$-functors $G_n : \mathcal{C} \to \mathcal{V}$ and $\mathcal{V}$-natural transformations $d_n^G : G_n \Rightarrow G_{n-1}$ for $n \in \mathbb{Z}$ with the property that $d^2 = 0$. This category is defined with morphisms $g : G \to G'$ being collections of $\mathcal{V}$-natural transformations $g_n : G_n \Rightarrow G'_n$ that commute with the differentials.

We are now in a position to prove the main result of the section.

THEOREM 5.4. *Let $\mathcal{V}$ be a closed symmetric monoidal Grothendieck category and $\mathcal{C}$ be a small $\mathcal{V}$-category. Then the category $\mathbf{Ch}[\mathcal{C}, \mathcal{V}]$ is naturally isomorphic to the category $[\mathcal{C}, \mathbf{Ch}(\mathcal{V})]$.*

PROOF. We split the proof into several steps.

Step 1. Given any $\mathbf{Ch}(\mathcal{V})$-functor $F \in [\mathcal{C}, \mathbf{Ch}(\mathcal{V})]$ we can associate a chain complex $G \in \mathbf{Ch}[\mathcal{C}, \mathcal{V}]$ to F in the following canonical way.

Firstly, we define the objects that constitute G as a collection of $\mathcal{V}$-functors $G_n : \mathcal{C} \to \mathcal{V}$ such that $G_n(c) := F(c)_n$. Further define the actions on morphisms of these G_n as maps $(G_n)_{a,b} : \mathcal{V}_{\mathcal{C}}(a,b) \to [G_n(a), G_n(b)]$ being equal to the n-th factor of the only non-trivial component of the map $F_{a,b}$, precisely the morphisms $(F_{a,b})_n : \mathcal{V}_{\mathcal{C}}(a,b) \to [F(a)_n, F(b)_n]$, for each $a, b \in \mathcal{C}$ and $n \in \mathbb{Z}$ (see p. 36).

We are able to see that G_n constitute valid $\mathcal{V}$-functors $\mathcal{C} \to \mathcal{V}$ because F is a $\mathbf{Ch}(\mathcal{V})$-functor if and only if

$$\begin{array}{ccc}
\mathbf{Ch}(\mathcal{V})_{\mathcal{C}}(a,b) \odot \mathbf{Ch}(\mathcal{V})_{\mathcal{C}}(b,c) & \xrightarrow{c_{\mathbf{Ch}(\mathcal{V})}} & \mathbf{Ch}(\mathcal{V})_{\mathcal{C}}(a,c) \\
\Big\downarrow {\scriptstyle F_{a,b} \odot F_{b,c}} & & \Big\downarrow {\scriptstyle F_{a,c}} \\
\underline{\mathrm{Hom}}(F(a), F(b)) \odot \underline{\mathrm{Hom}}(F(b), F(c)) & \xrightarrow{c_{\mathbf{Ch}(\mathcal{V})}} & \underline{\mathrm{Hom}}(F(a), F(c))
\end{array}$$

and

$$\begin{array}{ccc}
\varepsilon & \xrightarrow{u_a} & \mathbf{Ch}(\mathcal{V})_{\mathcal{C}}(a,a) \\
& {\scriptstyle u_{F(a)}} \searrow & \Big\downarrow {\scriptstyle F_{a,a}} \\
& & \underline{\mathrm{Hom}}(F(a), F(a))
\end{array}$$

commute in $\mathbf{Ch}(\mathcal{V})$ for all $a, b, c \in \mathcal{C}$. This reduces to the following diagrams

$$\begin{array}{ccc}
\mathcal{V}_{\mathcal{C}}(a,b) \otimes \mathcal{V}_{\mathcal{C}}(b,c) & \xrightarrow{c_{\mathcal{V}}} & \mathcal{V}_{\mathcal{C}}(a,c) \\
\Big\downarrow {\scriptstyle F_{a,b} \otimes F_{b,c}} & & \Big\downarrow {\scriptstyle F_{a,c}} \\
\prod_p [F(a)_p, F(b)_p] \otimes \prod_p [F(b)_p, F(c)_p] & \xrightarrow{c} & \prod_p [F(a)_p, F(c)_p]
\end{array}$$

and

$$\begin{array}{ccc}
e & \xrightarrow{u_a} & \mathcal{V}_{\mathcal{C}}(a,a) \\
& {\scriptstyle u_{F(a)}} \searrow & \Big\downarrow {\scriptstyle F_{a,a}} \\
& & \prod_p [F(a)_p, F(a)_p]
\end{array}$$

in $\mathcal{V}$ for all $a, b, c \in \mathcal{C}$ and every $p \in \mathbb{Z}$. Here c is the map determined by the collection of morphisms

$$c_{F(a)_p, F(b)_p, F(c)_p} \circ (\mathrm{pr}_{[F(a)_p, F(b)_p]} \otimes \mathrm{pr}_{[F(b)_p, F(c)_p]}),$$

with $p \in \mathbb{Z}$, as detailed in the previous section by (4.1). Therefore, we have that commutativity of those diagrams is equivalent to commutativity of the following diagrams in $\mathcal{V}$

$$\begin{array}{ccc}
\mathcal{V}_{\mathcal{C}}(a,b) \otimes \mathcal{V}_{\mathcal{C}}(b,c) & \xrightarrow{c_{\mathcal{V}}} & \mathcal{V}_{\mathcal{C}}(a,c) \\
\Big\downarrow {\scriptstyle (G_p)_{a,b} \otimes (G_p)_{b,c}} & & \Big\downarrow {\scriptstyle (G_p)_{a,c}} \\
[G_p(a), G_p(b)] \otimes [G_p(b), G_p(c)] & \xrightarrow{c_{\mathcal{V}}} & [G_p(a), G_p(c)]
\end{array}$$

and

$$\begin{array}{ccc}
e & \xrightarrow{u_a} & \mathcal{V}_{\mathcal{C}}(a,a) \\
& {\scriptstyle u_{G_p(a)}} \searrow & \Big\downarrow {\scriptstyle (G_p)_{a,a}} \\
& & [G_p(a), G_p(a)]
\end{array}$$

for all $a, b, c \in \mathcal{C}$ and every $p \in \mathbb{Z}$. We see that G_p are $\mathcal{V}$-functors.

commute in $\mathcal{V}$ for all $a, b, c \in \mathcal{C}$ and every $p \in \mathbb{Z}$. It follows that these diagrams commute if and only if

$$\begin{array}{ccc} \mathcal{V}_{\mathcal{C}}(a,b) \otimes \mathcal{V}_{\mathcal{C}}(b,c) & \xrightarrow{c_{\mathcal{V}}} & \mathcal{V}_{\mathcal{C}}(a,c) \\ \downarrow{\scriptstyle (G_p)_{a,b} \otimes (G_p)_{b,c}} & & \downarrow{\scriptstyle (G_p)_{a,c}} \\ [G_p(a), G_p(b)] \otimes [G_p(b), G_p(c)] & \xrightarrow{c_{\mathcal{V}}} & [G_p(a), G_p(c)] \end{array}$$

and

$$\begin{array}{ccc} e & \xrightarrow{u_a} & \mathcal{V}_{\mathcal{C}}(a,a) \\ & {\scriptstyle u_{G_p(a)}} \searrow & \downarrow{\scriptstyle (G_p)_{a,a}} \\ & & [G_p(a), G_p(a)] \end{array}$$

commute in $\mathcal{V}$ for all $a, b, c \in \mathcal{C}$ and every $p \in \mathbb{Z}$, which indeed commute as G_p are $\mathcal{V}$-functors.

Step 4. Now given any chain map in $\mathbf{Ch}[\mathcal{C}, \mathcal{V}]$, we associate a $\mathbf{Ch}(\mathcal{V})$-natural transformation in $[\mathcal{C}, \mathbf{Ch}(\mathcal{V})]$ in the following canonical way.

Consider a chain map $g : G \to G'$ with $G, G' \in \mathbf{Ch}[\mathcal{C}, \mathcal{V}]$. We can associate a $\mathbf{Ch}(\mathcal{V})$-natural transformation $f : F \Rightarrow F'$ with $F, F' \in [\mathcal{C}, \mathbf{Ch}(\mathcal{V})]$ being those functors associated to G and G' respectively. By Lemma 5.2 we can determine g at each component $n \in \mathbb{Z}$ by a family of maps $g_n(a) : G(a) \to G'(a) \in \mathbf{Ch}(\mathcal{V})$ such that for all $a, b \in \mathcal{C}$

$$\begin{array}{ccc} \mathcal{V}_{\mathcal{C}}(a,b) & \xrightarrow{(G_n)_{a,b}} & [G_n(a), G_n(b)] \\ {\scriptstyle (G'_n)_{a,b}}\downarrow & & \downarrow{\scriptstyle [\mathrm{id}, g_n(b)]} \\ [G'_n(a), G'_n(b)] & \xrightarrow{[g_n(a), \mathrm{id}]} & [G_n(a), G'_n(b)] \end{array}$$

is commutative. Thus we set $f(a)_p := g_p(a)$ and deduce that

$$\begin{array}{ccc} \mathcal{V}_{\mathcal{C}}(a,b) & \xrightarrow{F_{a,b}} & \prod_p [F(a)_p, F(b)_p] \\ {\scriptstyle F'_{a,b}}\downarrow & & \downarrow{\scriptstyle \prod_p [\mathrm{id}, f(b)_p]} \\ \prod_p [F'(a)_p, F'(b)_p] & \xrightarrow{\prod_p [f(a)_p, \mathrm{id}]} & \prod_p [F(a)_p, F'(b)_p] \end{array}$$

is commutative. However, in order to say that f is a map in $[\mathcal{C}, \mathbf{Ch}(\mathcal{V})]$, we must verify that all $f(a)$ belong to $\mathbf{Ch}(\mathcal{V})$ to claim that

$$\begin{array}{ccc} \mathbf{Ch}(\mathcal{V})_{\mathcal{C}}(a,b) & \xrightarrow{F_{a,b}} & \underline{\mathrm{Hom}}(F(a), F(b)) \\ {\scriptstyle F'_{a,b}}\downarrow & & \downarrow{\scriptstyle \underline{\mathrm{Hom}}(\mathrm{id}_a, f(b))} \\ \underline{\mathrm{Hom}}(F'(a), F'(b)) & \xrightarrow{\underline{\mathrm{Hom}}(f(a), \mathrm{id}_b)} & \underline{\mathrm{Hom}}(F(a), F'(b)) \end{array}$$

commutes. But this reduces to the fact that $f(a)_{p-1} \circ d_p^{F(a)} = (g_{p-1})(a) \circ d_p^G(a) = d_{p-1}^{G'}(a) \circ (g_p)(a) = d_{p-1}^{F'(a)} \circ f(a)_p$.

Conclusion. We have defined an association between objects and morphisms of the categories $\mathbf{Ch}[\mathcal{C}, \mathcal{V}]$ and $[\mathcal{C}, \mathbf{Ch}(\mathcal{V})]$, and further claim that it is functorial and an isomorphism of categories. Functoriality can be seen from Lemma 5.2 and the fact that composition of natural transformations is determined by the composition on each component. Clearly, this is an isomorphism of categories by the very construction, as required. □

Beke [**Be**, 3.13] and Hovey [**Hov**, 2.2] defined a proper cellular model structure on $\mathbf{Ch}(\mathcal{A})$ for every Grothendieck category $\mathcal{A}$, where cofibrations are the monomorphisms, and weak equivalences the quasi-isomorphisms. We also call it the *injective model structure*. Its fibrant objects are $\mathbf{K}$-injective complexes in the sense of Spaltenstein [**Sp**]. In particular, $\mathbf{Ch}[\mathcal{C},\mathcal{V}]$ has the injective model structure ($g : G \to G'$ in $\mathbf{Ch}[\mathcal{C},\mathcal{V}]$ is a quasi-isomorphism if and only if $g(a) : G(a) \to G'(a)$ is a quasi-isomorphism in $\mathbf{Ch}(\mathcal{V})$ for all $a \in \mathcal{C}$).

However, it is hard to deal with the injective model structure for some particular computations. Instead we want to transfer homotopy information from $\mathbf{Ch}(\mathcal{V})$ to $\mathbf{Ch}[\mathcal{C},\mathcal{V}]$ by using the identification $\mathbf{Ch}[\mathcal{C},\mathcal{V}] \cong [\mathcal{C},\mathbf{Ch}(\mathcal{V})]$ from the preceding theorem.

Suppose $\mathbf{Ch}(\mathcal{V})$ possesses a weakly finitely generated monoidal model structure in the sense of [**DRO1**] in which weak equivalences are the quasi-isomorphisms. Following [**DRO1**, Section 4] a morphism f in $[\mathcal{C},\mathbf{Ch}(\mathcal{V})]$ is a *pointwise fibration* if $f(c)$ is a fibration in $\mathbf{Ch}(\mathcal{V})$ for all $c \in \mathcal{C}$. It is a *cofibration* if it has the left lifting property with respect to all pointwise acyclic fibrations.

We have the following application of the preceding theorem.

THEOREM 5.5. *Let $\mathcal{V}$ be a closed symmetric monoidal Grothendieck category and $\mathcal{C}$ be a $\mathcal{V}$-category. Suppose $\mathbf{Ch}(\mathcal{V})$ is a weakly finitely generated monoidal model structure with respect to the tensor product $\odot$ of Theorem 3.2 and the monoid axiom holds for $\mathbf{Ch}(\mathcal{V})$. Then*

(1) *$\mathbf{Ch}[\mathcal{C},\mathcal{V}]$ with the classes of quasi-isomorphisms, cofibrations and pointwise fibrations defined above is a weakly finitely generated $\mathbf{Ch}(\mathcal{V})$-model category.*
(2) *$\mathbf{Ch}[\mathcal{C},\mathcal{V}]$ is a monoidal $\mathbf{Ch}(\mathcal{V})$-model category provided that $\mathcal{C}$ is a symmetric monoidal $\mathbf{Ch}(\mathcal{V})$-category. In this case the tensor product of $F, G \in \mathbf{Ch}[\mathcal{C},\mathcal{V}]$ is given by*

$$F \odot G = \int^{(a,b)\in\mathcal{C}\otimes\mathcal{C}} F(a) \odot G(b) \odot \mathbf{Ch}(\mathcal{V})_{\mathcal{C}}(a \otimes b, -).$$

Here $\mathbf{Ch}(\mathcal{V})_{\mathcal{C}}(a \otimes b, -)$ is regarded as a complex concentrated in zeroth degree. The internal Hom*-object is defined as*

$$\underline{\mathrm{Hom}}(F,G)(a) = \int_{b\in\mathcal{C}} \underline{\mathrm{Hom}}_{\mathbf{Ch}(\mathcal{V})}(F(b), G(a \otimes b)).$$

(3) *The pointwise model structure on $\mathbf{Ch}[\mathcal{C},\mathcal{V}]$ is right proper if $\mathbf{Ch}(\mathcal{V})$ is right proper, and left proper if $\mathbf{Ch}(\mathcal{V})$ is strongly left proper in the sense of* [**DRO1**, 4.6].

PROOF. In all statements we use Theorem 5.4. The first statement follows from [**DRO1**, 4.2]. The second statement follows from [**DRO1**, 4.4] and Day's Theorem [**Day**] for tensor products and internal Hom-objects. Finally, the third statement follows from [**DRO1**, 4.8]. □

EXAMPLE 5.6. Suppose $\mathcal{V} = \mathrm{Mod}\, R$ with R a commutative ring and $\mathcal{C} = \mathrm{mod}\, R$, the category of finitely presented R-modules. Then $\mathcal{C}$ and $\mathbf{Ch}(\mathrm{Mod}\, R)$ (together

PROOF. First note that $g : R^n \twoheadrightarrow A$ is left determined by I. One way to see this is that $\mathrm{im}(g,-)$ is equivalent to F_φ for some quantifier free formula φ [**Pre09**, 1.2.30] and thus its dual $D\varphi$ is freely realised in a projective module [**Pre09**, 1.2.29] and the dual of a projective is injective.

Now if $F \subseteq (R^n,-)$ and $F(C\oplus I) \subseteq \mathrm{im}(g\circ f, C\oplus I)$ then $FI \subseteq \mathrm{im}(g,I)$. Thus $F \subseteq \mathrm{im}(g,-)$. So F is a subfunctor of $\mathrm{im}(g,-)$. Let $F' \subseteq (A,-)$ be the inverse image of F under $(g,-)$. Now $F(C\oplus I) \subseteq \mathrm{im}(g\circ f, C\oplus I)$ implies $F'(C) \subseteq \mathrm{im}(f,C)$ since $(g,-)$ and hence (g,C) is an embedding. So $F' \subseteq \mathrm{im}(f,-)$. So $F \subseteq \mathrm{im}(g\circ f,-)$. □

We now finish this section by defining two canonical poset embeddings of the lattice of pp-definable subsets of finite length modules over an Artin algebra with the first a meet semi-lattice embedding and the second a join semi-lattice embedding.

If $\varphi \in \mathrm{pp}_R^k$ and $M \in \text{mod-}R$, then we will now write M determines φ to mean that M determines F_φ as a subfunctor of $(R^k,-)$.

LEMMA 2.7. *Let R be an Artin algebra and $M \in \text{mod-}R$. For any $\chi \in \mathrm{pp}_R^k$ there exist unique $\varphi, \psi \in \mathrm{pp}_R^k$ such that $\varphi(M) = \chi(M) = \psi(M)$, φ is freely realised in M^n for some $n \in \mathbb{N}$ and F_ψ is determined by M.*

PROOF. We want to show that there is a $\varphi \in \mathrm{pp}_R^k$ such that $\varphi(M) = \chi(M)$ and φ is freely realised in M^n for some $n \in \mathbb{N}$. We will define a finite sequence, indexed by i, of k-tuples $\overline{m}_i$ of elements from M^i such that $\overline{m}_i \in \chi(M^i)$ and if φ_i generates the pp-type of $\overline{m}_i$ then $\varphi_{i+1}(M) \supsetneq \varphi_i(M)$. Since M is finite endolength this process must terminate with $\overline{m}_n \in M^n$ and by details of the construction below, $\overline{m}_n$ will be such that $\varphi_n(M) = \chi(M)$.

Pick $\overline{m}_1 \in \chi(M)$. If $\chi(M) = \varphi_1(M)$ then we are done. Suppose that we have already defined $\overline{m}_i \in M^i$ and that $\chi(M) \supsetneq \varphi_i(M)$. Pick $\overline{m}' \in \chi(M)\backslash\varphi_i(M)$. Let $\overline{m}_{i+1} = m_i \oplus m'$. We have that $\overline{m}_{i+1} \in \chi(M)$ and that $\varphi_{i+1} \geq \varphi_i$ since $\overline{m}_i \in \varphi_{i+1}(M)$. Hence, since $m' \notin \varphi_i(M)$ and $m' \in \varphi_{i+1}(M)$, $\varphi_{i+1}(M) \supsetneq \varphi_i(M)$.

Now suppose that φ_1 and φ_2 are both freely realised in some power of M and that $\chi(M) = \varphi_1(M) = \varphi_2(M)$. Then M^* determines $D\varphi_1$ and $D\varphi_2$. Moreover $D\varphi_1(M^*) = D\varphi_2(M^*) = D\chi(M^*)$, so $D\varphi_1 = D\varphi_2$. Thus $\varphi_1 = \varphi_2$.

Let ψ be such that $D\psi$ is freely realised in $(M^*)^n$ for some $n \in \mathbb{N}$ and $D\psi(M^*) = D\chi(M^*)$. Then M determines ψ and $\psi(M) = \chi(M)$ as required. □

Let R be an Artin algebra and $M \in \text{mod-}R$. We define three order preserving maps based on the above lemma.

(1) Let $\rho_M : \mathrm{pp}^k(M) \to \mathrm{pp}_R^k$ take $\chi(M)$ to the unique $\varphi \in \mathrm{pp}_R^k$ such that $\varphi(M) = \chi(M)$ and φ is determined by M.
(2) Let $\lambda_M : \mathrm{pp}^k(M) \to \mathrm{pp}_R^k$ take $\chi(M)$ to the unique $\psi \in \mathrm{pp}_R^k$ such that $\psi(M) = \chi(M)$ and ψ is freely realised in M^n for some $n \in \mathbb{N}$.
(3) Let $\mu_M : \mathrm{pp}_R^k \to \mathrm{pp}^k(M)$ be the map taking $\chi \in \mathrm{pp}_R^k$ to $\chi(M)$.

We will use the same notation for the left module versions of these maps. Note that if $M \in \text{mod-}R$ (resp. $M \in R\text{-mod}$), by 1.11, the duality $D : \mathrm{pp}_R^k \to {}_R\mathrm{pp}^k$ (resp. $D : {}_R\mathrm{pp}^k \to \mathrm{pp}_R^k$) induces a duality $D_M : \mathrm{pp}_R^k(M) \to {}_R\mathrm{pp}^k(M^*)$ (resp. $D_M : {}_R\mathrm{pp}^k(M) \to \mathrm{pp}_R^k(M^*)$).

on each component. Clearly, this is an isomorphism of categories by the very construction, as required. □

Beke [**Be**, 3.13] and Hovey [**Hov**, 2.2] defined a proper cellular model structure on $\mathbf{Ch}(\mathcal{A})$ for every Grothendieck category $\mathcal{A}$, where cofibrations are the monomorphisms, and weak equivalences the quasi-isomorphisms. We also call it the *injective model structure*. Its fibrant objects are $\mathbf{K}$-injective complexes in the sense of Spaltenstein [**Sp**]. In particular, $\mathbf{Ch}[\mathcal{C},\mathcal{V}]$ has the injective model structure ($g : G \to G'$ in $\mathbf{Ch}[\mathcal{C},\mathcal{V}]$ is a quasi-isomorphism if and only if $g(a) : G(a) \to G'(a)$ is a quasi-isomorphism in $\mathbf{Ch}(\mathcal{V})$ for all $a \in \mathcal{C}$).

However, it is hard to deal with the injective model structure for some particular computations. Instead we want to transfer homotopy information from $\mathbf{Ch}(\mathcal{V})$ to $\mathbf{Ch}[\mathcal{C},\mathcal{V}]$ by using the identification $\mathbf{Ch}[\mathcal{C},\mathcal{V}] \cong [\mathcal{C},\mathbf{Ch}(\mathcal{V})]$ from the preceding theorem.

Suppose $\mathbf{Ch}(\mathcal{V})$ possesses a weakly finitely generated monoidal model structure in the sense of [**DRO1**] in which weak equivalences are the quasi-isomorphisms. Following [**DRO1**, Section 4] a morphism f in $[\mathcal{C},\mathbf{Ch}(\mathcal{V})]$ is a *pointwise fibration* if $f(c)$ is a fibration in $\mathbf{Ch}(\mathcal{V})$ for all $c \in \mathcal{C}$. It is a *cofibration* if it has the left lifting property with respect to all pointwise acyclic fibrations.

We have the following application of the preceding theorem.

THEOREM 5.5. *Let $\mathcal{V}$ be a closed symmetric monoidal Grothendieck category and $\mathcal{C}$ be a $\mathcal{V}$-category. Suppose $\mathbf{Ch}(\mathcal{V})$ is a weakly finitely generated monoidal model structure with respect to the tensor product $\odot$ of Theorem* 3.2 *and the monoid axiom holds for $\mathbf{Ch}(\mathcal{V})$. Then*

(1) $\mathbf{Ch}[\mathcal{C},\mathcal{V}]$ *with the classes of quasi-isomorphisms, cofibrations and pointwise fibrations defined above is a weakly finitely generated $\mathbf{Ch}(\mathcal{V})$-model category.*
(2) $\mathbf{Ch}[\mathcal{C},\mathcal{V}]$ *is a monoidal $\mathbf{Ch}(\mathcal{V})$-model category provided that $\mathcal{C}$ is a symmetric monoidal $\mathbf{Ch}(\mathcal{V})$-category. In this case the tensor product of $F, G \in \mathbf{Ch}[\mathcal{C},\mathcal{V}]$ is given by*
$$F \odot G = \int^{(a,b)\in\mathcal{C}\otimes\mathcal{C}} F(a) \odot G(b) \odot \mathbf{Ch}(\mathcal{V})_{\mathcal{C}}(a \otimes b, -).$$
Here $\mathbf{Ch}(\mathcal{V})_{\mathcal{C}}(a \otimes b, -)$ is regarded as a complex concentrated in zeroth degree. The internal Hom*-object is defined as*
$$\underline{\mathrm{Hom}}(F,G)(a) = \int_{b\in\mathcal{C}} \underline{\mathrm{Hom}}_{\mathbf{Ch}(\mathcal{V})}(F(b), G(a \otimes b)).$$
(3) *The pointwise model structure on $\mathbf{Ch}[\mathcal{C},\mathcal{V}]$ is right proper if $\mathbf{Ch}(\mathcal{V})$ is right proper, and left proper if $\mathbf{Ch}(\mathcal{V})$ is strongly left proper in the sense of* [**DRO1**, 4.6].

PROOF. In all statements we use Theorem 5.4. The first statement follows from [**DRO1**, 4.2]. The second statement follows from [**DRO1**, 4.4] and Day's Theorem [**Day**] for tensor products and internal Hom-objects. Finally, the third statement follows from [**DRO1**, 4.8]. □

EXAMPLE 5.6. Suppose $\mathcal{V} = \mathrm{Mod}\,R$ with R a commutative ring and $\mathcal{C} = \mathrm{mod}\,R$, the category of finitely presented R-modules. Then $\mathcal{C}$ and $\mathbf{Ch}(\mathrm{Mod}\,R)$ (together with the projective model structure) satisfies the assumptions of Theorem 5.5 and all statements are then true for $\mathbf{Ch}[\operatorname{mod} R, \operatorname{Mod} R]$. Since $[\operatorname{mod} R, \operatorname{Mod} R]$ is isomorphic to the category of generalized modules $\mathcal{C}_R = (\operatorname{mod} R, \mathrm{Ab})$ consisting of the additive functors from $\operatorname{mod} R$ to Abelian groups (see [**AG**, 6.1]), Theorem 5.5 recovers [**AG**, 6.3] stating similar model structures for $\mathbf{Ch}(\mathcal{C}_R)$.

6. Compact generators for the derived category

We consider the following situation when $\mathcal{V}$ is a closed symmetric monoidal Grothendieck category such that its derived category $\mathbf{D}(\mathcal{V})$ is compactly generated triangulated. We show that $\mathbf{D}[\mathcal{C}, \mathcal{V}]$ is also compactly generated in many reasonable cases with $\mathcal{C}$ a small $\mathcal{V}$-category.

EXAMPLE 6.1. (1) Given a commutative ring R, the category of R-modules is a closed symmetric monoidal Grothendieck category. Moreover, the derived category of R-modules $\mathbf{D}(\operatorname{Mod} R)$ is compactly generated triangulated. The compact generators are those complexes which are quasi-isomorphic to a bounded complex of finitely generated projective modules. Such complexes are called *perfect* complexes.

(2) Given a finite group G and a field k, $(\operatorname{Mod} kG, \otimes_k, k)$ is a closed symmetric monoidal Grothendieck category. The derived category $\mathbf{D}(\operatorname{Mod} kG)$ is compactly generated triangulated. Its compact objects are given by the perfect complexes.

(3) The category of Nisnevich sheaves with transfers $NSwT/F$ over a field F is a closed symmetric monoidal Grothendieck category. The derived category $\mathbf{D}(NSwT/F)$ is compactly generated triangulated. Its compact generators are given by complexes $\mathbb{Z}_{tr}(X)[n]$ (the sheaf $\mathbb{Z}_{tr}(X)$ concenrated in the nth degree), where X is an F-smooth algebraic variety (see, e.g., [**GP1**, p. 241]).

THEOREM 6.2. *Let $(\mathcal{V}, \otimes, e)$ be a closed symmetric monoidal Grothendieck category such that the derived category of $\mathcal{V}$ is a compactly generated triangulated category with compact generators $\{P_j\}_{j \in J}$. Further, suppose we have a small $\mathcal{V}$-category $\mathcal{C}$ and that any one of the following conditions is satisfied:*

1. *each P_j is $\mathbf{K}$-projective, in the sense of Spaltenstein [**Sp**];*
2. *for every $\mathbf{K}$-injective $Y \in \mathbf{Ch}[\mathcal{C}, \mathcal{V}]$ and every $c \in \mathcal{C}$, the complex $Y(c) \in \mathbf{Ch}(\mathcal{V})$ is $\mathbf{K}$-injective;*
3. *$\mathbf{Ch}(\mathcal{V})$ has a model structure, with quasi-isomorphisms being weak equivalences, such that for every injective fibrant complex $Y \in \mathbf{Ch}[\mathcal{C}, \mathcal{V}]$ the complex $Y(c)$ is fibrant in $\mathbf{Ch}(\mathcal{V})$.*

Then $\mathbf{D}[\mathcal{C}, \mathcal{V}]$ is a compactly generated triangulated category with compact generators $\{\mathcal{V}_{\mathcal{C}}(c, -) \oslash Q_j \mid c \in \mathcal{C}, j \in J\}$ where, if we assume either (1) or (2), $Q_j = P_j$ or if we assume (3) then $Q_j = P_j^c$ a cofibrant replacement of P_j.

PROOF. We write $(c, -)$ to denote $\mathcal{V}_{\mathcal{C}}(c, -)$. Assuming any of the three conditions, suppose $c \in \mathcal{C}, X \in \mathbf{Ch}[\mathcal{C}, \mathcal{V}]$ and take any Q_j. Clearly, if we denote the injective fibrant replacement of X by X_f (recall that every object is cofibrant in the injective model structure see [**Be**, **Hov**]), then

$$\mathbf{D}[\mathcal{C}, \mathcal{V}]\left((c, -) \oslash Q_j, X\right) \cong \mathbf{D}[\mathcal{C}, \mathcal{V}]\left((c, -) \oslash Q_j, X_f\right) \cong \mathbf{K}[\mathcal{C}, \mathcal{V}]\left((c, -) \oslash Q_j, X_f\right).$$

By the tensor hom adjunction in $\mathcal{V}$ and the Yoneda lemma, we have

$$\mathbf{K}[\mathcal{C}, \mathcal{V}]\left((c, -) \oslash Q_j, X_f\right) \cong \mathbf{K}(\mathcal{V})\left(Q_j, X_f(c)\right).$$

Next by assuming either (1) or (2), and the definition of **K**-projective (**K**-injective respectively) complexes we see

$$\mathbf{K}(\mathcal{V})\left(Q_j, X_f(c)\right) \cong \mathbf{D}(\mathcal{V})\left(Q_j, X_f(c)\right).$$

If, however, we assume (3) then Q_j is cofibrant, $X_f(c)$ is fibrant in $\mathbf{Ch}(\mathcal{V})$ and this natural isomorphism holds also.

Since the arrow $X(c) \to X_f(c)$ is a quasi-isomorphism, then

$$\mathbf{D}(\mathcal{V})\left(Q_j, X_f(c)\right) \cong \mathbf{D}(\mathcal{V})\left(Q_j, X(c)\right).$$

Hence we have established a natural isomorphism

$$\mathbf{D}[\mathcal{C},\mathcal{V}]\left((c,-)\oslash Q_j, X\right) \cong \mathbf{D}(\mathcal{V})\left(Q_j, X(c)\right).$$

With this isomorphism in hand the family $\{(c,-)\oslash Q_j \mid c\in\mathcal{C}, j\in J\}$ is a collection of compact generators can be verified as follows.

First, we verify that $\{(c,-)\oslash Q_j \mid c\in\mathcal{C}, j\in J\}$ is a family of generators for $\mathbf{D}[\mathcal{C},\mathcal{V}]$. Precisely, if $\mathbf{D}[\mathcal{C},\mathcal{V}]((c,-)\oslash Q_j, A) = 0$ for all $j\in J$ and $c\in\mathcal{C}$, then we must show that $A\cong 0$. Assume $\mathbf{D}[\mathcal{C},\mathcal{V}]((c,-)\oslash Q_j, A) = 0$ thus $\mathbf{D}(\mathcal{V})(Q_j, A(c)) = 0$ which implies $A(c)\cong 0$, for all $c\in\mathcal{C}$. We use the fact that $\{Q_j\}_J$ is a family of generators in $\mathbf{D}(\mathcal{V})$. Therefore A is pointwise acyclic and hence is acyclic itself, then $A\cong 0$ in $\mathbf{D}[\mathcal{C},\mathcal{V}]$ as required.

We now verify compactness, precisely we must demonstrate the following natural isomorphism

$$\mathbf{D}[\mathcal{C},\mathcal{V}]((c,-)\oslash Q_j, \bigoplus_i B_i) \cong \bigoplus_i \mathbf{D}[\mathcal{C},\mathcal{V}]((c,-)\oslash Q_j, B_i).$$

We have the following natural isomorphisms

$$\begin{aligned}
\mathbf{D}[\mathcal{C},\mathcal{V}]((c,-)\oslash Q_j, \bigoplus_i B_i) &\cong \mathbf{D}(\mathcal{V})(Q_j, \bigoplus_i B_i(c)) \\
&\cong \bigoplus_i \mathbf{D}(\mathcal{V})(Q_j, B_i(c)) \\
&\cong \bigoplus_i \mathbf{D}[\mathcal{C},\mathcal{V}]((c,-)\oslash Q_j, B_i).
\end{aligned}$$

Here we use the fact that direct sums commute with evaluation and our assumption about the compactness of Q_j. Hence, $\{(c,-)\oslash Q_j \mid c\in\mathcal{C}, j\in J\}$ is indeed a family of compact generators for $\mathbf{D}[\mathcal{C},\mathcal{V}]$. □

REMARK 6.3. Though conditions (1)-(2) of the preceding theorem have nothing to do with model structures, one should stress that condition (1) normally occurs whenever $\mathbf{Ch}(\mathcal{V})$ has a projective model structure with generating (trivial) cofibrations having finitely presented domains and codomains. Condition (2) is typical for the injective model structure on $\mathbf{Ch}(\mathcal{V})$, which always exists by [**Be**, **Hov**], and when $\mathcal{C} = \{*\}$, a singleton with $R := \mathcal{V}_{\mathcal{C}}(*,*)$ a flat ring object of $\mathcal{V}$ (i.e. the functor $R\otimes -$ is exact on $\mathcal{V}$). Finally, condition (3) is most common in practice. It often assumes intermediate model structures on $\mathbf{Ch}(\mathcal{V})$, i.e. model structures which are between the projective and injective model structures. This situation is often recovered from Theorem 5.5.

We conclude the paper with the following observation. Given a closed symmetric monoidal Grothendieck category $\mathcal{V}$ and a small symmetric monoidal $\mathcal{V}$-category $\mathcal{C}$, then $[\mathcal{C},\mathcal{V}]$ is also a closed symmetric monoidal Grothendieck $\mathcal{V}$-category by [**AG**, 4.2]. If $\mathbf{D}(\mathcal{V})$ has $\mathbf{K}$-projective compact generators $\{P_j\}_J$ then the proof of Theorem 6.2 shows that $\mathbf{D}[\mathcal{C},\mathcal{V}]$ has a family of $\mathbf{K}$-projective compact generators given by $\{(c,-)\oslash P_j \mid c\in\mathcal{C}, j\in J\}$. Thus we are able to iterate this process as follows. If we set $\mathcal{V}_1=[\mathcal{C},\mathcal{V}]$ and are given a small symmetric monoidal $\mathcal{V}_1$-category $\mathcal{C}_1$, we can conclude that $\mathbf{D}[\mathcal{C}_1,\mathcal{V}_1]$ is also compactly generated having $\mathbf{K}$-projective compact generators. We can then set $\mathcal{V}_2=[\mathcal{C}_1,\mathcal{V}_1]$ and repeat this procedure as many times as necessary to generate as many examples as we desire.

For instance, starting with $\mathcal{V}=\operatorname{Mod} R$, where R is commutative and $\mathcal{C}=\operatorname{mod} R$ (see Example 5.6), set $\mathcal{V}_1:=[\operatorname{mod} R,\operatorname{Mod} R]\cong\mathcal{C}_R$ and $\mathcal{C}_1:=\operatorname{fp}(\mathcal{V}_1)$, where $\operatorname{fp}(\mathcal{V}_1)$ consists of finitely presented objects of $\mathcal{V}_1$. Then $\mathcal{V}_2=[\operatorname{fp}(\mathcal{V}_1),\mathcal{V}_1]$ is a closed symmetric monoidal locally finitely presented Grothendieck category. Its finitely presented generators are given by $\mathcal{V}_1(a,-)\oslash c$, where $a,c\in\mathcal{C}_1$. We use here natural isomorphisms

$$\operatorname{Hom}_{\mathcal{V}_2}(\mathcal{V}_1(a,-)\oslash c,\varinjlim_I X_i)\cong\operatorname{Hom}_{\mathcal{V}_1}(c,\mathcal{V}_1(\mathcal{V}_1(a,-),\varinjlim_I X_i))\cong$$
$$\operatorname{Hom}_{\mathcal{V}_1}(c,\varinjlim_I X_i(a))\cong\varinjlim_I\operatorname{Hom}_{\mathcal{V}_1}(c,X_i(a))\cong\varinjlim_I\operatorname{Hom}_{\mathcal{V}_2}(\mathcal{V}_1(a,-)\oslash c,X_i)$$

and the fact that $\mathcal{C}_1$ is closed under tensor product in $\mathcal{V}_1$. Moreover, $\mathbf{D}(\mathcal{V}_2)=\mathbf{D}[\mathcal{C}_1,\mathcal{V}_1]$ is also compactly generated having $\mathbf{K}$-projective compact generators. Iterating this, we can define a closed symmetric monoidal locally finitely presented Grothendieck category $\mathcal{V}_n=[\operatorname{fp}(\mathcal{V}_{n-1}),\mathcal{V}_{n-1}]$ for all $n>1$. And then $\mathbf{D}(\mathcal{V}_n)=\mathbf{D}[\operatorname{fp}(\mathcal{V}_{n-1}),\mathcal{V}_{n-1}]$ is compactly generated having $\mathbf{K}$-projective compact generators.

References

[AG] H. Al Hwaeer and G. Garkusha, *Grothendieck categories of enriched functors*, J. Algebra **450** (2016), 204–241, DOI 10.1016/j.jalgebra.2015.09.052. MR3449691

[Be] T. Beke, *Sheafifiable homotopy model categories*, Math. Proc. Cambridge Philos. Soc. **129** (2000), no. 3, 447–475, DOI 10.1017/S0305004100004722. MR1780498

[BK] A. I. Bondal and M. M. Kapranov, *Framed triangulated categories* (Russian), Mat. Sb. **181** (1990), no. 5, 669–683, DOI 10.1070/SM1991v070n01ABEH001253; English transl., Math. USSR-Sb. **70** (1991), no. 1, 93–107. MR1055981

[Bor2] F. Borceux, *Handbook of categorical algebra 2*, Encyclopedia of Mathematics and its Applications, vol. 51, Cambridge University Press, Cambridge, 1994. Categories and structures. MR1313497

[Day] B. Day, *On closed categories of functors*, Reports of the Midwest Category Seminar, IV, Lecture Notes in Mathematics, Vol. 137, Springer, Berlin, 1970, pp. 1–38. MR0272852

[DRO1] B. I. Dundas, O. Röndigs, and P. A. Østvær, *Enriched functors and stable homotopy theory*, Doc. Math. **8** (2003), 409–488. MR2029170

[DRO2] B. I. Dundas, O. Röndigs, and P. A. Østvær, *Motivic functors*, Doc. Math. **8** (2003), 489–525. MR2029171

[GP1] G. Garkusha and I. Panin, *K-motives of algebraic varieties*, Homology Homotopy Appl. **14** (2012), no. 2, 211–264, DOI 10.4310/HHA.2012.v14.n2.a13. MR3007094

[GP2] G. Garkusha and I. Panin, *The triangulated category of K-motives* $DK_-^{\text{eff}}(k)$, J. K-Theory **14** (2014), no. 1, 103–137, DOI 10.1017/is014004013jkt263. MR3238259

[GP3] G. Garkusha and I. Panin, *On the motivic spectral sequence*, J. Inst. Math. Jussieu **17** (2018), no. 1, 137–170, DOI 10.1017/S1474748015000419. MR3742558

[Hov] M. Hovey, *Model category structures on chain complexes of sheaves*, Trans. Amer. Math. Soc. **353** (2001), no. 6, 2441–2457, DOI 10.1090/S0002-9947-01-02721-0. MR1814077

[HPS] M. Hovey, J. H. Palmieri, and N. P. Strickland, *Axiomatic stable homotopy theory*, Mem. Amer. Math. Soc. **128** (1997), no. 610, x+114, DOI 10.1090/memo/0610. MR1388895

[Il] L. Illusie, *Existence de ré solutions globales*, Sem. Geom. algebrique Bois Marie 1966/67, SGA VI, Lecture Notes in Mathematics 225, 1971, pp. 160-221.

[Kel] B. Keller, *On differential graded categories*, International Congress of Mathematicians. Vol. II, Eur. Math. Soc., Zürich, 2006, pp. 151–190. MR2275593

[R] E. Riehl, *Categorical homotopy theory*, New Mathematical Monographs, vol. 24, Cambridge University Press, Cambridge, 2014. MR3221774

[Sp] N. Spaltenstein, *Resolutions of unbounded complexes*, Compositio Math. **65** (1988), no. 2, 121–154. MR932640

[SV] A. Suslin and V. Voevodsky, *Bloch-Kato conjecture and motivic cohomology with finite coefficients*, The arithmetic and geometry of algebraic cycles (Banff, AB, 1998), NATO Sci. Ser. C Math. Phys. Sci., vol. 548, Kluwer Acad. Publ., Dordrecht, 2000, pp. 117–189. MR1744945

Department of Mathematics, Swansea University, Singleton Park, Swansea SA2 8PP, United Kingdom

Email address: g.garkusha@swansea.ac.uk

Department of Mathematics, Swansea University, Singleton Park, Swansea SA2 8PP, United Kingdom

Email address: darrenalexanderjones@gmail.com

Contemporary Mathematics
Volume **730**, 2019
https://doi.org/10.1090/conm/730/14709

Left determined morphisms and free realisations

Lorna Gregory

This paper is dedicated to Mike Prest on the occasion of his 65th birthday.

ABSTRACT. We investigate the connection between Prest's notion of the free realisation of a pp formula and Auslander's notion of determiners of functor and morphisms.

The aim of this note is to explain the connections between Auslander's notion of morphisms and subfunctors determined by objects introduced in [**Aus78**] and Prest's notion of free realisations of pp formulae introduced in [**Pre88**].

The concept of determiners of morphisms and subfunctors were largely ignored until recently. In the last 5-10 years, effort has been made to understand them (see for instance [**Rin13**], [**Rin12**], [**Kra13**]). On the other hand, the algebraic study of model theory of modules is unimaginable without the concept of free realisations of a pp formulae.

In 2.4 we explicitly describe the connection between determiners of functors defined by pp formulae and free realisations of pp formulae. This will give another proof, 2.5, of the existence of left determiners of morphisms between finitely presented modules for artin algebras. We then use determiners and free realisations to show that if $M \in \text{mod-}R$ and R is an artin algebra, then the lattice homomorphism $\text{pp}_R^k \to \text{pp}_R^k(M)$ which sends $\varphi \in \text{pp}_R^k$ to $\varphi(M) \in \text{pp}_R^k(M)$ has both a left and a right adjoint, both of which we explicitly describe.

Finally, in section 3, we will show that pushing the ideas from section 2 slightly harder actually gives a proof of the existence of minimal left determiners of morphisms between finitely presented modules for artin algebras.

1. Basic concepts

The material in this section about morphisms and functors determined by objects is from the introduction of [**Aus78**]. The material about pp formulae and free realisations can be found in [**Pre88**] and [**Pre09**].

If R is a ring then we write Mod-R (resp. R-Mod) for the category of right (resp. left) R-modules and mod-R (resp. R-mod) for the category of finitely presented

2010 *Mathematics Subject Classification.* Primary 03C60; Secondary 16G10.

The content of the paper was created while the author was a postdoc at the University of Manchester and prepared for publication while the author was a postdoc at the University of Camerino. The author acknowledges the support of EPSRC through Grant EP/K022490/1.

right (resp. left) R-modules. If $\mathcal{C}$ is an additive category then $(\mathcal{C}, \mathrm{Ab})$ will denote the category with objects the additive functors and morphisms natural transformations.

DEFINITION 1.1. Let $\mathcal{C}$ be an additive category and $G \in (\mathcal{C}, \mathrm{Ab})$ (resp. $G \in (\mathcal{C}^{\mathrm{op}}, \mathrm{Ab})$). We say that a subfunctor $H \subseteq G$ is **determined** by $C \in \mathcal{C}$ if for all subfunctors $H' \subseteq G$, $H' \subseteq H$ if and only if $H'C \subseteq HC$. We will call C a **determiner** for H.

We will say that C is a **minimal determiner** for H as a subfunctor of G, if C is a direct summand of all other determiners for H.

DEFINITION 1.2. Let $f : Y \to X$ be a morphism in an additive category $\mathcal{C}$.

(1) We say that f is **right determined** by $C \in \mathcal{C}$ provided that, for all $\beta : Z \to X$, if for all $\gamma : C \to Z$, $\beta\gamma$ factors through f then β factors through f.
(2) We say that f is **left determined** by $C \in \mathcal{C}$ provided that, for all $\beta : Y \to Z$, if for all $\gamma : Z \to C$, $\gamma\beta$ factors through f then β factors through f.

We will refer to C as a right (resp. left) **determiner** for f and say that C is a **minimal right (resp. left) determiner** for f if C is a direct summand of all other right (resp. left) determiner for f.

LEMMA 1.3. *Let $\mathcal{C}$ be an additive category.*

(1) *A morphism $f : X \to Y$ is right determined by C if and only if* $\mathrm{im}(-, f)$ *is determined by C as a subfunctor of $(-, Y)$.*
(2) *A morphism $f : Y \to X$ is left determined by C if and only if* $\mathrm{im}(f, -)$ *is determined by C as a subfunctor of $(Y, -)$.*

REMARK 1.4. A morphism $f : X \to Y \in \mathcal{C}$ is right determined by $C \in \mathcal{C}$ if and only if $f^{\mathrm{op}} : Y \to X \in \mathcal{C}^{\mathrm{op}}$ is left determined by $C \in \mathcal{C}^{\mathrm{op}}$.

Let R be an Artin algebra over a commutative Artinian ring S. The injective envelope I of $S/\mathrm{rad}(S)$ is a finitely presented injective cogenerator for Mod-S. If $M \in$ mod-R (resp. $M \in R$-mod) then $M^* := \mathrm{Hom}_S(M, I)$ is finitely presented as a left (resp. right) R-module and if $f : M \to N \in$ mod-R (resp. $f : M \to N \in R$-mod) then write $f^* := \mathrm{Hom}_S(f, I)$. The contravariant functor $\mathrm{Hom}_S(-, I)$: mod-$R \to R$-mod gives an equivalence of categories $(\text{mod-}R)^{op} \simeq R\text{-mod}$.

Specialising to the case where $\mathcal{C} :=$ mod-R (resp. $\mathcal{C} := R$-mod) and R is an Artin algebra, 1.4 shows that for $A, B \in$ mod-R, $f : A \to B$ is right determined by X if and only $f^* : B^* \to A^*$ is left determined by X^*.

Auslander showed in [**Aus78**], if R is an Artin algebra, then all morphisms $f : M \to N \in$ mod-R have minimal right and left determiners.

DEFINITION 1.5. Let R be a ring. A (right) **pp-n-formula** is a formula in the language $\mathcal{L}_R = (0, +, (\cdot r)_{r \in R})$ of (right) R-modules of the form

$$\exists \overline{y} (\overline{x}, \overline{y}) H = 0$$

where $\overline{x}$ is a n-tuple of variables and H is an appropriately sized matrix with entries in R.

If φ is a pp-n-formula and M is a module then we will write $\varphi(M)$ for the solution set of φ in M^n. Solutions sets of pp formulae have two important (and easily checked) properties. Firstly, the solution set of a pp-n-formula φ in a module

M is a subgroup of M^n under pointwise addition. Secondly, if $f : M \to N$ is a homomorphism of R-modules then the image of $\varphi(M)$ under f is contained in $\varphi(N)$. In this way, pp formulae give rise to functors in $(\text{mod-}R, \text{Ab})$ i.e. if φ is a pp formula then we define a functor $F_\varphi \in (\text{mod-}R, \text{Ab})$ which acts on objects by sending $M \in \text{mod-}R$ to $\varphi(M)$ and morphisms in the obvious way. If φ is a pp-n-formula, then F_φ is a subfunctor of the nth power of the forgetful functor $(R^n, -)$.

These properties also imply that if $M \in \text{Mod-}R$ and φ is a pp-n-formula then $\varphi(M)$ is closed under the diagonal action of $\text{End}(M)$ on M^n.

Note also that solution sets of pp formulae commute with direct sums. That is, if $\varphi \in \text{pp}_R^n$ and $L, M \in \text{Mod-}R$ then $\varphi(L \oplus M) = \varphi(L) \oplus \varphi(M)$.

After identifying pp-n-formulae φ, ψ such that $\varphi(M) = \psi(M)$ for all $M \in \text{Mod-}R$, the set of (equivalence classes of) pp-n-formulae becomes a lattice by setting $\varphi \leq \psi$ if and only if $\varphi(M) \subseteq \psi(M)$ for all $M \in \text{Mod-}R$. We will write pp_R^n for the lattice of right pp-n-formulae and ${}_R\text{pp}^n$ for the lattice of left pp-n-formulae. If $M \in \text{Mod-}R$ (resp. $M \in R\text{-Mod}$) then we will write $\text{pp}_R^n(M)$ (resp. ${}_R\text{pp}^n(M)$) for the lattice of all subgroups of M^n defined by right (resp. left) pp-n-formulae. This is just the quotient of pp_R^n (resp. ${}_R\text{pp}^n$) by the equivalence relation $\varphi \sim \psi$ if $\varphi(M) = \psi(M)$.

If R is an Artin algebra over a commutative Artinian ring S and M is a finitely presented R-module then all left $\text{End}(M)$-submodules of M^n are also S-submodules. Since M is a finite length as S-module, M^n is finite length as a left $\text{End}(M)$-module and hence $\text{pp}_R^n(M)$ is finite length.

If $\overline{m}$ is an n-tuple of elements from a module M then the **pp-type** of $\overline{m}$ is simply the set of pp-n-formulae φ such that $\overline{m} \in \varphi(M)$. If $M \in \text{mod-}R$ and $\overline{m}$ is an n-tuple of element from M then, [**Pre09**, 1.2.6], there exists $\varphi \in \text{pp}_R^n$ such that ψ is in the pp-type of $\overline{m}$ if and only if $\psi \geq \varphi$.

DEFINITION 1.6. Let $\varphi \in \text{pp}_R^n$. A **free realisation** of φ is a pair $(C, \overline{c})$ where $C \in \text{mod-}R$ and $\overline{c}$ is an n-tuple of elements from C with the property that the pp-type of $\overline{c}$ in C is generated by φ i.e. $\psi(\overline{c})$ holds in C if and only if $\varphi \leq \psi$.

PROPOSITION 1.7. [**Pre09**, 1.2.14,1.2.7] *Every pp formula φ has a free realisation. Moreover, if $(C, \overline{c})$ is a free realisation for φ and $\overline{m} \in \varphi(M)$ for some module M and tuple $\overline{m}$ of elements from M then there is a homomorphism $f : C \to M$ such that $f(\overline{c}) = \overline{m}$.*

If $\overline{c} \in C^n$ then we write $\overline{c} : R^n \to C$ for the map given by $\overline{r} \mapsto \overline{c} \cdot \overline{r}$. Equivalently, see [**Pre09**, 10.2.8], a free realisation is a pair $(C, \overline{c})$ where $C \in \text{mod-}R$ and $\overline{c}$ is an n-tuple of elements from C with the property that $\text{im}(\overline{c}, -) = F_\varphi$. This in fact implies that F_φ is finitely presented and that all finitely presented subfunctors of $(R^n, -)$ are of the form F_φ for some $\varphi \in \text{pp}_R^n$.

A free realisation $(C, \overline{c})$ of a pp formula φ is **minimal** if whenever $f \in \text{End}(C)$ and the pp-type of $\overline{c}$ and $f(\overline{c})$ are equal, f is an automorphism. Equivalently, $(\overline{c}, -) : (C, -) \to F_\varphi$ is a projective cover. Over an Artin algebra, or in fact over any Krull-Schmidt ring, all pp formulae have minimal free realisations [**Pre09**, 4.3.70].

LEMMA 1.8. [**Pre09**, 10.2.26] *Let $(C, \overline{c})$ be a minimal free realisation for a pp formula φ. If $(A, \overline{a})$ is a free realisation for φ then there is a split monomorphism such $g : C \to A$ such that $g(\overline{c}) = \overline{a}$.*

For each $n \in \mathbb{N}$, Prest defined a lattice anti-isomorphism $D : \mathrm{pp}_R^n \to {}_R\mathrm{pp}^n$ (see [**Pre09**, section 1.3.1] and [**Pre88**, 8.21]). As is standard, we denote its inverse ${}_R\mathrm{pp}^n \to \mathrm{pp}_R^n$ also by D. Since we will not need to explicitly take the dual of a pp formula here, we will not give its definition.

THEOREM 1.9. [**Her93**, 3.2][**Pre09**, 1.3.7] *Let $\overline{a}$ and $\overline{b}$ be n-tuples from $M \in$ Mod-R and $N \in R$-Mod respectively. Then $\overline{a} \otimes \overline{b} = 0$ if and only if there exists $\varphi \in \mathrm{pp}_R^n$ such that $\overline{a} \in \varphi(M)$ and $\overline{b} \in D\varphi(N)$.*

COROLLARY 1.10. *If $(C, \overline{c})$ is a free realisation of $\varphi \in \mathrm{pp}_R^n$ and $\overline{l}$ is an n-tuple for $L \in R$-Mod then $\overline{c} \otimes \overline{l} = 0$ if and only if $\overline{l} \in D\varphi(L)$.*

LEMMA 1.11. [**Pre09**, 1.3.13] *Let R be an Artin algebra, $M \in$ Mod-R and $\varphi, \psi \in \mathrm{pp}_R^n$. If $\psi(M) \subseteq \varphi(M)$ then $D\varphi(M^*) \subseteq D\psi(M^*)$.*

2. The relationship between free realisations and determiners

In this section we give a correspondence between determiners of pp formulae as subfunctors of $(R^n, -)$ and modules which free realise their dual. We then use free realisations and determiners to define a canonical meet-semi-lattice embedding $\rho_M : \mathrm{pp}_R^k(M) \to \mathrm{pp}_R^k$ and a canonical join-semi-lattice embedding $\lambda_M : \mathrm{pp}_R^k(M) \to \mathrm{pp}_R^k$ whenever R is an Artin algebra and $M \in$ mod-R.

Suppose that $\varphi \in \mathrm{pp}_R^k$ and $(M, \overline{m})$ is a free realisation for φ. Then $\varphi(M) \subseteq \psi(M)$ implies that $\overline{m} \in \psi(M)$. Hence $\varphi \leq \psi$. So $\varphi \leq \psi$ if and only if $\varphi(M) \subseteq \psi(M)$. The following lemma gives a partial converse to this.

LEMMA 2.1. *Let $M \in$ mod-R be such that $\mathrm{pp}_R^1(M)$ has the ascending chain condition. If $\varphi \in \mathrm{pp}_R^k$ is such that for all $\psi \in \mathrm{pp}_R^k$, $\psi(M) \supseteq \varphi(M)$ implies $\psi \geq \varphi$, then there exist $n \in \mathbb{N}$ and $\overline{m} \in M^n$ such that φ is freely realised by $(M^n, \overline{m})$.*

PROOF. Suppose that φ and M are as in the statement of this lemma. We will define a finite sequence, indexed by i, of k-tuples $\overline{m}_i$ of elements from M^i such that $\overline{m}_i \in \varphi(M^i)$ and if χ_i generates the pp-type of $\overline{m}_i$ then $\chi_{i+1}(M) \supsetneq \chi_i(M)$. Now, since $\mathrm{pp}_R^1(M)$ has the ascending chain condition, this process will terminate with $\overline{m}_n$ a k-tuple of elements from M^n and by the details of the construction below, $\overline{m}_n$ will be such that $\varphi(M) = \chi_n(M)$. So, by hypothesis and since $\varphi(M) = \chi_n(M)$, $\chi_n \geq \varphi$. Since $\overline{m}_n \in \varphi(M^n)$ and χ_n generates the pp-type of $\overline{m}_n$, we have that $\varphi \geq \chi_n$. Thus $\varphi = \chi_n$ and $(M^n, \overline{m}_n)$ freely realises φ.

Pick $\overline{m}_1 \in \varphi(M)$. If $\chi_1(M) = \varphi(M)$ then we are done. Supposing that we have already defined $\overline{m}_i \in M^i$ and $\chi_i(M) \subsetneq \varphi(M)$, pick $\overline{m}' \in \varphi(M) \backslash \chi_i(M)$. Let $\overline{m}_{i+1} = \overline{m}_i \oplus \overline{m}' \in M^{i+1}$. Since the solution sets of pp-formulae commute with direct sums, for all $\psi \in \mathrm{pp}_R^k$, $\overline{m}_{i+1} \in \psi(M^{i+1})$ if and only if $\overline{m}_i \in \psi(M^i)$ and $\overline{m}' \in \psi(M)$. Thus for all $\psi \in \mathrm{pp}_R^k$, $\overline{m}_{i+1} \in \psi(M^{i+1})$ implies $\psi \geq \chi_i$. Thus $\chi_{i+1} \geq \chi_i$. Since $\overline{m}' \in \chi_{i+1}(M)$ and $\overline{m}' \notin \chi_i(M)$, we have that $\chi_{i+1}(M) \supsetneq \chi_i(M)$ as required. □

If $M \in$ Mod-R is length e as a left End(M)-module then M^k is length ek as a left End(M)-module. Since for all $\varphi \in \mathrm{pp}_R^k$, $\varphi(M)$ is a left End(M)-submodule of M^k, $\mathrm{pp}_R^k(M)$ is at most length ek. Thus we have the following corollary to the proof of 2.1.

COROLLARY 2.2. *Let $M \in$ mod-R be of length e as a left End(M)-module. If $\varphi \in \mathrm{pp}_R^k$ is such that for all $\psi \in \mathrm{pp}_R^k$, $\psi(M) \supseteq \varphi(M)$ implies $\psi \geq \varphi$, then there exists $\overline{m} \in M^{ek}$ such that φ is freely realised by $(M^{ek}, \overline{m})$.*

COROLLARY 2.3. *Let R be an Artin algebra. A pp-k-formula φ over R is freely realised in M^n for some $n \in \mathbb{N}$ if and only if φ is freely realised in M^{ek} where e is the length of M as a left* $\mathrm{End}(M)$*-module.*

THEOREM 2.4. *Let R be an Artin algebra, φ a pp-k-formula and $M \in$* mod-R. *The following are equivalent:*

(1) *M^* determines $F_{D\varphi}$ as a subfunctor of $(R^k, -)$*
(2) *there is some $n \in \mathbb{N}$ and k-tuple $\overline{m}$ of elements from M^n such that $(M^n, \overline{m})$ freely realises φ.*

PROOF. (2)$\Rightarrow$ (1): Suppose that $\varphi \in \mathrm{pp}_R^k$ and $(M^n, \overline{m})$ is a free realisation of φ. If $\psi \in \mathrm{pp}_R^k$ then, since $(M^n, \overline{m})$ is a free realisation of φ, $\varphi \leq \psi$ if and only if $\overline{m} \in \psi(M^n)$. So $\varphi(M^n) \subseteq \psi(M^n)$ implies $\overline{m} \in \psi(M^n)$ and hence $\varphi \leq \psi$. Thus $\varphi \leq \psi$ if and only if $\varphi(M^n) \subseteq \psi(M^n)$ if and only if $\varphi(M) \subseteq \psi(M)$.

Taking duals, this gives that for all $\psi \in \mathrm{pp}_R^n$, $D\psi \leq D\varphi$ if and only if $D\psi(M^*) \subseteq D\varphi(M^*)$. Since, see [**Pre09**, 12.2.1], all subfunctors of $(R^n, -) \in (R\text{-mod}, \mathrm{Ab})$ are direct unions of functors of the form $F_{D\psi}$, this implies that for all $G \subseteq (R^n, -)$, $G \subseteq F_{D\varphi}$ if and only if $GM^* \subseteq F_{D\varphi}M^*$. That is $F_{D\varphi}$ is determined, as a subfunctor of $(R^n, -)$, by M^*.

(1)$\Rightarrow$ (2) Since M^* is a determiner for $F_{D\varphi}$, we have that for all $\psi \in \mathrm{pp}_R^k$, $D\psi(M^*) \subseteq D\varphi(M^*)$ if and only if $D\psi \leq D\varphi$. So $\psi(M) \supseteq \varphi(M)$ if and only if $\psi \geq \varphi$. Since M is a finitely presented module over an Artin algebra, it is of finite length as a left $\mathrm{End}(M)$-module. Thus, 2.1 implies there exists an $n \in \mathbb{N}$ and k-tuple $\overline{m}$ of elements from M^n such that $(M^n, \overline{m})$ freely realises φ. □

The following corollary explains how the existence of free realisations of pp formulae implies the existence of left determiners for morphisms between finitely presented modules over an Artin algebra.

COROLLARY 2.5. *Let R be an Artin algebra, $f : A \to B \in$* mod-R *and $\overline{a}$ a generating tuple for A. Let φ generate the pp-type of $f(\overline{a})$. If some tuple of elements in $C \in R$*-mod *freely realises $D\varphi$ then C^* is a left determiner for f.*

PROOF. Suppose that $\overline{a}$ is of length k and let $g : R^k \to A$ be the map defined by $g(\overline{r}) := \overline{a} \cdot \overline{r}$. Since $F_\varphi = \mathrm{im}(f \circ g, -)$, C^* is a determiner for $\mathrm{im}(f \circ g, -)$ as a subfunctor of $(R^n, -)$. Since g is an epimorphism, $(g, -) : (A, -) \to (R^n, -)$ is a monomorphism. Thus C^* is also a determiner for $\mathrm{im}(f, -)$ as a subfunctor of $(A, -)$. □

We now discuss the relationship between minimal free realisations of $\varphi \in \mathrm{pp}_R^k$ and minimal determiners of $F_{D\varphi}$. Suppose that $(C, \overline{c})$ is a minimal free realisation for φ and D is a minimal determiner of $F_{D\varphi}$ as a subfunctor of $(R^k, -)$. Then, by 1.8 and 2.4, C is a direct summand of $(D^*)^n$ for some $n \in \mathbb{N}$ and D is a direct summand of C^*. Thus C^* and D have the same indecomposable direct summands. Thus if $(C, \overline{c})$ is a minimal free realisation for φ and $C = C_1^{l_1} \oplus \ldots \oplus C_m^{l_m}$, where $C_i \cong C_j$ implies $i = j$, then $(C_1 \oplus \ldots \oplus C_m)^*$ is a minimal determiner for $F_{D\varphi}$.

The following lemma indicates how far 2.5 combined with minimal free realisations is from giving us minimal left determiners.

LEMMA 2.6. *Let R be an Artin algebra. Let $f : A \to B \in$* mod-R *and $g : R^n \twoheadrightarrow A$. If $C \in$* mod-R *left determines f then $C \oplus I$ left determines $g \circ f$ where I is an injective cogenerator for* mod-R.

PROOF. First note that $g : R^n \twoheadrightarrow A$ is left determined by I. One way to see this is that $\mathrm{im}(g, -)$ is equivalent to F_φ for some quantifier free formula φ [**Pre09**, 1.2.30] and thus its dual $D\varphi$ is freely realised in a projective module [**Pre09**, 1.2.29] and the dual of a projective is injective.

Now if $F \subseteq (R^n, -)$ and $F(C \oplus I) \subseteq \mathrm{im}(g \circ f, C \oplus I)$ then $FI \subseteq \mathrm{im}(g, I)$. Thus $F \subseteq \mathrm{im}(g, -)$. So F is a subfunctor of $\mathrm{im}(g, -)$. Let $F' \subseteq (A, -)$ be the inverse image of F under $(g, -)$. Now $F(C \oplus I) \subseteq \mathrm{im}(g \circ f, C \oplus I)$ implies $F'(C) \subseteq \mathrm{im}(f, C)$ since $(g, -)$ and hence (g, C) is an embedding. So $F' \subseteq \mathrm{im}(f, -)$. So $F \subseteq \mathrm{im}(g \circ f, -)$. □

We now finish this section by defining two canonical poset embeddings of the lattice of pp-definable subsets of finite length modules over an Artin algebra with the first a meet semi-lattice embedding and the second a join semi-lattice embedding.

If $\varphi \in \mathrm{pp}_R^k$ and $M \in \text{mod-}R$, then we will now write M determines φ to mean that M determines F_φ as a subfunctor of $(R^k, -)$.

LEMMA 2.7. *Let R be an Artin algebra and $M \in \text{mod-}R$. For any $\chi \in \mathrm{pp}_R^k$ there exist unique $\varphi, \psi \in \mathrm{pp}_R^k$ such that $\varphi(M) = \chi(M) = \psi(M)$, φ is freely realised in M^n for some $n \in \mathbb{N}$ and F_ψ is determined by M.*

PROOF. We want to show that there is a $\varphi \in \mathrm{pp}_R^k$ such that $\varphi(M) = \chi(M)$ and φ is freely realised in M^n for some $n \in \mathbb{N}$. We will define a finite sequence, indexed by i, of k-tuples $\overline{m}_i$ of elements from M^i such that $\overline{m}_i \in \chi(M^i)$ and if φ_i generates the pp-type of $\overline{m}_i$ then $\varphi_{i+1}(M) \supsetneq \varphi_i(M)$. Since M is finite endolength this process must terminate with $\overline{m}_n \in M^n$ and by details of the construction below, $\overline{m}_n$ will be such that $\varphi_n(M) = \chi(M)$.

Pick $\overline{m}_1 \in \chi(M)$. If $\chi(M) = \varphi_1(M)$ then we are done. Suppose that we have already defined $\overline{m}_i \in M^i$ and that $\chi(M) \supsetneq \varphi_i(M)$. Pick $\overline{m}' \in \chi(M) \backslash \varphi_i(M)$. Let $\overline{m}_{i+1} = m_i \oplus m'$. We have that $\overline{m}_{i+1} \in \chi(M)$ and that $\varphi_{i+1} \geq \varphi_i$ since $\overline{m}_i \in \varphi_{i+1}(M)$. Hence, since $m' \notin \varphi_i(M)$ and $m' \in \varphi_{i+1}(M)$, $\varphi_{i+1}(M) \supsetneq \varphi_i(M)$.

Now suppose that φ_1 and φ_2 are both freely realised in some power of M and that $\chi(M) = \varphi_1(M) = \varphi_2(M)$. Then M^* determines $D\varphi_1$ and $D\varphi_2$. Moreover $D\varphi_1(M^*) = D\varphi_2(M^*) = D\chi(M^*)$, so $D\varphi_1 = D\varphi_2$. Thus $\varphi_1 = \varphi_2$.

Let ψ be such that $D\psi$ is freely realised in $(M^*)^n$ for some $n \in \mathbb{N}$ and $D\psi(M^*) = D\chi(M^*)$. Then M determines ψ and $\psi(M) = \chi(M)$ as required. □

Let R be an Artin algebra and $M \in \text{mod-}R$. We define three order preserving maps based on the above lemma.

(1) Let $\rho_M : \mathrm{pp}^k(M) \to \mathrm{pp}_R^k$ take $\chi(M)$ to the unique $\varphi \in \mathrm{pp}_R^k$ such that $\varphi(M) = \chi(M)$ and φ is determined by M.
(2) Let $\lambda_M : \mathrm{pp}^k(M) \to \mathrm{pp}_R^k$ take $\chi(M)$ to the unique $\psi \in \mathrm{pp}_R^k$ such that $\psi(M) = \chi(M)$ and ψ is freely realised in M^n for some $n \in \mathbb{N}$.
(3) Let $\mu_M : \mathrm{pp}_R^k \to \mathrm{pp}^k(M)$ be the map taking $\chi \in \mathrm{pp}_R^k$ to $\chi(M)$.

We will use the same notation for the left module versions of these maps. Note that if $M \in \text{mod-}R$ (resp. $M \in R\text{-mod}$), by 1.11, the duality $D : \mathrm{pp}_R^k \to {}_R\mathrm{pp}^k$ (resp. $D : {}_R\mathrm{pp}^k \to \mathrm{pp}_R^k$) induces a duality $D_M : \mathrm{pp}_R^k(M) \to {}_R\mathrm{pp}^k(M^*)$ (resp. $D_M : {}_R\mathrm{pp}^k(M) \to \mathrm{pp}_R^k(M^*)$).

LEMMA 2.8. *Let R be an Artin algebra and $M \in$ mod-R.*

(i) *The functions ρ_M, λ_M, μ_M are order preserving.*
(ii) $\rho_M = D \circ \lambda_{M^*} \circ D_M$ *and* $\lambda_M = D \circ \rho_{M^*} \circ D_M$.
(iii) $\mu_M \circ \rho_M = Id_{\mathrm{pp}_R^k(M)}$ *and* $\mu_M \circ \lambda_M = Id_{\mathrm{pp}_R^k(M)}$
(iv) *For all $\psi \in \mathrm{pp}_R^k$, $\rho_M(\psi(M))$ is the largest element in*

$$\{\varphi \in \mathrm{pp}_R^k \mid \varphi(M) \subseteq \psi(M)\}$$

and $\lambda_M(\psi(M))$ is the smallest element in

$$\{\varphi \in \mathrm{pp}_R^k \mid \psi(M) \subseteq \varphi(M)\}.$$

(v) *If the ordered sets pp_R^k and $\mathrm{pp}_R^k(M)$ are viewed as categories in the usual way and λ_M, μ_M and ρ_M as functors then $\lambda_M \dashv \mu_M \dashv \rho_M$.*
(vi) *ρ_M is a meet-semi-lattice embedding and λ_M is a join-semi-lattice embedding.*

PROOF. (*i*) Suppose that $\varphi, \psi \in \mathrm{pp}_R^k$ and $\varphi(M) \subseteq \psi(M)$. Then $\varphi(M) = \rho_M(\varphi(M))(M) \subseteq \rho_M(\psi(M))(M) = \psi(M)$ and $\rho_M(\psi(M))$ is determined by M. So $\rho_M(\varphi(M)) \leq \rho_M(\psi(M))$. Thus ρ_M is order preserving. That λ_M is order preserving follows similarly and μ_M is order preserving by definition of the order on pp_R^k.
(*ii*) We show that $D \circ \rho_M = \lambda_{M^*} \circ D_M$. From this both equalities can be deduced. Suppose $\chi \in \mathrm{pp}_R^k$. Then $\rho_M(\chi(M))$ is determined by M and $\rho_M(\chi(M))(M) = \chi(M)$. So $D \circ \rho_M(\chi(M))(M^*) = D\chi(M^*)$ by 1.11 and $D \circ \rho_M(\chi(M))$ is freely realised in $(M^*)^n$ for some $n \in \mathbb{N}$ by 2.4. Thus $D \circ \rho_M(\chi(M)) = \lambda_{M^*}(D\chi(M^*)) = \lambda_{M^*} \circ D_M(\chi(M))$.
(*iv*) By definition $\rho_M(\psi(M))(M) = \psi(M) \subseteq \psi(M)$ and since $\rho_M(\psi(M))$ is determined by M, $\varphi(M) \subseteq \psi(M) = \rho_M(\psi(M))(M)$ implies $\varphi \leq \rho_M(\psi(M))$.
(*v*) It is well known and easy to check that if $g : P_1 \to P_2$ and $f : P_2 \to P_1$ are functors between preordered sets viewed as categories then g is right adjoint to f exactly if, for all $a \in P_2$ and $b \in P_1$, $f(a) \leq b$ if and only if $a \leq g(b)$.

Let $\varphi, \psi \in \mathrm{pp}_R^k$. If $\varphi \leq \rho_M(\psi(M))$ then $\mu_M(\varphi) \leq \psi(M)$ since $\mu_M \circ \rho_M = \mathrm{Id}_{\mathrm{pp}(M)}$. Suppose that $\mu_M(\varphi) \leq \psi(M)$ i.e. $\varphi(M) \subseteq \psi(M)$. Then $\rho_M(\psi(M))(M) = \psi(M)$ and $\rho_M(\psi(M))$ is determined by M. So $\varphi \leq \rho_M(\psi(M))$. Thus we have shown that ρ_M is right adjoint to μ_M.
(*vi*) That ρ_M and λ_M are embeddings of partially ordered sets is implied by (*iii*). Meets in preorders are products and joins in preorders are coproducts. So (*vi*) follows directly from (*v*) since right adjoints preserve limits and left adjoints preserve colimits. □

3. A proof of the existence of left determiners for morphisms

In this section we give a proof of the existence of minimal left determiners for morphisms in mod-R when R is an Artin algebra. This proof is inspired by 2.4 and we explain how the various steps correspond to those in 2.4.

As we have shown, the existence of minimal free realisations for pp formulae comes very close to implying the existence of minimal left determiners for morphisms in mod-R where R is an Artin algebra.

Let R be a ring and φ is a pp-n-formula. Suppose $f : R^n \to M \in \text{mod-}R$ and $\text{im}(f, -) = F_\varphi \subseteq (R^n, -)$. Then, by 1.10,

$$0 \longrightarrow F_{D\varphi} \longrightarrow R^n \otimes - \cong (R^n, -) \xrightarrow{f\otimes -} M \otimes -$$

is exact. A free realisation for $D\varphi(-)$ is just a map $\gamma : R^n \to N \in R\text{-mod}$ such that $(\gamma, -) : (N, -) \to (R^n, -)$ has image $F_{D\varphi(-)}$ and a free realisation is minimal if $(\gamma, -) : (N, -) \to (R^n, -)$ is a projective cover for its image $F_{D\varphi(-)}$.

THEOREM 3.1. *Let R be an Artin algebra. If $\gamma : (D, -) \to \ker(f \otimes -)$ is a projective cover then $C \in \text{mod-}R$ is a left determiner for f if and only if D^* is a direct summand of C^n for some $n \in \mathbb{N}$. That is, if $D = D_1^{l_1} \oplus \ldots \oplus D_m^{l_m}$, where $D_i \cong D_j$ implies $i = j$, then $(D_1 \oplus \ldots \oplus D_m)^*$ is a minimal left determiner for f.*

This statement follows from [**Kra13**, 3.13]. However, our proofs are somewhat more elementary and our motivation for the statement and proof have different roots.

In order to prove the above theorem we first prove 4 lemmas.

The first generalises elementary duality for pp formulae. The functor $\text{im}(g, -)$ plays the role of a pp formula φ and the functor $\ker(g \otimes -)$ plays the role of $D\varphi$.

LEMMA 3.2. *Let R be a ring. Let $f : A \to B$ and $g : A \to C$ be morphisms in* mod-R. *The following are equivalent:*

(1) $\text{im}(g, -) \subseteq \text{im}(f, -)$
(2) *there exists $h : B \to C$ such that $g = h \circ f$*
(3) $\ker(f \otimes -) \subseteq \ker(g \otimes -)$

PROOF. (1)⇒(2) Since $\text{im}(g, C) \subseteq \text{im}(f, C)$, $g \in \text{im}(f, C)$ and hence there exists $h : B \to C$ such that $g = h \circ f$.

(2)⇒(3) We have that $g \otimes - = (h \otimes -) \circ (f \otimes -)$.

(3)⇒(1) Let d be the Auslander-Gruson-Jensen dual as in [**Pre09**, section 10.3]. Since d is exact and $d(g \otimes -) = (g, -)$, the cokernel of $(f, -)$ factors through the cokernel of $(g, -)$. So $\text{im}(g, -) \subseteq \text{im}(f, -)$. □

The next lemma generalises the statement that if R is an Artin algebra, $\varphi, \psi \in \text{pp}_R^k$ and $M \in \text{mod-}R$ then $\varphi(M) \supseteq \psi(M)$ if and only if $D\psi(M^*) \supseteq D\varphi(M^*)$.

LEMMA 3.3. *Suppose that R is an Artin algebra. Let $f : A \to B$, $g : A \to C$ be morphisms in* mod-R *and let $D \in$* mod-R. *The following are equivalent:*

(1) $\text{im}(g, D) \subseteq \text{im}(f, D)$
(2) $\ker(f \otimes D^*) \subseteq \ker(g \otimes D^*)$

PROOF. Throughout, let $\overline{a}$ be a tuple of elements from A, $\overline{t} \in D^*$, ψ generate the pp-type of $f(\overline{a})$ and φ generate the pp-type of $g(\overline{a})$. Thus, by 1.10, $f(\overline{a}) \otimes \overline{t} = 0$ if and only if $\overline{t} \in D\psi(D^*)$ and $g(\overline{a}) \otimes \overline{t} = 0$ if and only if $\overline{t} \in D\varphi(D^*)$.
(1)⇒(2) Since $\text{im}(g, D) \subseteq \text{im}(f, D)$,

$$\varphi(D) = \{\delta g\overline{a} \mid \delta \in (C, D)\} \subseteq \{\gamma f\overline{a} \mid \gamma \in (B, D)\} = \psi(D).$$

Therefore, by 1.11, $D\psi(D^*) \subseteq D\varphi(D^*)$. Suppose that $\overline{a}\otimes\overline{t} \in \ker(f\otimes D^*) \subseteq A\otimes D^*$. Then $f(\overline{a}) \otimes t = 0$. So $\overline{t} \in D\psi(D^*)$. So $\overline{t} \in D\varphi(D^*)$. Thus $g(\overline{a}) \otimes \overline{t} = 0$, that is $\overline{a} \otimes \overline{t} \in \ker(g \otimes D^*)$.
(2)⇒(1) Let $\overline{a}$ generate A. By (2), $D\psi(D^*) \subseteq D\varphi(D^*)$. Thus

$$\varphi(D) = \{\delta g\overline{a} \mid \delta \in (C, D)\} \subseteq \{\gamma f\overline{a} \mid \gamma \in (B, D)\} = \psi(D).$$

So, for all $\delta \in (C, D)$, there is a $\gamma \in (B, D)$ such that $\delta g\overline{a} = \gamma f\overline{a}$. Since $\overline{a}$ generates A, $\delta g\overline{a} = \gamma f\overline{a}$ implies $\delta g = \gamma f$. Thus (1) holds. □

As we have already noted $(C, \overline{c})$ freely realises φ if and only if the map $(\overline{c}, -) : (C, -) \to (R^n, -)$ has image F_φ. So the following lemma replaces the use of free realisations.

LEMMA 3.4. *Let R be a ring. Let $f : A \to B$ be a morphism in* mod-R. *If $D \in R$-mod and $\gamma : (D, -) \to \ker(f \otimes -)$ is an epimorphism then for all $g : A \to C$, we have that $\ker(f \otimes -) \subseteq \ker(g \otimes -)$ if and only if $\ker(f \otimes D) \subseteq \ker(g \otimes D)$.*

PROOF. Let $i : \ker(g \otimes -) \to A \otimes -$ and $j : \ker(f \otimes -) \to A \otimes -$ be the kernels of $g \otimes -$ and $f \otimes -$ respectively. Since γ is an epimorphism and both i and j are monic, $j\gamma$ factors through i if and only if $\ker(f \otimes -) \subseteq \ker(g \otimes -)$.

We now show that $\ker(f \otimes D) \subseteq \ker(g \otimes D)$ implies $j \circ \gamma$ factors through i. Since $\ker(f \otimes D) \subseteq \ker(g \otimes D)$, there is a $c \in \ker(g \otimes D)$ such that $i_D(c) = (j\gamma)_D(1_D)$. Let $\pi \in ((D, -), \ker(g \otimes -))$ be such that $\pi_D(1_D) = c$. Now $(i\pi)_D(1_D) = (j\gamma)_D(1_D)$. So $i\pi = j\gamma$ as required. □

The following lemma is a generalisation of (1) ⇒ (2) in 2.4.

LEMMA 3.5. *Let R be an Artin algebra over a commutative Artinian ring S. If D^* is a left determiner for $f : A \to B \in$ mod-R then there exists an $n \in \mathbb{N}$ and an epimorphism $\gamma : (D^n, -) \to \ker(f \otimes -)$.*

PROOF. For any ring R, if $A \in$ mod-R then all finitely presented subobjects of $A \otimes -$ are of the form $\ker(g \otimes -)$ for some $g : A \to B \in$ mod-R. This follows from the fact that $(R\text{-mod}, \text{Ab})^{\text{fp}}$ has enough injectives and that all injectives are of the form $B \otimes -$ for some $B \in$ mod-R (see [**GJ81**, 5.5], [**Pre09**, 12.1.13]). Hence, all finitely presented subobjects of $A \otimes -$ are of the form $\ker \nu$ for some natural transformation $\nu : A \otimes - \to B \otimes -$. Further, since $h : A \otimes - \to B \otimes -$ is the zero morphism if and only if $h_R = 0$, all natural transformations $\nu : A \otimes - \to B \otimes -$ are of the form $g \otimes -$ for some $g : A \to B$.

Suppose that D^* is a left determiner for $f : A \to B \in$ mod-R. Combining 3.2 and 3.3, we have that for all $g : A \to C \in$ mod-R, $\ker(f \otimes -) \subseteq \ker(g \otimes -)$ if and only if $\ker(f \otimes D) \subseteq \ker(g \otimes D)$. Thus if F is any finitely presented subfunctor of $A \otimes -$, then $\ker(f \otimes -) \subseteq F$ if and only if $\ker(f \otimes D) \subseteq FD$. Since $A, B \in$ mod-R and $D \in R$-mod, $\ker(f \otimes D)$ is finitely generated as a module over S.

Let $a_1, ..., a_n$ generate $\ker(f \otimes D)$ over S. For each $1 \leq i \leq n$, let $\gamma_i : (D, -) \to \ker(f \otimes -)$ be such that $(\gamma_i)_D(1_D) = a_i$. Let $\gamma : (D^n, -) \to \ker(f \otimes -)$ be $(\gamma_1, ..., \gamma_n)$. Now the image of γ_D is the whole of $\ker(f \otimes D)$. Thus the image of γ is $\ker(f \otimes -)$ as required. □

Finally we are ready to prove 3.1.

PROOF OF 3.1. Lemma 3.4 shows that any projective $(D, -)$ which maps epimorphically onto $\ker(f \otimes -)$ has the property that $\ker(f \otimes -) \subseteq \ker(g \otimes -)$ if and only if $\ker(f \otimes D) \subseteq \ker(g \otimes D)$. Lemma 3.3 says that this is true if and only if $\text{im}(g, D^*) \subseteq \text{im}(f, D^*)$. Lemma 3.2 says that $\ker(f \otimes -) \subseteq \ker(g \otimes -)$ if and only if $\text{im}(g, -) \subseteq \text{im}(f, -)$. So D^* is a left determiner for f.

Suppose C^* is a left determiner for f. Then, by 3.5, there is an $n \in \mathbb{N}$ and epimorphism $\gamma : (C^n, -) \to \ker(f \otimes -)$. Since $(D, -)$ is a projective cover for

$\ker(f \otimes -)$, using general properties of projective covers, we have that $(D, -)$ is a direct summand of $(C^n, -)$. Therefore D is a direct summand of C^n. $\square$

COROLLARY 3.6. *Let R be an Artin algebra and $f : A \to B \in \text{mod-}R$. If $- \otimes D$ is an injective hull for coker$((f, -))$ then D^* is a left determiner for f and if C is a left determiner for f then D^* is a direct summand of C^n for some $n \in \mathbb{N}$.*

PROOF. Apply the Auslander-Gruson-Jensen dual d as in [**Pre09**, Section 10.3] to 3.1. $\square$

Acknowledgments

The content of this note was developed while attending Mike Prest's research group seminars while I was his postdoc in Manchester. I would like to thank him for introducing me to morphisms determined by objects and encouraging me to publish these results. I would also like to thank Sam Dean for reading a draft of this paper and making many helpful suggestions.

References

[Aus78] M. Auslander, *Functors and morphisms determined by objects*, Representation theory of algebras (Proc. Conf., Temple Univ., Philadelphia, Pa., 1976), Dekker, New York, 1978, pp. 1–244. Lecture Notes in Pure Appl. Math., Vol. 37. MR0480688

[GJ81] L. Gruson and C. U. Jensen, *Dimensions cohomologiques reliées aux foncteurs* $\varprojlim^{(i)}$ (French), Paul Dubreil and Marie-Paule Malliavin Algebra Seminar, 33rd Year (Paris, 1980), Lecture Notes in Math., vol. 867, Springer, Berlin-New York, 1981, pp. 234–294. MR633523

[Her93] I. Herzog, *Elementary duality of modules*, Trans. Amer. Math. Soc. **340** (1993), no. 1, 37–69, DOI 10.2307/2154545. MR1091706

[Kra13] H. Krause, *Morphisms determined by objects in triangulated categories*, Algebras, quivers and representations, Abel Symp., vol. 8, Springer, Heidelberg, 2013, pp. 195–207, DOI 10.1007/978-3-642-39485-0_9. MR3183886

[Pre88] M. Prest, *Model theory and modules*, London Mathematical Society Lecture Note Series, vol. 130, Cambridge University Press, Cambridge, 1988. MR933092

[Pre09] M. Prest, *Purity, spectra and localisation*, Encyclopedia of Mathematics and its Applications, vol. 121, Cambridge University Press, Cambridge, 2009. MR2530988

[Rin12] C. M. Ringel, *Morphisms determined by objects: the case of modules over Artin algebras*, Illinois J. Math. **56** (2012), no. 3, 981–1000. MR3161362

[Rin13] C. M. Ringel, *The Auslander bijections: how morphisms are determined by modules*, Bull. Math. Sci. **3** (2013), no. 3, 409–484, DOI 10.1007/s13373-013-0042-2. MR3128038

UNIVERSITY OF CAMERINO, SCHOOL OF SCIENCE AND TECHNOLOGIES, DIVISION OF MATHEMATICS, VIA MADONNA DELLE CARCERI 9, 62032 CAMERINO, ITALY

Email address: lorna.gregory@gmail.com

Contemporary Mathematics
Volume **730**, 2019
https://doi.org/10.1090/conm/730/14710

The universal abelian regular ring

Ivo Herzog and Sonia L'Innocente

Dedicated to Mike Prest on the occasion of his 65th birthday

ABSTRACT. Olivier's construction of a universal commutative regular ring is generalized to obtain for a noncommutative ring R the universal abelian regular R-ring $R \to \hat{R}$. This R-ring induces a homeomorphism from the spectrum of $\hat{R}$ to the constructible Cohn spectrum of R. An étale bundle is constructed over the Ziegler spectrum $\mathrm{Zg}^*(R)$, equipped with the Zariski topology, whose sheaf of sections is the sheafification of Prest's presheaf of definable scalars. When the bundle is restricted to $\mathrm{Zg}^*(\hat{R})$, which is homeomorphic to the spectrum of $\hat{R}$, one obtains an étale bundle whose R-ring of global sections is isomorphic to $\hat{R}$.

Let R be a commutative ring. The *universal commutative (von Neumann) regular* R-algebra is a ring morphism $R \to \hat{R}$ into a commutative regular ring that is universal with respect to this property: every ring morphism $f : R \to S$ to a commutative regular ring factors uniquely through $\hat{R}$ as in

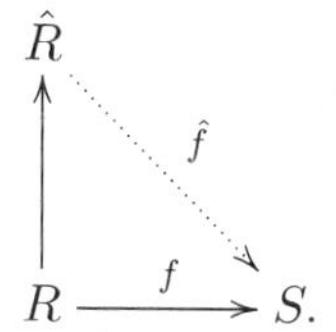

Olivier [**14**] proved the existence of the universal commutative regular R-algebra by the formal adjunction, for every element $r \in R$, of a *reflexive inverse:*

$$\hat{R} = R[y_r \mid r \in R]/(ry_r r - r, y_r r y_r - y_r \mid r \in R)$$

with $R \to \hat{R}$ given by the obvious morphism. The purpose of this article is to generalize Olivier's construction (Theorem 2.2) to the noncommutative case and study the universal abelian regular R-ring $R \to R^{\mathrm{ab}}$. One mimics the construction of Olivier, by formal adjunction of *group inverses*, defined as commuting reflexive inverses, iterated countably many times.

Wiegand [**22**, Thm 1] described $\hat{R}$ as the global sections of a sheaf on the Zariski spectrum $\mathrm{Spec}(R)$, equipped with the patch topology [**9**]. The universal abelian

2010 *Mathematics Subject Classification.* Primary 03C60, 16E50, 18E35.
Key words and phrases. Abelian regular ring, constructible Cohn spectrum, the sheaf of definable scalars, étale bundle.
The first author was supported by a Fulbright Distinguished Chair Grant and NSF Research Grant DMS 12-01523; the second by Italian PRIN 2012 and GNSAGA INdAM grants.

regular R-ring $R \to \hat{R}$ induces a homeomorphism $\mathrm{Spec}(\hat{R}) \to \widehat{\mathrm{Spec}}(R)$ between the spectrum of $\hat{R}$ and the Cohn spectrum $\widehat{\mathrm{Spec}}(R)$, equipped with the patch topology. There is a bijective correspondence $e \mapsto \mathcal{O}(e)$ between the idempotent elements of $\hat{R}$ and the clopen subsets of $\mathrm{Spec}(\hat{R})$. The Pierce sheaf [**15**, Thm 4.4] is a sheaf Γ of rings on $\mathrm{Spec}(\hat{R})$ with the property that $\Gamma[\mathcal{O}(e)] = e\hat{R}e$. In the last few sections of the paper, we invoke Prest's theory of the sheaf of definable scalars to show (Theorem 8.5) how the Pierce sheaf Γ is obtained by pullback of the sheaf of locally definable scalars on R along the homeomorphic embedding $\mathrm{Zg}^*(\hat{R}) \to \mathrm{Zg}^*(R)$ of Ziegler spectra, endowed with the Zariski topology.

Corollary 5.3 provides an axiomatization for the full subcategory $\hat{R}$-Mod $\subseteq$ R-Mod of modules over the universal abelian regular R-ring.

Throughout the paper, R will denote an associative ring with identity $1 \in R$. An R-ring is a morphism $f : R \to S$ of rings with domain R. An R-ring S may be thought of as a left R-module ${}_RS$ via the action $rs = f(r)s$. We may refer to the element $f(r)$ as $r \in S$. An R-field $R \to \Delta$ is an R-ring where Δ is a (not necessarily commutative) field. An R-algebra is an R-ring $f : R \to S$ with R commutative and the image of f contained in the center of S.

1. The group inverse

An element $r \in R$ is *regular* if there exists a $y \in R$ such that $ryr = r$; then y is called an *inner inverse* of r. Note that $y' = yry$ is also an inner inverse with the further property that $y'ry' = y'$; such an inner inverse is called a *reflexive inverse* of r and, if it commutes with r, a *group inverse* of r (see [**17**, §8.6]). This notion is important in the theory of *abelian regular rings*, which are rings in which every element has a group inverse. Equivalently, a regular ring R is abelian [**3**, §3] if and only if every idempotent $e \in R$ is central. If $r \in R$ is a unit, then the inverse of r in R is clearly a group inverse; and if a regular element $r \in R$ has a unique reflexive inverse then it must be a group inverse [**20**, Lemma 3.3]. A regular element $r \in R$ has at most one group inverse y, for if y' were another, then

$$y = y^2r = y^2r^2y' = y(yr^2)y' = yry'$$

and similarly $y' = yry'$. Here are some conditions equivalent to the existence of a group inverse.

PROPOSITION 1.1. *Let R be an associative ring. The following are equivalent for an element $r \in R$:*

(1) *r has a group inverse in R;*
(2) *there is an idempotent element $e \in R$ such that $rR = eR$ and $Rr = Re$;*
(3) *there exists a direct sum decomposition $R_R = rR \oplus \mathrm{r.ann}(r)$; and*
(4) *every left R-module ${}_RM$ admits a direct sum endodecomposition*[1] *$M = rM \oplus \mathrm{ann}_M(r)$.*

PROOF. (1) $\Rightarrow$ (2). Suppose that $ryr = r$ and $yry = y$, with $e = ry = yr$. Then $rR = eR$ and $Rr = Re$.
(2) $\Rightarrow$ (3). Since $Rr = Re$, we have that $\mathrm{r.ann}(r) = \mathrm{r.ann}(e) = (1 - e)R$.
(3) $\Rightarrow$ (4). Let $e \in R$ be the projection onto rR parallel to $\mathrm{r.ann}(r)$. Thus $rR = eR$ and $\mathrm{r.ann}(r) = (1 - e)R$, which imply that $M = eM \oplus (1 - e)M = rM \oplus \mathrm{ann}_M(r)$.

[1] A decomposition as a module over its endomorphism ring.

(4) $\Rightarrow$ (1). Take $M = {}_RR$ and let $e \in R$ be as in the previous implication. The equalities $rR = eR$ and $\mathrm{r.ann}(r) = (1-e)R$ imply that $r = er$, $r = re$ and that there exists a $y \in R$ such that $e = ry$; we may assume that $y = eye$. Then $ryr = er = r$ and $yry = ye = y$, so that y is a reflexive inverse of r. The idempotent $f = yr$ satisfies $r(1-f) = r - ryr = 0$, so that $1 - f \in (1-e)R$, whence $ry = e = ef = f = yr$. □

We will denote the group inverse of r, when it exists, by $\overline{r}$. Given an element $r \in R$, there exists an R-ring $R \to R\{\overline{r}\} := R\{y\}/I$ universal with respect to the property that r has a group inverse. It is obtained by freely adjoining a (usually noncentral) variable y and factoring out by the ideal $I = (ryr - r, yry - y, ry - yr)$ of relations satisfied by a group inverse. If $f : R \to S$ is any R-ring in which r has a group inverse, then there is a unique extension

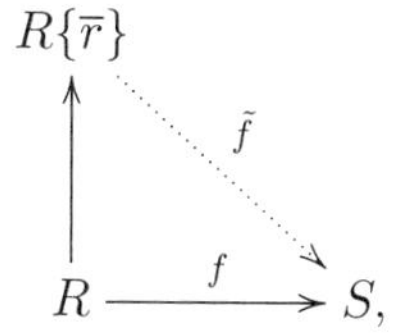

given by $\tilde{f} : \overline{r} = y + I \mapsto \overline{r}$. In particular, any ring morphism $g : R\{\overline{r}\} \to S'$ is determined by the image of $R \to R\{\overline{r}\}$, so that this universal R-ring is epic.

Prest [**16**, Cor 5.5.4] has shown that whenever $f : R \to S$ is an epic R-ring, then the full subcategory S-Mod $\subseteq R$-Mod, obtained by restriction of scalars along f, is axiomatizable in the language of left R-modules (see §3). For the epic R-ring $R \to R\{\overline{r}\}$, Condition (4) of Proposition 1.1 provides such an axiom. The formula $r|u := \exists v\ (u \doteq rv)$ defines in ${}_RM$ the endosubmodule rM, while the formula $ru \doteq 0$ defines $\mathrm{ann}_M(r)$. Thus we can express Condition (4) as follows.

PROPOSITION 1.2. *A left R-module M belongs to the subcategory $R\{\overline{r}\}$-Mod $\subseteq$ R-Mod if and only if*

$$M \models \{\forall u\ \exists v, w\ [(u \doteq v + w) \wedge r|v \wedge rw \doteq 0]\} \wedge \forall u\ [(ru \doteq 0 \wedge r|u) \to u \doteq 0].$$

2. Olivier's construction

Let us carry out the construction of the previous section for all $r \in R$ at once, by adjoining a group inverse for every element of R. An R-ring is obtained by adjoining noncentral variables y_r, one for every $r \in R$,

$$R \to R_1^{\mathrm{ab}} := R\{y_r \mid r \in R\}/I,$$

modulo the ideal $I = (ry_rr - r, y_rry_r - y_r, ry_r - y_rr \mid r \in R)$ generated by the relations that ensure each $y_r + I = \overline{r}$ is a group inverse of r. As above, the R-ring $R \to R_1^{\mathrm{ab}}$ is universal with respect to the property that every $r \in R$ obtains a group inverse. In particular, every abelian regular R-ring $f : R \to S$ factors uniquely, as above, through R_1^{ab},

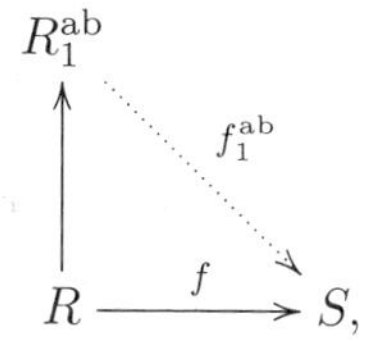

The universal property of the R-ring $R \to R_1^{\mathrm{ab}}$ ensures that it is also epic. As such, the full subcategory R_1^{ab}-Mod $\subseteq R$-Mod is axiomatizable and we obtain the following version of Proposition 1.2.

PROPOSITION 2.1. *A left R-module M belongs to the subcategory R_1^{ab}-Mod $\subseteq$ R-Mod if and only if for every $r \in R$,*

$$M \models \{\forall u\, \exists v, w\, [(u \doteq v + w) \wedge r|v \wedge rw \doteq 0]\} \wedge \forall u\, [(ru \doteq 0 \wedge r|u) \to u \doteq 0].$$

If $r \in R$ is a nilpotent element, then the existence of a group inverse in R_1^{ab} implies that $r = 0$ in R_1^{ab}. The kernel of $R \to R_1^{\mathrm{ab}}$ therefore contains the upper nilradical of R. Also note that if $e \in R$ is idempotent, then e will centralize R in R_1^{ab}, because the element $(1-e)re$ is nilpotent, for every $r \in R$. Thus $re = ere$ in R_1^{ab} and similarly $er = ere$.

The construction $R \to R_1^{\mathrm{ab}}$ is the first step in the noncommutative version of Olivier's construction. The process may be iterated to obtain a denumerable sequence $R = R_0^{\mathrm{ab}} \longrightarrow R_1^{\mathrm{ab}} \longrightarrow R_2^{\mathrm{ab}} \longrightarrow \cdots$, of ring morphisms defined recursively by $R_{n+1}^{\mathrm{ab}} := (R_n)_1^{\mathrm{ab}}$. Each of the compositions $R \to R_n^{\mathrm{ab}}$ is an epic R-ring and, therefore, so is the limit $R \to R^{\mathrm{ab}} := \varinjlim R_n^{\mathrm{ab}}$, which is abelian regular, because every element of R^{ab} has a group inverse. Indeed, if $r \in R^{\mathrm{ab}}$ is represented by some approximation $r_n \in R_n^{\mathrm{ab}}$, then the construction ensures that r_n obtains a group inverse in R_{n+1}^{ab}. Moreover, the abelian regular R-ring $R \to R^{\mathrm{ab}}$ is universal with respect to this property, because every abelian regular R-ring $f : R \to S$ factors uniquely through each of the $R \to R_n^{\mathrm{ab}}$ and therefore uniquely through $R \to R^{\mathrm{ab}}$.

THEOREM 2.2. *Every ring R admits a universal abelian regular R-ring $R \to R^{\mathrm{ab}}$.*

If a ring R is abelian regular to begin with, then $R \to R_1^{ab}$ is just the identity morphism, so that this noncommutative iterated version of Olivier's contruction halts at stage 0. More generally, if R_n^{ab} is abelian regular at some finite stage of the construction, then the process halts and $R^{\mathrm{ab}} = R_n^{\mathrm{ab}}$. It seems like an interesting general ring theoretic problem to determine the least n, if it exists, for which R_n^{ab} is abelian regular.

If R is a von Neumann regular ring (not necessarily abelian), let $N \subseteq R$ be the ideal generated by elements of the form $er(1-e)$, with e idempotent. These elements are nilpotent so that N is contained in the kernel of $R \to R_1^{\mathrm{ab}}$ and we obtain a morphism $R/N \to R_1^{\mathrm{ab}}$ of epic R-rings. On the other hand, R/N is abelian regular, by [**3**, Prop 3.3], so that we also have a morphisms $R_1^{\mathrm{ab}} \to R/N$. These morphisms of epic R-rings are therefore isomorphisms and we see that $R \to R_1^{\mathrm{ab}} = R/N$ is abelian regular.

COROLLARY 2.3. *There is a bijection $\mathcal{P} \mapsto R^{\mathrm{ab}}/\mathcal{P}$ between the prime ideals of R^{ab} and the epic R-fields $R \to R^{\mathrm{ab}} \to R^{\mathrm{ab}}/\mathcal{P}$. In particular, $R^{\mathrm{ab}} \neq 0$ if and only there exists a nonzero epic R-field.*

PROOF. If $\mathcal{P} \subseteq R^{\mathrm{ab}}$ is a prime ideal, then it is maximal and the quotient $R^{\mathrm{ab}}/\mathcal{P}$ is a field [**3**, Thm 3.2]. It follows that $R \to R^{\mathrm{ab}} \to \mathrm{R}^{\mathrm{ab}}/\mathcal{P}$ is an epic R-field. Conversely, every R-field $f : R \to \Delta$ is abelian regular, so that Theorem 2.2 implies that it factors through the universal abelian regular R-ring, with $f^{\mathrm{ab}} : \overline{r} \mapsto r^{-1} \in \Delta$ if $r \neq 0$ in Δ. Define $E_n(f) \subseteq \Delta$ by recursion on $n \geq 1$, so that $E_0(f) := \mathrm{Im}\, f$, the image of f, and E_{n+1} the subring of Δ generated by $E_n(f)$ and the elements r^{-1}, for nonzero $r \in E_n(f)$. A commutative diagram arises in the category of R-rings.

$$\begin{array}{ccccccccc}
R = R_0^{\mathrm{ab}} & \longrightarrow & R_1^{\mathrm{ab}} & \longrightarrow & R_2^{\mathrm{ab}} & \longrightarrow & \cdots & \longrightarrow & R^{\mathrm{ab}} \\
\downarrow{\scriptstyle f} & & \downarrow{\scriptstyle f_1} & & \downarrow{\scriptstyle f_2} & & & & \downarrow{\scriptstyle f^{\mathrm{ab}}} \\
E_0(f) & \longrightarrow & E_1(f) & \longrightarrow & E_2(f) & \longrightarrow & \cdots & \longrightarrow & E(f),
\end{array}$$

where $E(f) = \cup_n E_n(f)$ is the subfield of Δ generated by the image of f. The universal property of the ring morphisms $f_n : R_n^{\mathrm{ab}} \to E_n(f)$ ensures that every f_n is onto. It follows that all the R-rings in the diagram are epic and the limit $f^{\mathrm{ab}} : R^{\mathrm{ab}} \to \Delta$ is onto. Therefore, there exists a prime ideal $\mathcal{P} \subseteq R^{\mathrm{ab}}$ for which $\Delta = R^{\mathrm{ab}}/\mathcal{P}$. □

Corollary 2.3 implies that there is a canonical morphism of R-rings, $R^{\mathrm{ab}} \to \prod_{\mathcal{P}} R^{\mathrm{ab}}/\mathcal{P}$, to the product of all epic R-fields. This morphism is an embedding, because the kernel is the Jacobson radical $J(R^{\mathrm{ab}})$, which is 0 in a von Neumann regular ring.

The *inversion height* of an epic R-field $f : R \to \Delta$ is the least n for which $E_n(f) = E(f) = \Delta$. The construction of the universal abelian regular R-ring was inspired by the thesis of Javier Sanchez [**5**, **19**], who—building on the work of Fisher [**4**] and Reutenauer [**18**]—gives examples of epic R-fields of various inversion height.

3. Definable scalars of R-rings.

The language of left R-modules $\mathcal{L}(R) = (+, -, 0, r)_{r \in R}$ is the language $\mathcal{L}(\mathbb{Z}) = (+, -, 0)$ of abelian groups expanded to include unary function symbols, one for every $r \in R$, also denoted as r. If ${}_RM$ is a left R-module, the symbols of $\mathcal{L}(\mathbb{Z})$ are intended to interpret the underlying abelian group, while the unary function symbols $r \in R$ interpret the scalar action of $r \in R$ on that group. The axioms for a left R-module are expressible in $\mathcal{L}(R)$ by finitely many axiom schemata.

The nonlogical symbols of $\mathcal{L}(R)$ are precisely those needed to express a linear equation

$$r_1 u_1 + r_2 u_2 + \cdots + r_n u_n \doteq 0.$$

A system of m linear equations in n variables is a conjunction of such formulae

$$\bigwedge_{i=1}^{m} \sum_{j=1}^{n} r_{ij} u_j \doteq 0$$

and is denoted as usual by $A\mathbf{u}^t \doteq 0$, where $A = (r_{ij})$ is the associated $m \times n$ matrix of coefficients and $\mathbf{u}^t$ is the column vector of free variables given by the transpose of the n-tuple $\mathbf{u} = (u_1, u_2, \ldots, u_n)$.

A formula $\varphi(\mathbf{u})$ in $\mathcal{L}(R)$ is *positive primitive* (pp) if it is obtained from a system $(A, B)\begin{pmatrix} \mathbf{u}^t \\ \mathbf{v}^t \end{pmatrix} \doteq 0$ of linear equations by existential quantification over certain of the free variables, say $\mathbf{v}$. Thus A is an $m \times n$ matrix and B an $m \times k$ matrix, where $\mathbf{u} = (u_1, \ldots, u_n)$ and $\mathbf{v} = (v_1, \ldots, v_k)$, and the pp formula is displayed as

$$\varphi(\mathbf{u}) = \exists \mathbf{v}\ (A, B)\begin{pmatrix} \mathbf{u}^t \\ \mathbf{v}^t \end{pmatrix} \doteq 0$$

with the free variables $\mathbf{u}$ explicitly shown. We will mostly be interested in pp formulae in one and two (free) variables. A pp formula in one variable has the form

$$\psi(u) = \exists \mathbf{v}\ (A_0, A) \begin{pmatrix} u \\ \mathbf{v}^t \end{pmatrix} \doteq 0,$$

where A_0 is a column matrix with as many rows as there are in A. For a pp formula in two variables, we conform to the convention of Cohn [**1**, Ch 7] and write the formula as

$$\rho(u, v) = \exists \mathbf{w}\ (A_0, A_*, A_\infty) \begin{pmatrix} u \\ \mathbf{w}^t \\ v \end{pmatrix} \doteq 0,$$

where A_0 and A_∞ are now column vectors with the same number of rows as A_*.

Given a pp formula $\varphi(\mathbf{u})$ as above and left R-module ${}_RM$, an n-tuple $\mathbf{x} \in M^n$ satisfies $\varphi(\mathbf{u})$ in ${}_RM$, written ${}_RM \models \varphi(\mathbf{x})$, if there exists a k-tuple $\mathbf{y} \in M^k$ such that $(A, B) \begin{pmatrix} \mathbf{x}^t \\ \mathbf{y}^t \end{pmatrix} \doteq 0$. For example, consider an R-ring $f : R \to S$ as a left R-module and take a pp formula $\psi(u)$ in one variable as above. Then ${}_RS \models \psi(1)$ if and only if the nonhomogenous system $A\mathbf{v}^t = -A_0$ has a solution in S if and only if $A\mathbf{v}^t = A_0$ does.

If ${}_RM$ is a left R-module and $\varphi(\mathbf{u})$ a pp formula in n variables, the collection of n-tuples $\mathbf{x} \in M^n$ that satisfy $\varphi(\mathbf{u})$ in ${}_RM$ form the *pp definable subgroup*

$$\varphi(M) := \{\mathbf{x} \in M^n \mid M \models \varphi(\mathbf{x})\}$$

of M^n, closed under the action of the endomorphism ring $\mathrm{End}_R\ M$. For example, if $R \to \Delta$ is an R-field, then ${}_R\Delta$ is endosimple, so it must be that for every $\psi(u)$ in one variable, $\psi(\Delta) = \Delta$ or $\psi(\Delta) = 0$, depending on whether the homogeneous system $A\mathbf{v} = A_0$ is solvable in Δ or not.

For a pp formula $\rho(u, v)$ in two free variables and a left R-module ${}_RM$ there is the prospect that ρ defines the graph of a $\mathbb{Z}$-linear map $\rho : M \to M$,

$$(\dagger) \qquad M \models \forall u \exists! v\ \rho(u, v).$$

This happens precisely when the two pp formulae $\exists v \rho(u, v)$ and $\rho(0, v)$ in one variable define in M the pp definable subgroups M and 0, respectively. A function $\rho : M \to M$ that is pp definable in this way is called a *definable scalar* on M. Let us explain how the definable scalars on M form an R-ring $R \to R_M$. If $\rho, \sigma : M \to M$ are scalars on M defined by

$$\rho(u, v) = \exists \mathbf{w}\ (A_0, A_*, A_\infty) \begin{pmatrix} u \\ \mathbf{w}^t \\ v \end{pmatrix} \doteq 0 \quad \text{and} \quad \sigma(u, v) = \exists \mathbf{x}\ (B_0, D_*, B_\infty) \begin{pmatrix} u \\ \mathbf{x}^t \\ v \end{pmatrix} \doteq 0,$$

respectively, then $\rho - \sigma$ and $\rho \cdot \sigma$ are defined by the pp formulae

$$(\rho - \sigma)(u, v) = \exists \mathbf{y} \begin{pmatrix} B_0 & B_* & B_\infty & 0 & 0 \\ A_0 & 0 & A_\infty & A_* & A_\infty \end{pmatrix} \begin{pmatrix} u \\ \mathbf{y}^t \\ v \end{pmatrix} \doteq 0, \quad \text{and}$$

$$(\rho\sigma)(u, v) = \exists \mathbf{y} \begin{pmatrix} B_0 & B_* & B_\infty & 0 & 0 \\ 0 & 0 & A_0 & A_* & A_\infty \end{pmatrix} \begin{pmatrix} u \\ \mathbf{y}^t \\ v \end{pmatrix} \doteq 0.$$

These matrices can also be found on [**1**, p. 385]. The ring R_M of definable scalars is an R-ring, because the action of every $r \in R$ on M is itself pp definable, by the formula $(r, -1)\begin{pmatrix} u \\ v \end{pmatrix} \doteq 0$.

Consider an R-ring $f : R \to S$ as a left R-module and suppose that the pp formula $\rho(u, v)$ defines a scalar on ${}_RS$. Then the formula $\exists v\rho(u, v)$ defines in ${}_RS$ the entire endo-submodule S. But this occurs if and only if ${}_RS \models \exists v\rho(1, v)$, which is equivalent to the condition that the system $(A_*, A_\infty)\begin{pmatrix} \mathbf{w}^t \\ v \end{pmatrix} \doteq A_0$ has a solution in S. The other condition is that every solution to the corresponding homogeneous system $(A_*, A_\infty)\begin{pmatrix} \mathbf{w}^t \\ v \end{pmatrix} \doteq 0$ satisfies $v = 0$. Both of these conditions will certainly hold if the matrix (A_*, A_∞) is a square matrix, invertible in S. Cohn [**1**, p. 384] calls these scalars *admissible*.

Because the graph of ρ is an $\mathrm{End}_S(S_S)$-submodule of S^2, it follows that the scalar defined on ${}_RS$ is nothing more than multiplication by $s = \rho(1)$ on the left. Prest (cf. proof of [**16**, Thm 6.1.8]) has characterized the elements $s \in S$ whose left action is a definable scalar on ${}_RS$ as precisely those elements for which $s \otimes 1 = 1 \otimes s$ in $S \otimes_R S$. Schofield [**21**, Lemma 7.15] has shown that this is the R-ring given by the dominion of f, $R \to \mathrm{dom}(f) \subseteq S$. The *dominion* of an R-ring $f : R \to S$ is defined to be the maximal subring of S with the property that two ring morphisms from S that agree on the image of f agree on $\mathrm{dom}(f)$. Inspection of the matrices above shows that the admissible scalars form an R-subring $R^f(S)$ of the dominion; this is the *rational closure* [**1**, p. 382] of f.

THEOREM 3.1. *Let R be a ring and $f : R \to S = \prod_{\mathcal{P}} R^{\mathrm{ab}}/\mathcal{P}$ the product of all epic R-fields. The dominion of f is given by the universal abelian regular ring $R \to R^{\mathrm{ab}} \subseteq \prod_{\mathcal{P}} R^{\mathrm{ab}}/\mathcal{P}$.*

PROOF. The R-ring $R \to R^{\mathrm{ab}}$ is epic, so necessarily lies in the dominion of $R \to S = \prod_{\mathcal{P}} R^{\mathrm{ab}}/\mathcal{P}$. On the other hand, suppose that $s \in \mathrm{dom}(f)$. There is a pp formula $\rho(u, v)$ in the language $\mathcal{L}(R)$ that defines in ${}_RS$ the action given by left multiplication by s. This implies that ${}_RS \models \rho(1, s)$ and s is the unique element of S for which this satisfaction holds. In particular, ${}_RS \models \exists v\ \rho(1, v)$. The result follows if we can establish that R^{ab} is a pure left R-submodule of S. For then ${}_RR^{\mathrm{ab}} \models \exists v\ \rho(1, v)$, and the witness must be s. This puts s in R^{ab}. But the inclusion $R^{\mathrm{ab}} \subseteq {}_RS$ may be seen as an inclusion in R^{ab}-Mod, where it is necessarily a pure monomorphism, because R^{ab} is von Neumann regular ring. Since $R \to R^{\mathrm{ab}}$ is epic, it is also a pure monomorphism in R-Mod. □

Theorem 3.1 gives a characterization of the elements of $R^{\mathrm{ab}} \subseteq \prod\{\Delta \mid R \to \Delta \text{ epic}\}$. A pp formula $\rho(u, v)$ defines a scalar on the product $\prod \Delta$ if and only if it defines a scalar $\rho(\Delta) : \Delta \to \Delta$ on every epic R-field. It follows that the elements of R^{ab} are those of the form $(\rho(\Delta)(1_\Delta))_\Delta$, where ρ is such a pp formula.

4. The commutative case

Let us show that if R is a commutative ring, then $R_1^{\mathrm{ab}} = \hat{R}$ is the *commutative* universal regular R-algebra.

LEMMA 4.1. *If C is a commutative ring and R is a C-algebra, then so is every epic R-ring $f : R \to S$.*

PROOF. Because $f : R \to S$ is epic, the dominion of f is S, and the left action of every $s \in S$ is a definable scalar on ${}_RS$. The graph of this action is closed under the endomorphisms of ${}_RS$. It follows that the left action of s commutes with that of C. □

The lemma implies that if R is commutative, then every epic R-field $f : R \to k$ is commutative. This is easy to see, because the center of k is a field containing the image of f. In fact, the kernel of f is a prime ideal $\mathcal{P} \subseteq R$ and $k = k(\mathcal{P})$ is the field of fractions of $R/\mathcal{P}$. This implies that $E_1(f) = k(\mathcal{P})$ and that the extension

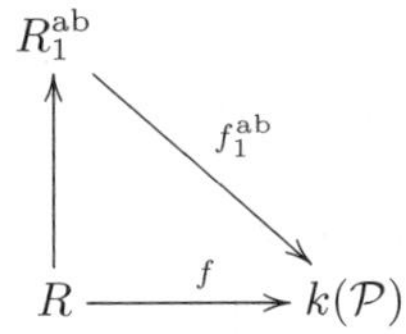

is onto.

THEOREM 4.2. *If R is a commutative ring, then $R_1^{\mathrm{ab}} = \hat{R}$ is the universal commutative regular R-algebra.*

PROOF. To show that R_1^{ab} is commutative, first observe that if R is commutative, then so is every epic R-ring of the form $R \to R\{\overline{r}\}$ constructed in §1. This is clear, because the lemma implies that $R\{\overline{r}\}$ is an R-algebra generated by the element $\overline{r}$. So if we adjoin one group inverse at a time, $R \to R\{\overline{r}\} \to R\{\overline{r}\}\{\overline{s}\}$, a commutative R-algebra is obtained. The universal property of $R\{\overline{r}\}\{\overline{s}\}$ ensures that there is morphism $R\{\overline{r}\}\{\overline{s}\} \to R_1^{\mathrm{ab}}$ that respects the group inverses of r and s. It follows that any two group inverses of elements in R commute in R_1^{ab}. As these group inverses generate the R-algebra R_1^{ab}, it must be commutative. The universal properties of $R \to R_1^{\mathrm{ab}}$ and $R \to \hat{R}$ now ensure that these two R-algebras are isomorphic. To see that $R_1^{\mathrm{ab}} = \hat{R}$ is regular, look at the proof of [**10**, Prop 4.2] which implies that every element of $\hat{R}$ has a reflexive inverse. It works even if R is not reduced, as the hypothesis there assumes. □

Theorem 4.2 implies that the iterated noncommutative variant of Olivier's construction is a generalization of his commutative construction. **It justifies the notation $R \to \hat{R} := R^{\mathrm{ab}}$ that we will use from now on.**

5. The lattice of pp definable subgroups

In general, we consider two pp formulae $\varphi(u)$ and $\psi(u)$ equivalent if for every left R-module ${}_RM$, $\varphi(M) = \psi(M)$, or, what amounts to the same thing, that the sentence $\varphi(u) \leftrightarrow \psi(u)$ is provable from the axioms for a left R-module. The lattice of pp formulae $\psi(u)$ in one variable is denoted by $\mathbb{L}(R, 1)$. This is a modular lattice where the infimum is given by $(\psi_1 \wedge \psi_2)(u) = \psi_1(u) \wedge \psi_2(u)$ and the supremum by

$$(\psi_1 + \psi_2)(u) = \exists v\ [\psi_1(u - v) \wedge \psi_2(v)].$$

There is a maximum element $u \doteq u$ in $\mathbb{L}(R, 1)$ and a minimum element $u \doteq 0$. A morphism $f : R \to S$ of rings induces a morphism of languages $\mathcal{L}(f) : \mathcal{L}(R) \to \mathcal{L}(S)$, which induces the obvious morphism $\mathbb{L}(f, 1) : \mathbb{L}(R, 1) \to \mathbb{L}(S, 1)$ of pp lattices. Similar considerations hold for the lattice $\mathbb{L}(R, n)$ of pp formulae in n variables.

If ${}_RM$ is a left R-module, the pp definable subgroups $\psi(M) \subseteq M$ represent the elements of the quotient lattice $\mathbb{L}(R, 1) \twoheadrightarrow \mathbb{L}(R, 1)_M$, $\psi(u) \mapsto \psi(M)$, modulo the congruence given by equivalence relative to M, $\varphi(M) = \psi(M)$. If $R \to R_M$ is the ring of definable scalars on M, then M can be considered as a left module over R_M and every pp definable subgroup $\psi(M)$ in $\mathbb{L}(R_M, 1)_M$ is pp definable over R. This is expressed by the commutative diagram

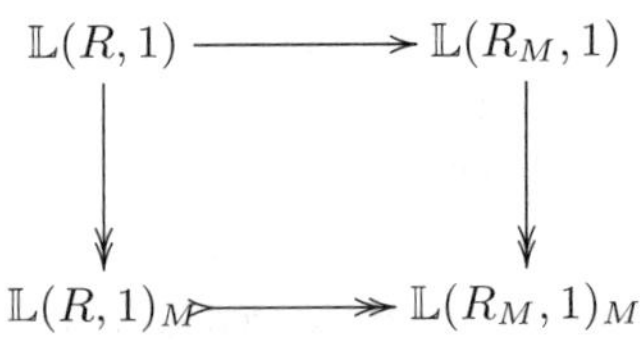

of bounded modular lattices, where the bottom horizontal arrow is an isomorphism. These considerations hold even if the number of variables is $n \geq 1$ and the case $n = 2$ includes an isomorphism $R_M \to (R_M)_M$ of the respective R-rings of definable scalars.

If R is a regular ring, then every pp formula $\psi(u)$ in one variable is equivalent to one of the form $e|u$ for some idempotent $e \in R$. Furthermore, the localization $\mathbb{L}(R,1) \to \mathbb{L}(R,1)_R$, $e|u \mapsto eR$ is an isomorphism. The lattice $\mathbb{L}(R,1)$ is therefore isomorphic to the lattice of principal right ideals of R. A bounded complemented modular lattice $\mathbb{L}$ is said to be *coordinatized* by a given von Neumann regular ring R, if $\mathbb{L}$ is isomorphic to the lattice $\mathbb{L}(R,1)_R$ of principal right ideals of the ring R. More generally, if M is a left R-module over a regular ring, then the vertical arrow in the diagram

$$\begin{array}{ccc} \mathbb{L}(R,1) & \longrightarrow & \mathbb{L}(R_M,1) \\ & \searrow & \downarrow \\ & & \mathbb{L}(R,1)_M \end{array}$$

is an isomorphism of lattices, so that $R_M = R/\operatorname{ann}(M)$ coordinatizes $\mathbb{L}(R,1)_M$. The following proposition provides a kind of converse.

PROPOSITION 5.1. *Let R be an associative ring and M a left R-module for which $\mathbb{L}(R,1)_M$ is complemented. Then the vertical arrow in the diagram*

$$\begin{array}{ccc} \mathbb{L}(R,1) & \longrightarrow & \mathbb{L}(R_M,1) \\ & \searrow & \downarrow \\ & & \mathbb{L}(R,1)_M \end{array}$$

is an isomorphism and $R \to R_M$ is a regular epic R-ring that coordinatizes $\mathbb{L}(R,1)_M$.

PROOF. This is proved in [**7**, Prop 9] or in [**16**, p. 28], the salient point being that if $\psi(u)$ is the complement of $\varphi(u)$ in $\mathbb{L}(R,1)_M$, i.e., that $\varphi(M) \oplus \psi(M) = M$, then there is a pp definable projection $E(\varphi) \in R_M$ onto $\varphi(M)$ parallel to $\psi(M)$; it is defined by the pp formula $E(\varphi)(u,v) = \varphi(v) \wedge \psi(u-v)$. □

Abelian regular rings can also be characterized by the property that every element in the lattice $\mathbb{L}(R,1)_R$ of principal right ideals has a unique complement.

THEOREM 5.2. *The following are equivalent for a left R-module ${}_RM$:*

(1) *$M \in \hat{R}$-Mod;*
(2) *the R-ring $R \to R_M$ of definable scalars is abelian regular; and*
(3) *every pp definable subgroup $\psi(M) \in \mathbb{L}(R,1)_M$ has a unique complement.*

PROOF. (1) $\Leftrightarrow$ (2). If M belongs to $\hat{R}$-Mod, then $R_M = \hat{R}_M = \hat{R}/\operatorname{ann}_{\hat{R}}(M)$ is abelian regular. Conversely, if $R \to R_M$ is abelian regular, then it factors through the universal abelian regular R-ring $R \to \hat{R}$, and M acquires the structure of a left $\hat{R}$-module. (2) $\Leftrightarrow$ (3). If the R-ring $R \to R_M$ of definable scalars is abelian regular, the considerations above imply that $\mathbb{L}(R,1)_M = \mathbb{L}(R_M,1)_M$, which is coordinatized by the abelian regular ring R_M. Conversely, if the lattice $\mathbb{L}(R,1)_M$ is complemented, then R_M is a regular ring that coordinatizes it. If these complements are unique, then R_M must be abelian. □

Important examples of complemented lattices have been explored in the context of the universal enveloping algebra U associated to the Lie algebra $\mathrm{sl}(2,k)$ (of all 2×2 traceless matrices with entries in a field k). Prest and the second author [**12**] investigated the infinite-dimensional Verma modules over U and proved that the ring of definable scalars is (non-abelian) regular. The second author and Macintyre [**11**] examined the decidability issue for the theory of finite dimensional U-modules and brought out a connection to fundamental problems in the diophantine geometry of curves. Finally, Herzog's analysis [**7**] of the theory of finite dimensional representations of U was generalized by the authors [**8**] for quantized universal enveloping algebras U_q, (often called quantum groups) when the parameter q is not a root of unity. Coming back to our framework, note that condition (2) of Theorem 5.2 can be used to axiomatize the elementary class $\hat{R}$-Mod $\subseteq R$-Mod in a way that is analogous to Propositions 1.2 and 2.1.

COROLLARY 5.3. *A module M belongs to $\hat{R}$-Mod if and only if for every definable scalar $\rho(u,v) \in R_M$,*

$$M \models \{\forall u\, \exists v, w\, [(u \doteq v + w) \wedge \exists u'\, \rho(u',v) \wedge \rho(w,0)]\} \wedge \forall u\, [(\rho(u,0) \wedge \exists v \rho(v,u)) \to u \doteq 0].$$

PROOF. The sentence expresses the condition that $M = \rho M \oplus \mathrm{ann}_M(\rho)$. Let $E \in R_M$ be the pp definable projection onto $\rho(M)$ parallel to $\mathrm{ann}_M(\rho)$. It follows that $EM = \rho M$, and therefore $E\rho = \rho$, and that $\rho(1-E) = 0$, or $\rho = \rho E$. Let us note that the restriction $\rho|_{EM} : EM \to EM$ is invertible: for every $y \in \rho M = EM$, there is a unique $x \in \rho M = EM$ such that $\rho x = y$. For if $x' \in EM$ were another, then $\rho(x-x') = 0$ and $x - x' = E(x-x')$. Because $\mathrm{ann}_M(\rho) \cap EM = 0$, we have that $x - x' = 0$. This observation shows that the group inverse of $\rho \in R_M$ is defined by the pp formula

$$\overline{\rho}(u,v) := \exists u', v'\, [\rho(v,u') \wedge \rho(u-u',0) \wedge \rho(v',v)],$$

because its restriction to EM is the inverse of $\rho|_{EM} : EM \to EM$, while the restriction to $(1-E)M = \mathrm{ann}_M(\rho)$ is 0. □

The sentences given in Corollary 5.3 form a system of axioms for $\hat{R}$-Mod, because the condition that $\rho(u,v)$ define a scalar on ${}_RM$ is expressible by (†) in the language $\mathcal{L}(R)$. A nicer system of axioms could be given if we can find, for every φ an explicit form for a pp formula $\varphi^\perp$ that defines in M the unique complement of $\varphi(M)$ in $\mathbb{L}(R,1)_M$. The axiom schema would then be of the form $\varphi(M) \oplus \varphi^\perp(M) = M$.

It would suffice to find such a $\varphi^\perp$ for the R-ring $R \to T = \prod_{\mathcal{P}} \hat{R}/\mathcal{P}$, because $\mathbb{L}(R,1)_T$ is coordinatized by $\hat{R}$. Given a pp formula $\varphi(u)$, the task therefore is to find a complementary pp formula $\varphi^\perp(u)$ such that for every epic R-field Δ, $\varphi^\perp(\Delta) = \Delta$ if and only if $\varphi(\Delta) = 0$. More concretely, Condition (3) of Theorem 5.2 implies that for every nonhomogeneous system $A\mathbf{v}^t \doteq A_0$, there exists a complementary system $B\mathbf{w}^t \doteq B_0$ such that for every epic R-field Δ, the first has a solution in Δ if and only if the second does not. It would be interesting to find an explicit form of the second, obtained constructively from the first, because it could be used as the starting point of the present theory.

This is possible when the ring R is commutative. We make implicit use of the Prest dual [**16**, §1.3.1] to see that a complementary system is given by the nonhomogeneous system $A^t\mathbf{w}^t \doteq 0 \wedge A_0^t\mathbf{w}^t \doteq 1$. For if k is an epic R-field, then the system $A\mathbf{v}^t \doteq A_0$ has a solution in k if and only if A_0 belongs to the column space of $A = (A_1, \dots, A_k)$, $A_0 = \sum_{i=1}^k c_i A_i$, which holds if and only if $\mathbf{w}A = 0$ implies $\mathbf{w}A_0 = 0$. The complementary system $\mathbf{w}A = 0 \wedge \mathbf{w}A_0 = 1$ is thus obtained, but the scalars of R act on the right.

6. The constructible Cohn spectrum

Denote by $\mathrm{Spec}(R)$ the Cohn spectrum [**1**, p. 410] of a ring R. The points of $\mathrm{Spec}(R)$ are the epic R-fields $R \to \Delta$, with a basis of quasi-compact open subsets given by

$$\mathcal{O}(A) := \{\Delta \mid A \text{ is invertible in } \Delta\},$$

as A ranges over the square matrices with entries in R. If R is commutative, then the points of $\mathrm{Spec}(R)$ are in bijective correspondence with the Zariski spectrum $\mathcal{P} \mapsto k(\mathcal{P})$. If $r \in R$, then $\mathcal{O}(r) = \{\mathcal{P} \mid r \notin \mathcal{P}\}$ corresponds to a basic Zariski open subset, so that the Cohn topology is at least as fine as the Zariski topology. On the other hand, if A is a square matrix over R, then $\mathcal{O}(A) = \mathcal{O}(\det(A))$ so that the Cohn topology is no finer than the Zariski topology. It follows that the Cohn spectrum $\mathrm{Spec}(R)$ is a generalization of the Zariski spectrum to the noncommutative case.

Like the Zariski spectrum, the Cohn spectrum satisfies the axioms [**1**, Exercise 13, p. 141] for a spectral topological space, in the sense of Hochster [**9**]. The rule $R \mapsto \mathrm{Spec}(R)$ is a contravariant functor from the category of rings to that of spectral topological spaces: if $f : R \to S$ is a morphism of rings, and $S \to \Delta$ is an epic S-field, then $R \to S \to \Delta$ is an R-field, whose image generates the epic R-field $R \to \mathrm{Spec}(f)(\Delta)$. The map $\mathrm{Spec}(f) : \mathrm{Spec}(S) \to \mathrm{Spec}(R)$ is continuous; if A is a square matrix over R, then $\mathrm{Spec}(f)^{-1}\mathcal{O}(A) = \mathcal{O}(f(A))$ in $\mathrm{Spec}(S)$.

There is another notion of spectrum $\mathrm{JSpec}(R)$ whose points are the prime ideals of the ring R, equipped with the Jacobson topology: the closed sets are given by $V(I) := \{\mathcal{P} \mid \mathcal{P} \supseteq I\}$, as I ranges over the ideals of R. If R is abelian regular, then the epic R-rings correspond to maximal ideals, $\mathcal{P} \mapsto R/\mathcal{P}$ and every prime ideal is maximal, so there is a bijective function $\mathrm{ann} : \mathrm{Spec}(R) \to \mathrm{JSpec}(R)$, $\Delta \mapsto \mathrm{ann}(\Delta)$. By [**3**, Prop 3.12], the topological space $\mathrm{JSpec}(R)$ is a totally disconnected compact space, with a clopen basis given by $\mathcal{O}(e) = \{\mathcal{P} \mid e \notin \mathcal{P}\}$, where e ranges over the (central) idempotent elements of R. This implies that ann is a continuous bijection from a quasi-compact space to a Hausdorff space. By an exercise in Topology, it must be a homeomorphism, and we have established the following.

Proposition 6.1. *If R is abelian regular, then the Cohn spectrum* $\mathrm{Spec}(R)$ *is a totally disconnected compact space with a clopen basis given by $\mathcal{O}(e)$, as e ranges over the idempotent elements in R.*

Because the Cohn spectrum is spectral, we may introduce the patch topology on it, as in [**9**]. This space is the *constructible* Cohn spectrum of R, denoted by $\widehat{\mathrm{Spec}}(R)$; an open basis is given by the boolean combinations of quasi-compact open subsets of $\mathrm{Spec}(R)$.

Theorem 6.2. (cf. [**22**, Thm 1]) *The universal abelian regular R-ring $R \to \hat{R}$ induces a homeomorphism* $\mathrm{Spec}(\hat{R}) \to \widehat{\mathrm{Spec}}(R)$, $\mathcal{P} \mapsto \hat{R}/\mathcal{P}$, *of constructible Cohn spectra.*

Proof. The induced function $\mathcal{P} \mapsto \hat{R}/\mathcal{P}$ is a bijection, by Corollary 2.3. The preimage in $\mathrm{Spec}(\hat{R})$ of a basic open subset $\mathcal{O}(A)$ in $\mathrm{Spec}(R)$ is quasi-compact, and therefore clopen in $\mathrm{Spec}(\hat{R})$, which implies that $\mathrm{Spec}(\hat{R}) \to \widehat{\mathrm{Spec}}(R)$, $\mathcal{P} \mapsto \hat{R}/\mathcal{P}$, is a bijective continuous function between compact Hausdorff spaces. By that same exercise from Topology, it is a homeomorphism. □

7. The Ziegler spectrum

The first author noticed [**6**, Thm 9.5] that the constructible Cohn spectrum $\widehat{\mathrm{Spec}}(R)$ of a ring R may be identified with the closed subspace $\mathrm{Zg}_1(R)$ of endosimple points of the Ziegler spectrum $\mathrm{Zg}(R)$. In this section, we observe that this embedding is natural, induced by the universal abelian regular R-ring.

The Ziegler spectrum $\mathrm{Zg}(R)$ of a ring R is the space [**16**, **23**] whose points are given by indecomposable pure injective left R-modules, topologized so that a basis of open subsets is given by

$$\mathcal{O}(\varphi/\psi) := \{U \in \mathrm{Zg}(R) \mid \varphi(U)/\psi(U) \neq 0\},$$

as $\psi \leq \varphi$ range over $\mathbb{L}(R, 1)$. The quasi-compact open subsets of this topology have the form $\mathcal{O}(\varphi/\psi)$, as $\psi \leq \varphi$ range over the various $\mathbb{L}(R, n)$, $n \geq 1$.

A module ${}_RU$ is *endosimple* if it is simple as a module over its endomorphism ring $\mathrm{End}_R U$. An endosimple module U is pure injective [**16**, Cor 4.4.24] so that every such indecomposable module represents a point $U \in \mathrm{Zg}(R)$. For example, every epic R-field $R \to \Delta$ becomes, by restriction of scalars, an indecomposable endosimple left R-module and it is shown in [**6**, §9] how every indecomposable endosimple left R-module arises in this way. The subspace $\mathrm{Zg}_1(R)$ of endosimple points of $\mathrm{Zg}(R)$ forms a closed subset [**16**, Example 5.1.16]. If we endow $\mathrm{Zg}_1(R)$ with the relative subspace topology, then the quasi-compact open subsets of $\mathrm{Zg}_1(R)$ are also closed, for suppose that $\psi \leq \varphi$ in $\mathbb{L}(R,1)$ and $\Delta \in \mathcal{O}(\varphi/\psi)$. Then $\varphi(\Delta) = \Delta$ and $\psi(\Delta) = 0$, which implies that in $Zg_1(R)$,

$$\mathcal{O}(\varphi/\psi)^c = \mathcal{O}(u \doteq u/\varphi(u)) \cup \mathcal{O}(\psi(u)/u \doteq 0)$$

is also open. Now every quasi-compact open subset is a finite union of such open subsets and therefore has an open complement. It follows that the Ziegler topology restricted to $\mathrm{Zg}_1(R)$ is its own patch topology.

If R is abelian regular, then the points of the Ziegler spectrum are given by the endosimple modules $R/\mathcal{P}$, as $\mathcal{P}$ ranges over the prime, i.e., maximal, ideals. To see that every indecomposable pure injective left R-module is of this form, note that if M is indecomposable, then every idempotent $e \in R$ induces a decomposition $M = eM \oplus (1-e)M$ so that either e or $1-e$ belongs to $\mathrm{ann}(M)$. This implies that $\mathrm{ann}(M) = \mathcal{P}$ is a prime ideal and that $M = R/\mathcal{P}$. Thus $\mathrm{Zg}(R) = \mathrm{Zg}_1(R)$ is totally disconnected.

PROPOSITION 7.1. *If R is abelian regular, then $\mathrm{Zg}(R) = \mathrm{Zg}_1(R)$ and the function* $\mathrm{ann} : \mathrm{Zg}(R) \to \mathrm{Spec}(R)$, $\Delta \mapsto \mathrm{ann}(\Delta)$, *is a homeomorphism.*

PROOF. The first equality has been established. To prove that ann is a homeomorphism, it suffices to verify that it is continuous, because it is then a bijective continuous function of compact Hausdorff spaces. The preimage of a basic open subset $\mathcal{O}(e) \subseteq \mathrm{Spec}(R)$ consists of those Δ for which $e\Delta = \Delta$. This is the Ziegler open subset $\mathcal{O}(e|u/u \doteq 0)$. □

If $f : R \to S$ is an epic R-ring and $U \in \mathrm{Zg}(S)$ is an indecomposable pure injective left S-module, then the left R-module ${}_RU$ obtained by restriction of scalars is also indecomposable pure injective, ${}_RU \in \mathrm{Zg}(R)$. The function ${}_SU \mapsto {}_RU$ is then a homeomorphic embedding $\mathrm{Zg}(f) : \mathrm{Zg}(S) \to \mathrm{Zg}(R)$ onto a closed subset. These ideas are clearly explained in [**16**, §5.5.1]. For the universal abelian regular R-ring $R \to \hat{R}$, these consideration take on the following form.

THEOREM 7.2. *The universal abelian regular R-ring $R \to \hat{R}$ induces a homeomorphism*

$$\widehat{\mathrm{Spec}}(R) = \mathrm{Zg}(\hat{R}) \to \mathrm{Zg}_1(R) \subseteq \mathrm{Zg}(R)$$

from the constructible Cohn spectrum of R to the closed subset of endosimple points in the Ziegler spectrum.

PROOF. The only claim that requires a comment is that $\mathrm{Zg}(\hat{R})$ is homeomorphic to the constructible Cohn spectrum. By Proposition 7.1, it is homeomorphic to $\mathrm{Spec}(\hat{R})$, which is homeomorphic to $\widehat{\mathrm{Spec}}(R)$, by Theorem 6.2. □

8. The étale bundle of definable scalars

The theory of the Pierce spectrum [**15**, Thm 4.4] (but see [**2**]) shows how to present an abelian regular ring R as the ring of global sections of a sheaf over $\mathrm{Spec}(R)$. In this section, we indicate how this sheaf is an instance of Prest's notion of the sheaf of locally definable scalars over the Ziegler spectrum $\mathrm{Zg}^*(R)$ - of an arbitrary ring - endowed with the Zariski topology. The universal abelian R-ring $\iota : R \to \hat{R}$ is then presented as the the ring of global sections of the inverse image sheaf along the homeomorphic embedding $\mathrm{Zg}^*(\iota) : \mathrm{Zg}^*(\hat{R}) \to \mathrm{Zg}^*(R)$ of the constructible Cohn spectrum.

The Ziegler spectrum $\mathrm{Zg}(R)$ of a ring R satisfies enough of the axioms of a spectral topological space so that a dual topology may be given, whose basic open subsets are the complements $\mathcal{O}(\varphi/\psi)^c$, $\psi \leq \varphi \in \mathbb{L}(R,n)$, of the quasi-compact open subsets of $\mathrm{Zg}(R)$. This is explained nicely in [**16**, §5.6] and this topology on $\mathrm{Zg}(R)$ is called the Zariski topology and denoted by $\mathrm{Zg}^*(R)$. In general, it is not so clear what the quasi-compact open subsets of $\mathrm{Zg}^*(R)$ are, but it is obvious that every clopen Ziegler subset is also clopen in the the Zariski topology. If R is abelian regular, Proposition 7.1 implies that $\mathrm{Zg}^*(R) = \mathrm{Zg}(R)$.

If ${}_RU \in \mathrm{Zg}(R)$, then the endomorphism ring $\mathrm{End}_R\, U$ is local. If R is commutative, there is a continuous function $\mathcal{P} : \mathrm{Zg}^*(R) \to \mathrm{Spec}(R)$, given by

$$U \mapsto \mathcal{P}(U) := \{r \in R \mid r : U \to U \text{ is not invertible}\}.$$

To verify continuity, let $\mathcal{O}(r) = \{\mathcal{P} \mid r \notin \mathcal{P}\} \subseteq \mathrm{Spec}(R)$ be a basic open subset and note that

$$\mathcal{P}^{-1}(\mathcal{O}(r)) = \mathcal{O}(u \doteq u/\exists v\ (rv \doteq u))^c \ \cap\ \mathcal{O}(rv \doteq 0/v \doteq 0)^c$$

is open in $\mathrm{Zg}^*(R)$. This is a feature of the Zariski topology that the Ziegler topology does not enjoy.

More generally, let $\rho(u, v)$ be a pp formula in two variables. Then ρ defines a scalar on every point in the Zariski open subset

$$\mathcal{O}_{\mathrm{Zar}}(\rho(u, v)) := \mathcal{O}(u \doteq u/\exists v\ \rho(u, v))^c \ \cap\ \mathcal{O}(\rho(0, v)/v \doteq 0)^c$$

of $\mathrm{Zg}^*(R)$. Thus $U \in \mathcal{O}_{\mathrm{Zar}}(\rho(u, v))$ if and only if $\rho(U) \in R_U$. For example, if $r \in R$, then $\mathcal{O}_{\mathrm{Zar}}(rv \doteq u)$ is $\mathcal{P}^{-1}(\mathcal{O}(r))$, which consists of those indecomposable pure injective modules ${}_RU$ on which r^{-1} is a definable scalar. Let $\mathrm{Bun}(R) := \dot{\bigcup} \{R_U \mid U \in \mathrm{Zg}^*(R)\}$ be the disjoint union of the R_U and define $p : \mathrm{Bun}(R) \to \mathrm{Zg}^*(R)$ to be the function whose fiber over U is $p^{-1}\{U\} = R_U$. There is a commutative diagram

$$\begin{array}{ccc} & & \mathrm{Bun}(R) \\ & \overset{\mathrm{Ev}(\rho)}{\nearrow} & \big\downarrow p \\ \mathcal{O}_{\mathrm{Zar}}(\rho(u, v)) & \xrightarrow{\ \subseteq\ } & \mathrm{Zg}^*(R), \end{array}$$

where $\mathrm{Ev}(\rho)(U) := \rho(U) \in R_U$. The function p acquires the structure of a *topological bundle* [**13**, §II.4] over $\mathrm{Zg}^*(R)$ if we endow $\mathrm{Bun}(R)$ with the coarsest topology for which p and all the maps $\mathrm{Ev}(\rho) : \mathcal{O}_{\mathrm{Zar}}(\rho(u, v)) \to \mathrm{Bun}(R)$, $\rho(u, v) \in \mathbb{L}(R, 2)$, are open continuous.

PROPOSITION 8.1. *The topological bundle* $p : \mathrm{Bun}(R) \to \mathrm{Zg}^*(R)$ *is an étale bundle, with a subbasis of open subsets for* $\mathrm{Bun}(R)$ *given by the images* $\mathrm{Im}\ \mathrm{Ev}(\rho)$, $\rho(u, v) \in \mathbb{L}(R, 2)$, *and preimages* $p^{-1}(\mathcal{O})$, *as* $\mathcal{O}$ *ranges over a basis for* $\mathrm{Zg}^*(R)$. *A basis of open subsets is given by finite intersections of these:* $\bigcap_i \mathrm{Im}\ \mathrm{Ev}(\rho_i) \cap p^{-1}(\mathcal{O})$.

PROOF. Let us first check that the given subsets are open. If $\mathrm{Ev}(\rho) : \mathcal{O}_{\mathrm{Zar}}(\rho) \to \mathrm{Bun}(R)$ is open, then its image must be open and if $p : \mathrm{Bun}(R) \to \mathrm{Zg}^*(R)$ is continuous, then all of the preimages $p^{-1}(\mathcal{O})$, with $\mathcal{O} \subseteq \mathrm{Zg}^*(R)$ are open.

Conversely, let us show that for any topology containing this open basis, the maps $\mathrm{Ev}(\rho)$ and p are open continuous. All the subsets $p^{-1}(\mathcal{O})$ are open in $\mathrm{Bun}(R)$, so that p is continuous. To see that p is open, let us see that the image under p of a basic open subset $\bigcap_i \mathrm{Im}\ \mathrm{Ev}(\rho)_i \cap p^{-1}(\mathcal{O}) \subseteq \mathrm{Bun}(R)$ is open. To that end, let $\rho(U)$ belong to this basic open subset. It means that $U \in \mathcal{O}$ and that for all i, $\rho(U) = \rho_i(U)$. Consider the Zariski open subset $\mathcal{O}' \subseteq \mathcal{O}$ given by

$$\mathcal{O}' = \bigcap_i \mathcal{O}_{\mathrm{Zar}}(\rho_i) \ \cap\ \bigcap_{i<j} \mathcal{O}(u \doteq u/(\rho_i - \rho_j)(u, 0))^c \ \cap\ \mathcal{O}.$$

Then $U \in \mathcal{O}' \subseteq p[\bigcap_i \mathrm{Im}\ \mathrm{Ev}(\rho)_i \cap p^{-1}(\mathcal{O})]$ and $\mathcal{O}'$ is the open subset of $\mathcal{O}$ on which all of the ρ_i's define the same scalar.

Now let us prove that $\mathrm{Ev}(\rho)$ is continuous. It is injective, so it suffices to prove that for every open subset $\mathrm{Im}\ \mathrm{Ev}(\sigma) \cap p^{-1}(\mathcal{O})$ in the subbasis, $\mathrm{Ev}(\rho)^{-1}[\mathrm{Im}\ \mathrm{Ev}(\sigma) \cap p^{-1}(\mathcal{O})]$ is open in $\mathcal{O}_{\mathrm{Zar}}(\rho)$. Well, the elements of $\mathrm{Im}\ \mathrm{Ev}(\rho) \cap \mathrm{Im}\ \mathrm{Ev}(\sigma) \cap p^{-1}(\mathcal{O})$ are those of the form $\rho(U)$, where $U \in \mathcal{O}$ and $\rho(U) = \sigma(U)$. As above, the preimage of that is the open subset

$$\mathcal{O}_{\mathrm{Zar}}(\rho) \cap \mathcal{O}_{\mathrm{Zar}}(\sigma) \cap \mathcal{O}(u \doteq u/(\rho - \sigma)(u, 0))^c \cap \mathcal{O}.$$

That $\mathrm{Ev}(\rho)$ is open follows from the fact that its image is open in $\mathrm{Bun}(R)$ and that $p|_{\mathrm{Im}\ \mathrm{Ev}(\rho)}$ is a continuous inverse.

Finally, recall that a topological bundle $p : E \to X$ is *étale* if for every point $e \in E$, there is an open neighborhood $e \in \mathcal{O} \subseteq E$ such that the restriction $p|_{\mathcal{O}} : \mathcal{O} \to X$ is a homeomorphism onto an open subset $p(\mathcal{O}) \subseteq X$. Such a neighborhood$(\mathcal{O}, e)$ is called *étale.* Because the section $\mathrm{Ev}(\rho)$ on $\mathcal{O}_{\mathrm{Zar}}(\rho)$ is a homeomorphism onto an open image, it is clear that if $\rho(U) \in R_U$, then the image $(\mathrm{Im}\ \mathrm{Ev}(\rho), \rho(U))$ is an étale neighborhood. □

There is a natural correspondence [**13**, Cor II.6.3] between étale bundles and sheaves over a topological space B. For the étale bundle $p : \mathrm{Bun}(R) \to \mathrm{Zg}^*(R)$, the corresponding sheaf is the *sheaf* Def_R *of sections.* It assigns to an open subset $\mathcal{O} \subseteq \mathrm{Zg}^*(R)$ the R-ring $R \to \mathrm{Def}_R(\mathcal{O})$ of continuous maps $s : \mathcal{O} \to \mathrm{Bun}(R)$ for which the diagram

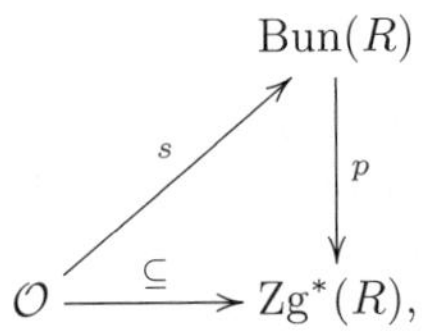

commutes. It is clear that the ring $\mathrm{Def}_R(\mathrm{Zg}^*(R))$ of global sections carries an R-ring structure, because the action of every $r \in R$ gives a global scalar. The R-ring structure on $\mathrm{Def}_R(\mathcal{O})$ is then inherited via the restriction map $R \to \mathrm{Def}_R(\mathrm{Zg}^*(R)) \to \mathrm{Def}_R(\mathcal{O})$.

A section $s \in \mathrm{Def}_R(\mathcal{O})$ is *definable* if there is a pp formula $\rho(u, v)$ such that $\mathcal{O} \subseteq \mathcal{O}_{\mathrm{Zar}}(\rho)$ and $s = \mathrm{Ev}(\rho)|_{\mathcal{O}}$. Prest [**16**, Chs 6 & 14] defines the notion of a *presheaf-on-a-basis* of definable scalars, which assigns to a basic open subset $\mathcal{O} \subseteq \mathrm{Zg}^*(R)$ the R-ring of definable sections on $\mathcal{O}$. The sheaf Def_R on $\mathrm{Zg}^*(R)$ is the sheafification [**13**, Thm 11.2.3] of this presheaf; the following proposition shows that it is the *sheaf of locally definable scalars* on $\mathrm{Zg}^*(R)$.

PROPOSITION 8.2. *If $\mathcal{O} \subseteq \mathrm{Zg}^*(R)$ is open, then $s \in \mathrm{Def}_R(\mathcal{O})$ if and only if for every point $U \in \mathcal{O}$, there is an open neighborhood $U \in \mathcal{O}' \subseteq \mathcal{O}$ such that $s|_{\mathcal{O}'}$ is definable.*

PROOF. The collation property of the sheaf Def_R ensures that every section s of the kind described belongs to $\mathrm{Def}_R(\mathcal{O})$. On the other hand, let $s \in \mathrm{Def}_R(\mathcal{O})$ and pick $\rho(U) \in R_U$ in the image $\mathrm{Im}\ s$. The intersection of an étale neighborhood $(\mathrm{Im}\ \mathrm{Ev}(\rho), \rho(U))$ with the image $\mathrm{Im}\ s$ of s is itself an étale neighborhood of $\rho(U)$. Then $\mathcal{O}' = p(\mathrm{Im}\ \mathrm{Ev}(\rho) \cap \mathrm{Im}\ s) \subseteq \mathcal{O}$ is a neighborhood such that $s|_{\mathcal{O}'} = \mathrm{Ev}(\rho)|_{\mathcal{O}'}$, as required. □

If the ring R is abelian regular, then, according to the next proposition, the sheaf Def_R over $\mathrm{Zg}^*(R)$ coincides with the sheaf Γ over $\mathrm{Spec}(R)$ introduced in [**2**, **15**].

PROPOSITION 8.3. *If R is abelian regular, then* $\mathrm{Bun}(R) = \dot{\bigcup} \{R/\mathcal{P} \mid \mathcal{P} \in \mathrm{Spec}(R)\}$, *and for a quasi-compact open subset $\mathcal{O}(e) \subseteq \mathrm{Zg}^*(R)$,*

$$\mathrm{Def}_R(\mathcal{O}(e)) = eRe.$$

In particular, the ring R is represented as the R-ring of global sections $R \to \mathrm{Def}_R(\mathrm{Zg}^*(R))$.

PROOF. If R is abelian regular, then $\mathrm{Zg}(R) = \mathrm{Zg}^*(R)$ is the space whose points are the division rings $\Delta = R/\mathcal{P}$, $\mathcal{P} \in \mathrm{Spec}(R)$, and a basis of quasi-compact open subsets is given by $\mathcal{O}(e) = \{\Delta \mid e\Delta = \Delta\}$, with $e \in R$ idempotent. Each $R \to \Delta$ is an epic R-ring, so that its dominion is given by $R_\Delta = \Delta$; the topological space $\mathrm{Bun}(R)$ is therefore the disjoint union of these Δ.

Let us first show that the statement holds for every definable section $s \in \mathrm{Def}_R(\mathcal{O}(e))$. So suppose that $\rho(u, v)$ defines a scalar on every division ring $\Delta = R/\mathcal{P}$ such that $e\Delta = \Delta$. Then it defines a definable scalar on the R-ring $f : R \to \prod_{e\Delta=\Delta} \Delta$, and hence belongs to the dominion of f. The kernel of f is $(1 - e)R$, so that the dominion of f is the dominion of $eRe = R/(1-e)R \to \prod_{e\Delta=\Delta} \Delta$. But $\mathrm{Spec}(eRe) = \mathcal{O}(e)$, so that the claim follows from Theorem 3.1.

The foregoing considerations imply that every definable section $s \in \mathrm{Def}_R(\mathcal{O}(e))$ extends to a global section of the form ere. This global section ere has the property that $ere|_{\mathcal{O}(e)} = s$, while $ere|_{\mathcal{O}(1-e)} = 0$. If $t \in \mathcal{O}(e)$, then Proposition 8.2 implies that there is a finite partition $\mathcal{O}(e) = \dot\bigcup_i \mathcal{O}(e_i)$ and that $t|_{\mathcal{O}(e_i)}$ is the restriction of a global scalar $e_i r_i e_i \in e_i R e_i$. Then t is the restriction of $\sum_i e_i r_i e_i \in eRe$ to $\mathcal{O}(e)$, as required. $\square$

If $f : R \to S$ is an epic R-ring, then the induced homeomorphic embedding $\mathrm{Zg}(f) : \mathrm{Zg}(S) \to \mathrm{Zg}(R)$ is also continuous with respect to the Zariski topology $\mathrm{Zg}(f) = \mathrm{Zg}^*(f) : \mathrm{Zg}^*(S) \to \mathrm{Zg}^*(R)$. The action of every element $s \in S$ on a left S-module ${}_SM$ is a definable scalar over R. So if $U \in \mathrm{Zg}^*(S)$ is an indecomposable pure injective, then $S_U = R_U$ and we get the obvious map $\mathrm{Bun}(f) : \mathrm{Bun}(S) \to \mathrm{Bun}(R)$ of étale bundles.

PROPOSITION 8.4. *The morphism of étale bundles given by*

$$\begin{array}{ccc} \mathrm{Bun}(S) & \xrightarrow{\mathrm{Bun}(f)} & \mathrm{Bun}(R) \\ \Big\downarrow{\scriptstyle p_S} & & \Big\downarrow{\scriptstyle p_R} \\ \mathrm{Zg}^*(S) & \xrightarrow{\mathrm{Zg}^*(f)} & \mathrm{Zg}^*(R), \end{array}$$

where $p_S = p_R|_{\mathrm{Bun}(S)}$, *is a pullback diagram.*

PROOF. It suffices to verify that the topology on $\mathrm{Bun}(S) \subseteq \mathrm{Bun}(R)$ is the relative subspace topology, so we just need to check that an open subset from the subbasis of $\mathrm{Bun}(S)$ given in Proposition 8.1 is open in the relative subspace topology. The topology on $\mathrm{Zg}^*(S) \subseteq \mathrm{Zg}^*(R)$ is given by the relative subspace topology, so that the Zariski open subsets of $\mathrm{Zg}^*(S)$ have the form $\mathcal{O} \cap \mathrm{Zg}^*(S)$, where $\mathcal{O} \subseteq \mathrm{Zg}^*(R)$ is Zariski open. Together with the fact that $p_S = p_R|_{\mathrm{Bun}(S)}$, these considerations imply

$$p_S^{-1}(\mathcal{O} \cap \mathrm{Zg}^*(S)) = p_R^{-1}(\mathcal{O}) \cap \mathrm{Bun}(S).$$

The fact that the action of S is definable over R, implies that if $\rho(u, v)$ is a pp formula over S, then there is a pp formula $\rho'(u, v)$ over R such that $\mathcal{O}_{\mathrm{Zar}}(\rho) = \mathcal{O}_{\mathrm{Zar}}(\rho') \cap \mathrm{Zg}^*(S)$ and the section $\mathrm{Ev}(\rho) : \mathcal{O}_{\mathrm{Zar}}(\rho) \to \mathrm{Bun}(S)$ is the restriction of $\mathrm{Ev}(\rho') : \mathcal{O}_{\mathrm{Zar}}(\rho') \to \mathrm{Bun}(R)$ to $\mathrm{Zg}^*(S)$. It follows that $\mathrm{Im}\ \mathrm{Ev}(\rho) = \mathrm{Im}\ \mathrm{Ev}(\rho') \cap \mathrm{Bun}(S)$. $\square$

If $f : R \to S$ is an epic R-ring, then the correspondence [**13**, Cor II.6.3] between sheaves and topological bundles, implies that the sheaf of locally definable scalars on $\mathrm{Zg}^*(S)$ is given by the pullback along $\mathrm{Zg}^*(f)$ of the sheaf of locally definable scalars on $\mathrm{Zg}^*(R)$. This is also referred to as the inverse image sheaf [**13**, §II.9],

$$\mathrm{Def}_S = [\mathrm{Zg}^*(f)]^*(\mathrm{Def}_R).$$

Let us combine the contents of the previous two propositions for the universal abelian regular R-ring.

THEOREM 8.5. *Let R be a ring with universal abelian regular R-ring $R \to \hat{R}$. The sheaf* $\mathrm{Def}_{\hat{R}}$ *of locally definable scalars over the constructible Cohn spectrum* $\widehat{\mathrm{Spec}}(R) = \mathrm{Zg}^*(\hat{R})$ *is obtained by pullback of the sheaf* Def_R *along the homeomorphic embedding* $\mathrm{Zg}^*(\hat{R}) \to \mathrm{Zg}^*(R)$. *It has the property that for every quasi-compact open subset* $\mathcal{O}(e)$, $\mathrm{Def}_{\hat{R}}(\mathcal{O}(e)) = e\hat{R}e$.

Acknowledgement

We thank Pace Nielsen for his critical reading of an earlier version and numerous helpful suggestions.

References

[1] P. M. Cohn, *Free rings and their relations*, 2nd ed., London Mathematical Society Monographs, vol. 19, Academic Press, Inc. [Harcourt Brace Jovanovich, Publishers], London, 1985. MR800091

[2] J. Dauns and K. H. Hofmann, *The representation of biregular rings by sheaves*, Math. Z. **91** (1966), 103–123, DOI 10.1007/BF01110158. MR0186693

[3] K. R. Goodearl, *Von Neumann regular rings*, 2nd ed., Robert E. Krieger Publishing Co., Inc., Malabar, FL, 1991. MR1150975

[4] J. L. Fisher, *Embedding free algebras in skew fields*, Proc. Amer. Math. Soc. **30** (1971), 453–458, DOI 10.2307/2037715. MR0281750

[5] D. Herbera and J. Sánchez, *Computing the inversion height of some embeddings of the free algebra and the free group algebra*, J. Algebra **310** (2007), no. 1, 108–131, DOI 10.1016/j.jalgebra.2006.11.020. MR2307784

[6] I. Herzog, *The Ziegler spectrum of a locally coherent Grothendieck category*, Proc. London Math. Soc. (3) **74** (1997), no. 3, 503–558, DOI 10.1112/S002461159700018X. MR1434441

[7] I. Herzog, *Pseudo-finite dimensional representations of* $\mathrm{sl}(2,k)$, Selecta Math. (N.S.) **7** (2001), no. 2, 241–290, DOI 10.1007/PL00001402. MR1860015

[8] I. Herzog and S. L'Innocente, *The nonstandard quantum plane*, Ann. Pure Appl. Logic **156** (2008), no. 1, 78–85, DOI 10.1016/j.apal.2008.06.009. MR2474442

[9] M. Hochster, *Prime ideal structure in commutative rings*, Trans. Amer. Math. Soc. **142** (1969), 43–60, DOI 10.2307/1995344. MR0251026

[10] J. F. Kennison, *Structure and costructure for strongly regular rings*, J. Pure Appl. Algebra **5** (1974), 321–332, DOI 10.1016/0022-4049(74)90041-3. MR0371987

[11] S. L'Innocente and A. MacIntyre, *Towards decidability of the theory of pseudo-finite dimensional representations of* $\mathrm{sl}_2(k)$. *I*, Andrzej Mostowski and foundational studies, IOS Press, Amsterdam, 2008, pp. 235–260. MR2422688

[12] S. L'Innocente and M. Prest, *Rings of definable scalars of Verma modules*, J. Algebra Appl. **6** (2007), no. 5, 779–787, DOI 10.1142/S0219498807002430. MR2355619

[13] S. Mac Lane and I. Moerdijk, *Sheaves in geometry and logic: A first introduction to topos theory*, Universitext, Springer-Verlag, New York, 1994. Corrected reprint of the 1992 edition. MR1300636

[14] Olivier, J.-P., Anneaux absolutment plats universels et épimorphismes à buts réduits, Séminaire Samuel, Algèbre commutative **2**(6) (1967-1968), 1-12.

[15] R. S. Pierce, *Modules over commutative regular rings*, Memoirs of the American Mathematical Society, No. 70, American Mathematical Society, Providence, R.I., 1967. MR0217056

[16] M. Prest, *Purity, spectra and localisation*, Encyclopedia of Mathematics and its Applications, vol. 121, Cambridge University Press, Cambridge, 2009. MR2530988

[17] K. P. S. Bhaskara Rao, *The theory of generalized inverses over commutative rings*, Algebra, Logic and Applications, vol. 17, Taylor & Francis, Ltd., London, 2002. MR1919967

[18] C. Reutenauer, *Inversion height in free fields*, Selecta Math. (N.S.) **2** (1996), no. 1, 93–109, DOI 10.1007/BF01587940. MR1403353

[19] Sanchez, J., *Localization: On Division Rings and Tilting Modules,* Doctoral Dissertation, UAB, Bellaterra, 2008.

[20] T. R. Savage, *Generalized inverses in regular rings*, Pacific J. Math. **87** (1980), no. 2, 455–466. MR592745

[21] A. H. Schofield, *Representation of rings over skew fields*, London Mathematical Society Lecture Note Series, vol. 92, Cambridge University Press, Cambridge, 1985. MR800853

[22] R. Wiegand, *Modules over universal regular rings*, Pacific J. Math. **39** (1971), 807–819. MR0379472

[23] M. Ziegler, *Model theory of modules*, Ann. Pure Appl. Logic **26** (1984), no. 2, 149–213, DOI 10.1016/0168-0072(84)90014-9. MR739577

The Ohio State University at Lima, Lima, OH 45804 USA
Email address: herzog.23@osu.edu

Università di Camerino, Camerino, ITALY
Email address: sonia.linnocente@unicam.it

Contemporary Mathematics
Volume **730**, 2019
https://doi.org/10.1090/conm/730/14711

A characterisation of τ-tilting finite algebras

Lidia Angeleri Hügel, Frederik Marks, and Jorge Vitória

ABSTRACT. We prove that a finite dimensional algebra is τ-tilting finite if and only if it does not admit large silting modules. Moreover, we show that for a τ-tilting finite algebra A there is a bijection between isomorphism classes of basic support τ-tilting (that is, finite dimensional silting) modules and equivalence classes of ring epimorphisms $A \longrightarrow B$ with $\mathrm{Tor}_1^A(B,B) = 0$. It follows that a finite dimensional algebra is τ-tilting finite if and only if there are only finitely many equivalence classes of such ring epimorphisms.

Introduction

Adachi, Iyama and Reiten [**AIR**] defined support τ-tilting modules in the context of finite dimensional algebras, motivated by the study of mutation in cluster theory. Their work has inspired the notion of a silting module, which was introduced in [**AMV1**] with the aim of extending the concept of support τ-tilting modules to arbitrary rings. In this broader context, however, it is necessary to allow for large modules. For example, over a commutative ring A there are no non-trivial finitely presented silting modules, but there is a rich supply of large silting modules parametrising the Gabriel localisations of A, see [**AH**] for details. Similarly, over finite dimensional algebras some phenomena only become visible when leaving the finite dimensional world. For example, as illustrated in [**AMV2**], certain posets associated to the algebra, such as the poset of noncrossing partitions considered in [**IT**], can be completed to lattices by dropping finiteness assumptions. Hence, it is often useful to pass to large modules – provided they do exist.

A well-known result due to Auslander states that a finite dimensional algebra does not admit large indecomposable modules if and only if it has finite representation type. In this note, we will ask about the existence of large silting modules, that is, infinite dimensional silting modules which are not obtained trivially as direct sums of finite dimensional ones. We will prove that a finite dimensional algebra does not admit large silting modules if and only if it is τ-tilting finite. This class of algebras introduced in [**DIJ**] encompasses representation-finite algebras, local

2010 *Mathematics Subject Classification.* Primary 16G20; Secondary 16S85, 16S90.
Key words and phrases. τ-tilting finite algebra, silting module, ring epimorphism.
The authors acknowledge funding from the Project "Ricerca di Base 2015" of the University of Verona. Additionally, the third named author acknowledges support from the Department of Computer Sciences of the University of Verona in the earlier part of this project, as well as from the Engineering and Physical Sciences Research Council of the United Kingdom, grant number EP/N016505/1, in the later part of this project.

algebras, and preprojective algebras of Dynkin type [**Mi**]. An algebra A is called τ-tilting finite if there are only finitely many isomorphism classes of basic support τ-tilting (that is, finite dimensional silting) modules. We show that the latter correspond bijectively to equivalence classes of ring epimorphisms $A \longrightarrow B$ with $\mathrm{Tor}_1^A(B,B) = 0$. As a consequence, we obtain that a finite dimensional algebra A is τ-tilting finite if and only if there are only finitely many equivalence classes of such ring epimorphisms.

Notation. Throughout this note, let A be a (unital) ring, Mod-A the category of right A-modules, and mod-A its subcategory of finitely presented modules. We denote by Proj-A and proj-A the class of all projective and of all finitely generated projective right A-modules, respectively. Furthermore, we write $\mathsf{D}(A)$ for the derived category of Mod-A. Given a module $M \in$ Mod-A, we denote by Add(M) the class of all modules which are isomorphic to direct summands of direct sums of copies of M, by Gen(M) the class of all M-generated modules, i. e. all modules N admitting a surjection $M^{(I)} \longrightarrow N$ for some set I, and we set $\mathrm{gen}(M) = \mathrm{Gen}(M) \cap \text{mod-}A$. Moreover, we define

$$M^{\circ} = \{X \in \text{Mod-}A \mid \mathrm{Hom}_A(M,X) = 0\}$$

$$M^{\perp_1} = \{X \in \text{Mod-}A \mid \mathrm{Ext}_A^1(M,X) = 0\}.$$

We fix a ground field $\mathbb{K}$, and all algebras considered in this note are $\mathbb{K}$-algebras.

1. Silting modules and ring epimorphisms

1.1. Silting modules. We first introduce silting modules over a ring A. Given a set Σ of morphisms in Proj-A, we define the subcategory

$$\mathcal{D}_\Sigma := \{X \in \text{Mod-}A \mid \mathrm{Hom}_A(\sigma, X) \text{ is surjective for all } \sigma \in \Sigma\}.$$

Note that, if we identify a morphism σ with its cone in $\mathsf{D}(A)$, then we can write

$$\mathcal{D}_\Sigma = \{X \in \text{Mod-}A \mid \mathrm{Hom}_{\mathsf{D}(A)}(\sigma, X[1]) = 0 \text{ for all } \sigma \in \Sigma\}.$$

If Σ contains just one element σ, we write $\mathcal{D}_\sigma$.

Definition 1.1. [**AMV1**] We say that an A-module T is

- *partial silting* if it admits a projective presentation $P \xrightarrow{\sigma} Q \longrightarrow T \longrightarrow 0$ such that $\mathcal{D}_\sigma$ is a torsion class containing T;
- *silting* if it admits a projective presentation $P \xrightarrow{\sigma} Q \longrightarrow T \longrightarrow 0$ such that $\mathcal{D}_\sigma = \mathrm{Gen}(T)$.

We say that T is (partial) silting *with respect to* σ. The torsion class Gen(T) generated by a silting module T is called a *silting class.* Two silting modules T and T' are said to be *equivalent* if they generate the same silting class or, equivalently, $\mathrm{Add}(T) = \mathrm{Add}(T')$.

By [**AMV1**, Theorem 3.12] every partial silting module T_1 with respect to a presentation σ_1 can be completed to a silting module T, called the *Bongartz completion* of T_1, such that $\mathrm{Gen}(T) = \mathcal{D}_{\sigma_1}$. It was also shown in [**AMV1**] that every silting module T induces an exact sequence of the form

$$A \xrightarrow{f} T_0 \longrightarrow T_1 \longrightarrow 0 \tag{1.1}$$

where T_0 and T_1 are in Add(T) and f is a left Add(T)-approximation.

Recall that a module T is *tilting* if it is silting with respect to a projective resolution $0 \longrightarrow P \xrightarrow{\sigma} Q \longrightarrow T \longrightarrow 0$, or equivalently, if $T^{\perp_1} = \mathrm{Gen}(T)$. In this case, the module T_1 in (1.1) is a partial silting module (actually, partial tilting). We do not know, however, whether this is always true. In this subsection we explore some cases that allow us to conclude that T_1 is partial silting.

We say that an A-module M admits a *presilting presentation* if there are P and Q in Proj-A and an exact sequence

$$P \xrightarrow{\sigma} Q \longrightarrow M \longrightarrow 0$$

such that $M^{(I)} \in \mathcal{D}_\sigma$ or, equivalently, $\mathrm{Hom}_{\mathsf{D}(A)}(\sigma, \sigma^{(I)}[1]) = 0$, for all sets I.

EXAMPLE 1.2. (1) It follows from the definition above that every partial silting module admits a presilting presentation.

(2) If a module M admits both a minimal projective presentation p and a presilting presentation, then p is a presilting presentation. Indeed, p must be a summand of the presilting presentation, making it presilting as well.

PROPOSITION 1.3. *Let T_0 be an A-module, and let $A \xrightarrow{f} T_0 \longrightarrow T_1 \longrightarrow 0$ be an* $\mathrm{Add}(T_0)$*-approximation sequence. If T_0 admits a presilting presentation, then $T = T_0 \oplus T_1$ is a silting A-module, and T_1 is a partial silting module with respect to a projective presentation σ_1 such that* $\mathrm{Gen}(T) = \mathcal{D}_{\sigma_1}$.

PROOF. Let σ_0 be a presilting presentation of T_0. Then, the A-module map f can be lifted to a morphism of chain complexes $A \longrightarrow \sigma_0$. Consider the induced triangle in the derived category $\mathsf{D}(A)$

$$A \longrightarrow \sigma_0 \longrightarrow \sigma_1 \longrightarrow A[1].$$

By construction, σ_1 is a projective presentation of T_1. We show that $T = T_0 \oplus T_1$ is a silting A-module with respect to the presentation $\sigma_0 \oplus \sigma_1$, that is, $\mathrm{Gen}(T) = \mathcal{D}_{\sigma_0 \oplus \sigma_1}$. First, observe that $\mathcal{D}_{\sigma_0 \oplus \sigma_1} = \mathcal{D}_{\sigma_1}$. The inclusion $\mathcal{D}_{\sigma_0 \oplus \sigma_1} \subseteq \mathcal{D}_{\sigma_1}$ is clear. For the converse, apply the functor $\mathrm{Hom}_{\mathsf{D}(A)}(-, X[1])$ for X in $\mathcal{D}_{\sigma_1}$ to the triangle above. Since $\mathrm{Hom}_{\mathsf{D}(A)}(A, X[1]) = 0 = \mathrm{Hom}_{\mathsf{D}(A)}(\sigma_1, X[1])$, we also get $\mathrm{Hom}_{\mathsf{D}(A)}(\sigma_0, X[1]) = 0$ or, in other words, $X \in \mathcal{D}_{\sigma_0}$. Hence, we conclude that X lies in $\mathcal{D}_{\sigma_0} \cap \mathcal{D}_{\sigma_1} = \mathcal{D}_{\sigma_0 \oplus \sigma_1}$.

It remains to show that $\mathrm{Gen}(T) = \mathrm{Gen}(T_0) = \mathcal{D}_{\sigma_1}$. Using that f is a left $\mathrm{Add}(T_0)$-approximation, we see that an A-module X belongs to $\mathrm{Gen}(T_0)$ if and only if the map $\mathrm{Hom}_A(f, X) : \mathrm{Hom}_A(T_0, X) \longrightarrow \mathrm{Hom}_A(A, X)$ is surjective. This amounts to the surjectivity of the map $\pi : \mathrm{Hom}_{\mathsf{D}(A)}(\sigma_0, X) \longrightarrow \mathrm{Hom}_{\mathsf{D}(A)}(A, X)$ which is obtained by applying the functor $\mathrm{Hom}_{\mathsf{D}(A)}(-, X)$ to the triangle above. Considering the exact sequence

$$\mathrm{Hom}_{\mathsf{D}(A)}(\sigma_0, X) \xrightarrow{\pi} \mathrm{Hom}_{\mathsf{D}(A)}(A, X) \to \mathrm{Hom}_{\mathsf{D}(A)}(\sigma_1, X[1]) \to \mathrm{Hom}_{\mathsf{D}(A)}(\sigma_0, X[1]),$$

it follows immediately that $\mathrm{Gen}(T_0) \supseteq \mathcal{D}_{\sigma_1}$. Moreover, since σ_0 is a presilting presentation of T_0, the class $\mathcal{D}_{\sigma_0}$ contains $\mathrm{Add}(T_0)$ and therefore also $\mathrm{Gen}(T_0)$. So, if $X \in \mathrm{Gen}(T_0)$, then $\mathrm{Hom}_{\mathsf{D}(A)}(\sigma_0, X[1]) = 0$ and π is surjective, hence $\mathrm{Hom}_{\mathsf{D}(A)}(\sigma_1, X[1]) = 0$. This proves that $\mathrm{Gen}(T_0) \subseteq \mathcal{D}_{\sigma_1}$. □

COROLLARY 1.4. *If T is a silting module such that every module in* $\mathrm{Add}(T)$ *admits a minimal projective presentation, then any module T_1 appearing in a left* $\mathrm{Add}(T)$*-approximation sequence of the form (1.1) is partial silting with respect to a projective presentation σ_1 such that* $\mathcal{D}_{\sigma_1} = \mathrm{Gen}(T)$.

PROOF. By Proposition 1.3, it is enough to show that T_0 admits a presilting presentation. Since T_0 lies in $\mathrm{Add}(T)$, it is a summand of $T^{(I)}$, for some set I. Hence, the minimal projective presentation of T_0 is a summand of the minimal projective presentation of $T^{(I)}$. Since T has a presilting presentation, it follows from Example 1.2(2) that the minimal projective presentation of $T^{(I)}$ is presilting and, therefore, so is the minimal projective presentation of T_0. □

There is a different way of interpreting the statements above. Recall from [**Wei**] that if T is a silting module with respect to σ, then there is a triangle

$$A \xrightarrow{\phi} \sigma_0 \longrightarrow \sigma_1 \longrightarrow A[1] \tag{1.2}$$

in $\mathsf{D}(A)$, where σ_0 and σ_1 lie in $\mathrm{Add}(\sigma)$ and ϕ is a left $\mathrm{Add}(\sigma)$-approximation of A. Note that applying the functor $H^0(-)$ to the triangle above yields a left $\mathrm{Add}(T)$-approximation sequence in Mod-A as in (1.1) for which $T_1 := H^0(\sigma_1)$ is partial silting, by Proposition 1.3. In fact, if every module in $\mathrm{Add}(T)$ admits a minimal projective presentation, then every exact sequence as in (1.1) can be lifted to a triangle as in (1.2). This is precisely Corollary 1.4.

DEFINITION 1.5. Let T be a silting module with respect to σ. If the map ϕ in the triangle (1.2) above can be chosen left-minimal (i. e. any $h \in \mathrm{End}_{\mathsf{D}(A)}(\sigma_0)$ with $hf = f$ is an isomorphism), then T is said to be a *minimal* silting module.

Using that left-minimal maps are essentially unique, we can associate to every minimal silting module T a canonical partial silting module $T_1 := H^0(\mathsf{cone}(\phi))$. This assignment will become useful later in this note.

REMARK 1.6. If A is (right) hereditary or perfect, then a silting module T with respect to σ is minimal if and only if there is an exact sequence as in (1.1) for which the left $\mathrm{Add}(T)$-approximation f can be chosen left-minimal, compare with [**AMV2**, Definition 5.4].

Indeed, it is true for any ring A that a left-minimal $\mathrm{Add}(\sigma)$-approximation of A in $\mathsf{D}(A)$ induces a left-minimal $\mathrm{Add}(T)$-approximation of A in Mod-A. To see this, given a left-minimal $\mathrm{Add}(\sigma)$-approximation $\phi : A \longrightarrow \sigma_0$, we check that for any endomorphism g of $H^0(\sigma_0)$ fulfilling $gH^0(\phi) = H^0(\phi)$ there is an endomorphism $\tilde{g}$ of σ_0 in $\mathsf{D}(A)$ such that $H^0(\tilde{g}) = g$ and $\tilde{g}\phi = \phi$. The minimality of ϕ then guarantees that $\tilde{g}$ is an isomorphism and, hence, so is g.

For the converse, we first need to assure that a left $\mathrm{Add}(T)$-approximation $f : A \longrightarrow T_0$ in Mod-A can be lifted to a left $\mathrm{Add}(\sigma)$-approximation $\phi : A \longrightarrow \sigma_0$ in $\mathsf{D}(A)$. Here, we use that if A is (right) hereditary or perfect, then H^0 induces a dense functor $\mathrm{Add}(\sigma) \longrightarrow \mathrm{Add}(T)$ (see also Corollary 1.4). Secondly, observe that we can choose ϕ to be left-minimal provided that f is left-minimal. Indeed, in the hereditary (respectively, perfect) case we can choose ϕ with σ_0 being a monomorphic projective presentation (respectively, a minimal projective presentation) of T_0.

1.2. Ring epimorphisms.

We first recall some notions and basic results.

DEFINITION 1.7. A ring homomorphism $f : A \longrightarrow B$ is a *ring epimorphism* if it is an epimorphism in the category of rings with unit, or equivalently, if the functor given by restriction of scalars $f_* : \text{Mod-}B \longrightarrow \text{Mod-}A$ is fully faithful.

Two ring epimorphisms $f_1 : A \longrightarrow B_1$ and $f_2 : A \longrightarrow B_2$ are said to be *equivalent* if there is an isomorphism of rings $h : B_1 \longrightarrow B_2$ such that $f_2 = h \circ f_1$. We then say that f_1 and f_2 lie in the same *epiclass* of A.

DEFINITION 1.8. A full subcategory $\mathcal{X}$ of Mod-A is called *bireflective* if the inclusion functor $\mathcal{X} \longrightarrow$ Mod-A admits both a left and right adjoint or, equivalently, if it is closed under products, coproducts, kernels and cokernels. A full subcategory $\mathcal{W}$ of mod-A is said to be *wide* if it is closed under kernels, cokernels and extensions.

THEOREM 1.9. (1) [**GdP, GL**] *There is a bijection between*

- *epiclasses of ring epimorphisms $A \longrightarrow B$;*
- *bireflective subcategories of* Mod-A

mapping a ring epimorphism $A \longrightarrow B$ to the essential image $\mathcal{X}_B$ of its restriction functor.

(2) [**Sch**, Theorem 4.8] *The bijection in (1) restricts to a bijection between*
   - *epiclasses of ring epimorphisms $A \longrightarrow B$ with $\operatorname{Tor}_1^A(B,B) = 0$;*
   - *bireflective subcategories closed under extensions in* Mod-A.

(3) [**I**, Theorem 1.6.1] *If A is a finite dimensional algebra, mapping a ring epimorphism $A \longrightarrow B$ to $\mathcal{X}_B \cap$ mod-A defines a bijection between*
   - *epiclasses of ring epimorphisms $A \longrightarrow B$ with B finite dimensional and $\operatorname{Tor}_1^A(B,B) = 0$;*
   - *functorially finite wide subcategories of* mod-A.

We will be particularly interested in ring epimorphisms that arise from partial silting modules. Given a set Σ of maps in Proj-A, we define the subcategory

$$\mathcal{X}_\Sigma := \{X \in \text{Mod-}A \mid \operatorname{Hom}_A(\sigma, X) \text{ is bijective for all } \sigma \in \Sigma\}.$$

Similar to before, identifying a morphism σ with its cone in $\mathsf{D}(A)$, we can write

$$\mathcal{X}_\Sigma = \{X \in \text{Mod-}A \mid \operatorname{Hom}_{\mathsf{D}(A)}(\sigma, X[1]) = 0 = \operatorname{Hom}_{\mathsf{D}(A)}(\sigma, X) \text{ for all } \sigma \in \Sigma\}.$$

If Σ contains just one element σ, we write $\mathcal{X}_\sigma$.

If T_1 is a partial silting module with respect to a projective presentation σ_1, then we have the inclusions $\operatorname{Gen}(T_1) \subseteq \mathcal{D}_{\sigma_1} \subseteq {T_1}^{\perp_1}$, and $(\operatorname{Gen}(T_1), T_1^\circ)$ is a torsion pair. Let T be the Bongartz completion of T_1 with $\operatorname{Gen}(T) = \mathcal{D}_{\sigma_1}$. The subcategory

$$\mathcal{X}_{\sigma_1} = \mathcal{D}_{\sigma_1} \cap \operatorname{Coker}(\sigma_1)^\circ = \operatorname{Gen}(T) \cap T_1^\circ$$

is bireflective and extension-closed by [**AMV2**, Proposition 3.3]. Thus, it can be realised as $\mathcal{X}_B$ for a ring epimorphism $A \longrightarrow B$ with $\operatorname{Tor}_1^A(B,B) = 0$ as in Theorem 1.9(2), and by [**AMV2**, Theorem 3.5] the ring B is isomorphic to an idempotent quotient of the endomorphism ring of T.

DEFINITION 1.10. A ring epimorphism arising from a partial silting module as above will be said to be a *silting ring epimorphism*.

The following result provides a large supply of silting ring epimorphisms.

THEOREM 1.11. *Let Σ be a set of morphisms in* proj-A.

(1) [**Sch**, Theorem 4.1] *There is a ring homomorphism $f_\Sigma : A \longrightarrow A_\Sigma$, called the* universal localisation *of A at Σ, such that*
   (i) *f_Σ is* Σ-inverting, *i.e. $\sigma \otimes_A A_\Sigma$ is an isomorphism for every σ in Σ.*
   (ii) *f_Σ is* universal Σ-inverting, *i.e. every Σ-inverting ring homomorphism $f : A \longrightarrow B$ factors uniquely through f_Σ.*
(2) [**MS2**, Theorem 6.7] *The ring homomorphism $f_\Sigma \colon A \longrightarrow A_\Sigma$ is a silting ring epimorphism.*

Note that $\mathcal{X}_\Sigma$ is precisely the bireflective subcategory corresponding to f_Σ under the bijection in Theorem 1.9, see [**Sch**, p. 52 and 57].

2. From torsion classes to abelian subcategories

For any torsion class $\mathcal{T}$ in Mod-A, or more generally, in an abelian category $\mathcal{A}$, we consider the following subcategory of $\mathcal{A}$ studied in [**IT**]

$$\mathfrak{a}(\mathcal{T}) := \{X \in \mathcal{T} \mid \text{ if } (g : Y \longrightarrow X) \in \mathcal{T}, \text{ then } \operatorname{Ker}(g) \in \mathcal{T}\}. \tag{2.1}$$

PROPOSITION 2.1. [**IT**, Proposition 2.12] *The subcategory* $\mathfrak{a}(\mathcal{T})$ *of* $\mathcal{A}$ *is closed under kernels, cokernels, and extensions.*

In general, however, $\mathfrak{a}(\mathcal{T})$ is not a bireflective subcategory, not even in case $\mathcal{T}$ is a tilting class, as the following example shows.

EXAMPLE 2.2. Let A be the Kronecker algebra. It is well-known that the class $\mathcal{T}$ of all A-modules without indecomposable preprojective summands is a tilting class, that is, $\mathcal{T} = \operatorname{Gen}(L)$ for a tilting module L, called the Lukas tilting module. We will show that $\mathfrak{a}(\mathcal{T})$ is not closed under coproducts.

We first claim that $\mathfrak{a}(\mathcal{T})$ contains all regular modules. By Proposition 2.1, it is enough to check that every simple regular (that is, quasisimple) module S lies in $\mathfrak{a}(\mathcal{T})$. Moreover, since A is hereditary, we know from [**IT**, Proposition 2.15] that

$$\mathfrak{a}(\mathcal{T}) = \{X \in \mathcal{T} \mid \text{ if } (g : Y \longrightarrow X) \in \mathcal{T} \text{ is surjective, then } \operatorname{Ker}(g) \in \mathcal{T}\}.$$

Certainly, $S \in \mathcal{T} = \operatorname{Gen}(L)$. Assume that there is a surjection $g : X \longrightarrow S$ with X in $\operatorname{Gen}(L)$ and $K := \operatorname{Ker}(g)$ not lying in $\operatorname{Gen}(L)$. Since $\operatorname{Gen}(L)$ is the class of all modules without indecomposable preprojective summands, K must have an indecomposable preprojective summand P. Consider the pushout diagram

$$\begin{array}{ccccccccc}
0 & \longrightarrow & K & \longrightarrow & X & \xrightarrow{g} & S & \longrightarrow & 0 \\
 & & \downarrow{\scriptstyle \pi} & & \downarrow{\scriptstyle \pi'} & & \| & & \\
0 & \longrightarrow & P & \longrightarrow & X' & \xrightarrow{g'} & S & \longrightarrow & 0
\end{array}$$

with π the canonical projection onto P. The module X' is finite dimensional. It has no non-zero preprojective summands because π' is surjective and thus X' lies in $\operatorname{Gen}(L)$. Further, it has no non-zero preinjective summands, because they would have to be summands of $\operatorname{Ker}(g') = P$. But then X' is regular, and $\operatorname{Ker}(g')$ must be regular as well, a contradiction.

Now assume that $\mathfrak{a}(\operatorname{Gen}(L))$ is closed under coproducts. Then all direct limits of regular modules, in particular all Prüfer modules, would belong to $\mathfrak{a}(\operatorname{Gen}(L))$. But this is not possible: for any Prüfer module S_∞ there is a short exact sequence

$$0 \longrightarrow A \longrightarrow A_S \longrightarrow S_\infty{}^{(2)} \longrightarrow 0$$

where A_S is the universal localisation of A at the minimal projective presentation of S, and $A_S \in \operatorname{Gen}(L)$ since the category of A_S-modules identifies with the subcategory $S^\circ \cap S^{\perp_1}$ of Mod-A, which cannot contain indecomposable preprojective modules. For details on universal localisations and tilting modules over the Kronecker algebra we refer to [**AS**].

On the other hand, if $\mathcal{T}$ is a silting class generated by a minimal silting module, then $\mathfrak{a}(\mathcal{T})$ is bireflective. Indeed, the following proposition expresses the abelian category $\mathfrak{a}(\operatorname{Gen}(T))$ as a subcategory that is also closed under products and coproducts.

PROPOSITION 2.3. [**AMV2**, Remark 5.7] *Let T be a minimal silting module with associated partial silting module T_1. Then we have that*

$$\mathfrak{a}(\mathrm{Gen}(T)) = \mathrm{Gen}(T) \cap T_1^\circ.$$

To summarise, given a minimal silting module T, we can associate to it a canonical partial silting module and, therefore, a silting ring epimorphism, and it turns out that the corresponding bireflective subcategory is precisely $\mathfrak{a}(\mathrm{Gen}(T))$. The following corollary collects these findings.

COROLLARY 2.4. *There is a commutative diagram as follows.*

$$\begin{array}{ccc}
\left\{\begin{array}{c}\text{Equivalence classes}\\ \text{of minimal}\\ \text{silting } A\text{-modules}\end{array}\right\} & \xrightarrow{\quad\alpha\quad} & \left\{\begin{array}{c}\text{Epiclasses of ring}\\ \text{epimorphisms } A \to B\\ \text{with } \mathrm{Tor}_1^A(B,B)=0\end{array}\right\} \\
 & \searrow^{\mathfrak{a}} \qquad \swarrow_{\epsilon} & \\
 & \left\{\begin{array}{c}\text{Bireflective}\\ \text{extension-closed}\\ \text{subcategories}\\ \text{of Mod-}A\end{array}\right\} &
\end{array}$$

The map α assigns to the minimal silting module T the associated silting ring epimorphism. The map ϵ is the bijection from Theorem 1.9*(1) and $\mathfrak{a}$ is the map defined in (*2.1*) for the abelian category* Mod-A.

Over a finite dimensional algebra, the triangle above restricts to the level of finite dimensional modules.

THEOREM 2.5. [**MS1**] *If A is a finite dimensional algebra, there is a commutative diagram of injections as follows.*

$$\begin{array}{ccc}
\left\{\begin{array}{c}\text{Equivalence classes}\\ \text{of finite dimensional}\\ \text{silting } A\text{-modules}\end{array}\right\} & \xrightarrow{\quad\alpha\quad} & \left\{\begin{array}{c}\text{Epiclasses of ring}\\ \text{epimorphisms } A \to B\\ \text{with } \mathrm{Tor}_1^A(B,B)=0,\\ B \text{ finite dimensional}\end{array}\right\} \\
 & \searrow^{\mathfrak{a}} \qquad \swarrow_{\epsilon} & \\
 & \left\{\begin{array}{c}\text{Functorially}\\ \text{finite wide}\\ \text{subcategories}\\ \text{of mod-}A\end{array}\right\} &
\end{array}$$

Here, the map α is as above, ϵ is the bijection from Theorem 1.9*(3), and $\mathfrak{a}$ is the map defined in (*2.1*) for the abelian category* mod-A. *Moreover, ring epimorphisms in the image of the assignment α are universal localisations.*

PROOF. Observe that every finite dimensional silting module T is minimal. If T_1 is the associated partial silting module (with respect to a projective presentation σ_1), then it follows from the very construction that the associated silting ring epimorphism is the universal localisation of A at $\{\sigma_1\}$. By [**AMV2**, Theorem 3.5], the algebra $A_{\{\sigma_1\}}$ is finite dimensional. In particular, α is well-defined. Moreover, by

[**MS1**, Lemma 3.8], the functorially finite torsion class $\operatorname{gen}(T)$ gives rise to the functorially finite wide subcategory $\mathfrak{a}(\operatorname{gen}(T)) = \operatorname{gen}(T) \cap T_1^\circ = \mathcal{X}_{\sigma_1} \cap \operatorname{mod}\text{-}A$. Therefore, the above diagram commutes. Finally, $\mathfrak{a}$ is injective by [**MS1**, Proposition 3.9]. □

It was shown in [**Asai**] that the maps α and $\mathfrak{a}$ in Theorem 2.5 are surjective only once we add further assumptions. We will pass to such a situation in the last section of this note.

3. From abelian subcategories to torsion classes

The injectivity of the map $\mathfrak{a}$ in Theorem 2.5 is proved in [**MS1**] by providing a left inverse map. Our next aim is to find an analogous map for large modules. Proceeding as in [**MS1**], we will try to recover a given torsion class $\mathcal{T}$ from $\mathfrak{a}(\mathcal{T})$ by considering the class of modules admitting filtrations by modules in $\operatorname{Gen}(\mathfrak{a}(\mathcal{T}))$.

Definition 3.1. Given a module M and an ordinal number μ, we call an ascending chain $\mathcal{M} = (M_\lambda, \lambda \leq \mu)$ of submodules of M a *filtration* of M, if $M_0 = 0$, $M_\mu = M$, and $M_\nu = \bigcup_{\lambda<\nu} M_\lambda (= \varinjlim_{\lambda<\nu} M_\lambda)$ for each limit ordinal $\nu \leq \mu$. Moreover, given a class of modules $\mathcal{C}$, we call $\mathcal{M}$ a *$\mathcal{C}$-filtration* of M, provided that each of the consecutive factors $M_{\lambda+1}/M_\lambda$ ($\lambda < \mu$) is isomorphic to a module from $\mathcal{C}$. A module M admitting a $\mathcal{C}$-filtration is said to be *$\mathcal{C}$-filtered*, and the class of all $\mathcal{C}$-filtered modules is denoted by $\operatorname{Filt}(\mathcal{C})$.

Lemma 3.2. *Let $\mathcal{C}$ be a full subcategory of* Mod-A *closed under coproducts and quotients. Then* $\operatorname{Filt}(\mathcal{C})$ *is a torsion class in* Mod-A.

Proof. We verify that $\operatorname{Filt}(\mathcal{C})$ is closed under coproducts, extensions and quotients. The closure under coproducts and extensions is easy. Indeed, since coproducts are exact in Mod-A and $\mathcal{C}$ is coproduct-closed, the coproduct of filtrations with factors in $\mathcal{C}$ is still a $\mathcal{C}$-filtration. On the other hand, given a short exact sequence of the form

$$0 \longrightarrow X \longrightarrow Y \longrightarrow Z \longrightarrow 0$$

with X and Z in $\operatorname{Filt}(\mathcal{C})$, with filtrations given by, respectively, $(X_\lambda, \lambda \leq \mu)$ and $(Z_\omega, \omega \leq \nu)$ for some ordinals μ and ν. For each Z_ω, we can choose a subobject Y_ω of Y such that $Y_\omega/X \cong Z_\omega$. We claim that $(W_\theta)_{\theta \leq \mu+\nu}$ given by $W_\theta = X_\theta$ whenever $\theta \leq \mu$ and $W_\theta = Y_\theta$ whenever $\mu < \theta \leq \mu + \nu$ is a $\mathcal{C}$-filtration of Y. It suffices to observe that $W_{\mu+1}/W_\mu = W_{\mu+1}/X = Y_1/X \cong Z_1$ which lies in $\mathcal{C}$.

It remains to show that $\operatorname{Filt}(\mathcal{C})$ is closed under quotients. Given a $\mathcal{C}$-filtration $(X_\lambda : \lambda \leq \mu)$ of X and a surjection $f\colon X \longrightarrow Y$, we can consider the restrictions $f_\lambda\colon X_\lambda \longrightarrow Y$ for each $\lambda \leq \mu$. Clearly, $(\operatorname{Im}(f_\lambda) : \lambda \leq \mu)$ is a filtration of Y since, for any limit ordinal $\nu \leq \mu$, we have

$$\bigcup_{\lambda<\nu} \operatorname{Im}(f_\lambda) = f(\bigcup_{\lambda<\nu} X_\lambda) = f(X_\nu) = \operatorname{Im}(f_\nu).$$

We show that indeed $(\operatorname{Im}(f_\lambda) : \lambda \leq \mu)$ is a $\mathcal{C}$-filtration of Y. In fact, for each $\lambda < \mu$, we have the following diagram.

$$\begin{array}{ccccccccc}
0 & \longrightarrow & X_\lambda & \longrightarrow & X_{\lambda+1} & \longrightarrow & X_{\lambda+1}/X_\lambda & \longrightarrow & 0 \\
 & & \downarrow{\scriptstyle \tilde{f}_\lambda} & & \downarrow{\scriptstyle \tilde{f}_\lambda} & & \downarrow{\scriptstyle g} & & \\
0 & \longrightarrow & \operatorname{Im}(f_\lambda) & \longrightarrow & \operatorname{Im}(f_{\lambda+1}) & \longrightarrow & \operatorname{Im}(f_{\lambda+1})/\operatorname{Im}(f_\lambda) & \longrightarrow & 0
\end{array}$$

Since all vertical maps are surjective and $\mathcal{C}$ is closed under quotients, it follows that $\mathrm{Im}(f_{\lambda+1})/\mathrm{Im}(f_\lambda)$ lies in $\mathcal{C}$, as wanted. $\square$

If $\mathcal{C}$ is a class of modules, we use the abbreviation $\mathrm{FiltGen}(\mathcal{C}) = \mathrm{Filt}(\mathrm{Gen}(\mathcal{C}))$. By the lemma above, this is the smallest torsion class in Mod-A containing $\mathcal{C}$.

LEMMA 3.3. *Let $\mathcal{C}$ be a full subcategory of* Mod-A *closed under kernels, cokernels, coproducts and extensions. Then $\mathcal{C}$ is closed under subobjects in* $\mathrm{FiltGen}(\mathcal{C})$.

PROOF. Let C be in $\mathcal{C}$ and $f : Y \longrightarrow C$ be an injection with Y in $\mathrm{FiltGen}(\mathcal{C})$. Suppose that Y lies in $\mathrm{Gen}(\mathcal{C})$. Since $\mathcal{C}$ is closed under coproducts, there is a surjection $g : W \longrightarrow Y$ for some W in $\mathcal{C}$. Hence, Y is isomorphic to the image of the map $fg \colon W \longrightarrow C$. Since $\mathcal{C}$ is closed under kernels and cokernels, Y lies in $\mathcal{C}$.

Now we consider a $\mathrm{Gen}(\mathcal{C})$-filtration $(Y_\lambda : \lambda \leq \mu)$ of Y and we proceed by transfinite induction. Let $\lambda < \mu$ and suppose Y_λ lies in $\mathcal{C}$. Since $\mathcal{C}$ is closed under cokernels, C/Y_λ lies in $\mathcal{C}$ as well (we can assume without loss of generality that f is an inclusion map). Hence $Y_{\lambda+1}/Y_\lambda$, which lies in $\mathrm{Gen}(\mathcal{C})$ and is a subobject of C/Y_λ, also lies in $\mathcal{C}$. Since $\mathcal{C}$ is extension-closed, we conclude that $Y_{\lambda+1}$ lies in $\mathcal{C}$. Finally, let $\nu \leq \mu$ be a limit ordinal, and assume that Y_λ lies in $\mathcal{C}$ for any $\lambda < \nu$. Since $\mathcal{C}$ is closed under coproducts and cokernels, it is also closed under direct limits and, therefore, $Y_\nu = \varinjlim_{\lambda<\nu} Y_\lambda$ lies in $\mathcal{C}$. $\square$

The following result can be viewed as a large version of [**MS1**, Proposition 3.3].

PROPOSITION 3.4. *Let $\mathcal{C}$ be a full subcategory of* Mod-A *closed under kernels, cokernels, coproducts and extensions. Then we have* $\mathfrak{a}(\mathrm{FiltGen}(\mathcal{C})) = \mathcal{C}$.

PROOF. Let us first show that $\mathcal{C}$ is contained in $\mathfrak{a}(\mathrm{FiltGen}(\mathcal{C}))$. Consider a map $f : X \longrightarrow C$ with X in $\mathrm{FiltGen}(\mathcal{C})$ and C in $\mathcal{C}$. By the previous lemma, without loss of generality we may assume f to be surjective. We have to show that $K := \mathrm{Ker}(f)$ lies in $\mathrm{FiltGen}(\mathcal{C})$. If X lies in $\mathrm{Gen}(\mathcal{C})$, then there is W in $\mathcal{C}$ and a surjection $g : W \longrightarrow X$ inducing the following diagram.

$$\begin{array}{ccccccccc}
0 & \longrightarrow & L & \xrightarrow{a} & W & \xrightarrow{fg} & C & \longrightarrow & 0 \\
 & & \downarrow{\scriptstyle b} & & \downarrow{\scriptstyle g} & & \| & & \\
0 & \longrightarrow & K & \longrightarrow & X & \xrightarrow{f} & C & \longrightarrow & 0
\end{array}$$

Since g is a surjection, so is b, and since $\mathcal{C}$ is closed under kernels, L lies in $\mathcal{C}$. Hence, K lies in $\mathrm{Gen}(\mathcal{C})$. Now consider a $\mathrm{Gen}(\mathcal{C})$-filtration $(X_\lambda : \lambda \leq \mu)$ of X. It is easy to see that $(K_\lambda : \lambda \leq \mu)$, where K_λ is the kernel of the restriction $f_\lambda : X_\lambda \longrightarrow C$ of f, is a filtration of $K = \mathrm{Ker}(f)$. We show that this is in fact a $\mathrm{Gen}(\mathcal{C})$-filtration. Let $\lambda < \mu$ and consider the following commutative diagram

$$\begin{array}{ccccccccc}
0 & \longrightarrow & X_\lambda & \xrightarrow{a} & X_{\lambda+1} & \xrightarrow{p} & X_{\lambda+1}/X_\lambda & \longrightarrow & 0 \\
 & & \downarrow{\scriptstyle \tilde{f}_\lambda} & & \downarrow{\scriptstyle f_{\lambda+1}} & & \downarrow{\scriptstyle b} & & \\
0 & \longrightarrow & Im(f_\lambda) & \longrightarrow & C & \xrightarrow{d} & C/Im(f_\lambda) & \longrightarrow & 0
\end{array}$$

with $\tilde{f}_\lambda$ being the natural factorisation of f_λ via its image. By Lemma 3.3, we have that $\mathrm{Im}(f_\lambda)$ lies in $\mathcal{C}$. Therefore, so does $C/\mathrm{Im}(f_\lambda)$. By the previous argument, using that $X_{\lambda+1}/X_\lambda$ lies in $\mathrm{Gen}(\mathcal{C})$, we get that $\mathrm{Ker}(b) \cong K_{\lambda+1}/K_\lambda$ lies in $\mathrm{Gen}(\mathcal{C})$.

It remains to show that $\mathfrak{a}(\text{FiltGen}(\mathcal{C})) \subseteq \mathcal{C}$. Take Y in $\mathfrak{a}(\text{FiltGen}(\mathcal{C}))$. Then Y lies in FiltGen($\mathcal{C}$), and we proceed by transfinite induction as before. If Y lies in Gen($\mathcal{C}$), then there is a surjection $f : C \longrightarrow Y$ with C in $\mathcal{C}$ (as $\mathcal{C}$ is closed under coproducts). Since Y lies in $\mathfrak{a}(\text{FiltGen}(\mathcal{C}))$, the kernel $K := \text{Ker}(f)$ lies in FiltGen($\mathcal{C}$), and even in $\mathcal{C}$ by Lemma 3.3. Then Y lies in $\mathcal{C}$ as $\mathcal{C}$ is closed under cokernels. Now suppose that Y has a Gen($\mathcal{C}$)-filtration $(Y_\lambda : \lambda \leq \mu)$. Let $\lambda < \mu$ and suppose that Y_λ lies in $\mathcal{C}$. Then $\mathfrak{a}(\text{FiltGen}(\mathcal{C}))$, which contains $\mathcal{C}$ and is clearly a subcategory closed under subobjects in FiltGen($\mathcal{C}$), must contain both Y_λ and $Y_{\lambda+1}$, and therefore also $Y_{\lambda+1}/Y_\lambda$ by Proposition 2.1. But $Y_{\lambda+1}/Y_\lambda$ lies in Gen($\mathcal{C}$) and, hence, it lies in $\mathcal{C}$ by the argument above. Since $\mathcal{C}$ is closed under extensions, we infer that also $Y_{\lambda+1}$ belongs to $\mathcal{C}$. Finally, let $\nu \leq \mu$ be a limit ordinal and assume that all Y_λ with $\lambda < \nu$ belong to $\mathcal{C}$. Since $\mathcal{C}$ is closed under coproducts and cokernels, it follows that $Y_\nu = \varinjlim_{\lambda<\nu} Y_\lambda$ also lies in $\mathcal{C}$, thus finishing the proof. □

We remark that under our assumptions above $\mathcal{C}$ is not necessarily bireflective.

EXAMPLE 3.5. Let A be the Kronecker algebra, and let $\mathbf{t}$ denote the class of all finite dimensional indecomposable regular modules. Take $\mathcal{C}$ to be the direct limit closure of add($\mathbf{t}$). This is a full subcategory of Mod-A closed under kernels, cokernels, coproducts and extensions which contains all Prüfer modules and lies in the torsion class $\text{Gen}(\mathbf{t}) = \text{Gen}(\mathcal{C}) = \text{FiltGen}(\mathcal{C})$. The corresponding torsionfree class $\mathbf{t}^o$ is the class of all modules cogenerated by the generic module G. For details we refer to [**RR**]. Notice that G is a summand of any countable product of copies of a Prüfer module. This shows that $\mathcal{C}$ it is not closed under products (so it is not bireflective).

We now aim at a large version of [**MS1**, Proposition 3.9]. Let us first recall the following result.

LEMMA 3.6. [**CB1**, (4.4) Lemma] *Let A be a right coherent ring, and let $(\mathcal{X}, \mathcal{Y})$ be a torsion pair in* mod-A. *Then the classes $\mathcal{T} := \varinjlim \mathcal{X}$ and $\mathcal{F} := \varinjlim \mathcal{Y}$ consisting of all direct limits of modules in $\mathcal{X}$ and $\mathcal{Y}$, respectively, form a torsion pair $(\mathcal{T}, \mathcal{F})$ in* Mod-A *and, furthermore, $\mathcal{T} = \text{Gen}(\mathcal{X})$ and $\mathcal{F} = \mathcal{X}^o$.*

PROPOSITION 3.7. *Let A be a finite dimensional algebra and let T be a finite dimensional silting module. Then* $\text{FiltGen}(\mathfrak{a}(\text{Gen}(T))) = \text{Gen}(T)$.

PROOF. By construction we have $\text{FiltGen}(\mathfrak{a}(\text{Gen}(T))) \subseteq \text{Gen}(T)$. For the reverse inclusion, we observe that $\text{Gen}(T) \cap \text{mod-}A$ coincides with the class gen(T) of all finitely T-generated modules, which is a functorially finite torsion class in mod-A by [**AIR**, Theorem 2.7]. It follows from Lemma 3.6 that $\text{Gen}(T) = \varinjlim(\text{gen}(T))$, and by [**MS1**, Proposition 3.9] we know that gen(T) consists of all modules with a finite filtration by quotients of modules in $\mathfrak{a}(\text{gen}(T))$ (here $\mathfrak{a}$ is the map defined in (2.1) for the abelian category mod-A). If we prove that $\mathfrak{a}(\text{gen}(T))$ is contained in $\mathfrak{a}(\text{Gen}(T))$ (here $\mathfrak{a}$ is now defined for the abelian category Mod-A), then gen(T) will be contained in $\text{FiltGen}(\mathfrak{a}(\text{Gen}(T)))$ and, hence, so will $\text{Gen}(T) = \varinjlim \text{gen}(T)$, concluding our proof. So, let us show that $\mathfrak{a}(\text{gen}(T)) \subseteq \mathfrak{a}(\text{Gen}(T))$. Take a module X in $\mathfrak{a}(\text{gen}(T))$. Of course X lies also in Gen(T). Consider a map $g : Y \longrightarrow X$ with Y in Gen(T). Then there is a direct system $(Y_i)_{i \in I}$ in gen(T) such that $Y = \varinjlim Y_i$, and the induced maps $g_i : Y_i \longrightarrow X$ form a direct system with $\text{Ker}(g) = \varinjlim \text{Ker}(g_i)$. Since by assumption $\text{Ker}(g_i)$ belongs to gen(T) for each i, we conclude that Ker(g) lies in Gen(T). □

We obtain a large version of Theorem 2.5 together with an explicit description of the image of $\mathfrak{a}$.

COROLLARY 3.8. *If A is a finite dimensional algebra, there is a commutative diagram of injections as follows.*

$$\begin{array}{ccc}
\left\{\begin{array}{c}\textit{Equivalence classes}\\ \textit{of finite dimensional}\\ \textit{silting } A\textit{-modules}\end{array}\right\} & \xrightarrow{\qquad\alpha\qquad} & \left\{\begin{array}{c}\textit{Epiclasses of ring}\\ \textit{epimorphisms } A\to B\\ \textit{with } \mathrm{Tor}_1^A(B,B)=0\end{array}\right\} \\
 & {\scriptstyle\mathfrak{a}}\searrow \qquad \swarrow{\scriptstyle\epsilon} & \\
 & \left\{\begin{array}{c}\textit{Bireflective}\\ \textit{extension-closed}\\ \textit{subcategories}\\ \textit{of } \mathrm{Mod}\text{-}A\end{array}\right\} &
\end{array}$$

The map α assigns to the minimal silting module T the associated silting ring epimorphism. The map ϵ is the bijection from Theorem 1.9*(1) and $\mathfrak{a}$ is the map defined in (*2.1*) for the abelian category* Mod-A. *Moreover, the image of $\mathfrak{a}$ consists of the bireflective extension-closed subcategories $\mathcal{C}$ in* Mod-A *for which* FiltGen($\mathcal{C}$) *is generated by a finite dimensional silting module.*

PROOF. By Proposition 3.7, the map $\mathfrak{a}$ yields an injection from equivalence classes of finite dimensional silting modules to bireflective extension-closed subcategories of Mod-A. The description of the image follows from Proposition 3.4. □

Note that, if all torsion classes in Mod-A are generated by a finite dimensional silting module, then $\mathfrak{a}$ is a bijection. Let us turn to algebras with such property.

4. τ-tilting finite algebras

In this section, A will be a finite dimensional algebra and we denote by τ the Auslander–Reiten translate. Recall the following definitions.

DEFINITION 4.1. [**AIR**] A finite dimensional A-module M is said to be *τ-rigid* if $\mathrm{Hom}_A(M, \tau M) = 0$, and it is called *$\tau$-tilting* if, in addition, the number of non-isomorphic indecomposable direct summands of M coincides with the number of isomorphism classes of simple A-modules. Finally, M is said to be *support τ-tilting*, if it is τ-tilting over A/AeA for an idempotent $e \in A$.

The next theorem provides a characterisation of algebras that have only finitely many τ-tilting modules, up to additive equivalence. Recall that a finite dimensional A-module is said to be a *brick* if its endomorphism ring is a division ring.

THEOREM 4.2. [**DIJ**] *The following statements are equivalent for a finite dimensional algebra A.*

(1) *There are only finitely many isomorphism classes of indecomposable τ-rigid A-modules.*
(2) *There are only finitely many isomorphism classes of basic τ-tilting A-modules.*
(3) *There are only finitely many isomorphism classes of bricks in* mod-A.
(4) *There are only finitely many torsion classes in* mod-A.
(5) *Every torsion class in* mod-A *is functorially finite.*

DEFINITION 4.3. A finite dimensional algebra A is called *τ-tilting finite* if it satisfies the equivalent conditions of Theorem 4.2.

EXAMPLE 4.4. Let A be a preprojective algebra of Dynkin type. Then A is τ-tilting finite by [**Mi**]. Indeed, the isomorphism classes of basic support τ-tilting A-modules correspond bijectively to the elements of the Weyl group associated to the Dynkin graph. It is also not hard to check that every brick in mod-A has no non-trivial self-extensions and, hence, its dimension vector needs to be a positive root of the underlying Dynkin graph. In particular, up to isomorphism, there are only finitely many bricks in mod-A.

EXAMPLE 4.5. The following example is due to [**R**]. Let Q be the quiver

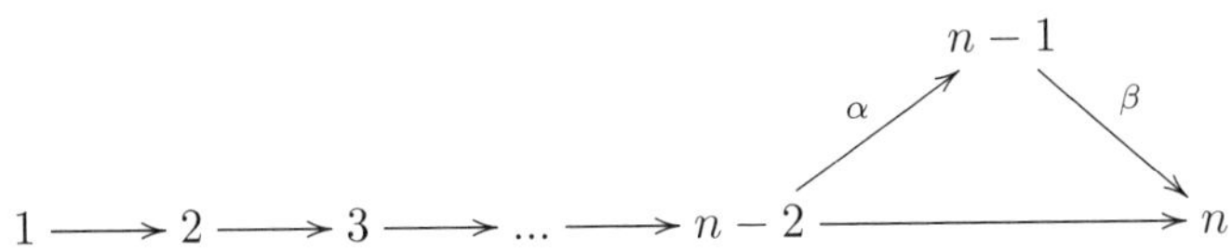

and let A be the finite dimensional algebra given by Q modulo the relation $\beta\alpha$. Then A has global dimension two. It was observed in [**R**] that for $n \geq 9$ the algebra A is of wild representation type and that, independently of n, there are only finitely many bricks in mod-A, up to isomorphism. In particular, A is τ-tilting finite. Note that, by choosing n large enough and adding relations to the ideal above, we can construct further τ-tilting finite algebras of wild representation type and of any finite global dimension.

Since support τ-tilting modules over A are precisely the finite dimensional silting modules [**AMV1**, Proposition 3.15], it is only natural to ask whether any torsion class in Mod-A whose intersection with mod-A is functorially finite is a silting torsion class. The following lemma answers this question positively. For its proof, recall that a subcategory of Mod-A is said to be *definable* if it is closed under products, direct limits and pure submodules (X is a *pure submodule* of Y if the inclusion of X into Y remains injective after tensoring it with any left A-module).

LEMMA 4.6. *Let $\mathcal{T}$ be a torsion class in* Mod-A *such that $\mathcal{T} \cap$* mod-A *is functorially finite. Then there is a finite dimensional silting module T such that $\mathcal{T} =$* Gen(T).

PROOF. If $\mathcal{T} \cap \text{mod-}A$ is a functorially finite torsion class in mod-A, it has the form gen(T) for a finite dimensional silting module T, by [**AIR**, Theorem 2.7]. Then the direct limit closure $\varinjlim \text{gen}(T)$ coincides with Gen(T) by Lemma 3.6, and it is obviously contained in $\mathcal{T}$. Conversely, since every module X is a pure submodule of the direct product of all its finitely generated quotients, see [**CB2**, 2.2, Example 3], any module X in $\mathcal{T}$ lies in the smallest definable subcategory containing $\mathcal{T} \cap \text{mod-}A = \text{gen}(T)$. Since silting classes are definable by [**AMV1**, Corollary 3.5], it is then clear that the smallest definable subcategory containing gen(T) is Gen(T), thus proving that $\mathcal{T} = \text{Gen}(T)$. □

We are now ready for the main results of this note. Observe that the diagram of injections from Theorem 2.5 becomes a diagram of bijections if we assume our algebra A to be τ-tilting finite, see [**MS1**, Corollary 3.11]. In particular, it follows that an algebra A is τ-tilting finite if and only if there are only finitely many epiclasses of ring epimorphisms $A \longrightarrow B$ with $\text{Tor}_1^A(B, B) = 0$ and such that B is again a finite dimensional algebra. The following result shows that an analogous statement holds true if we allow for large modules and ring epimorphisms $A \longrightarrow B$ for which B is no longer assumed to be a finite dimensional algebra.

THEOREM 4.7. *If A is a τ-tilting finite algebra, there is a commutative diagram of bijections as follows.*

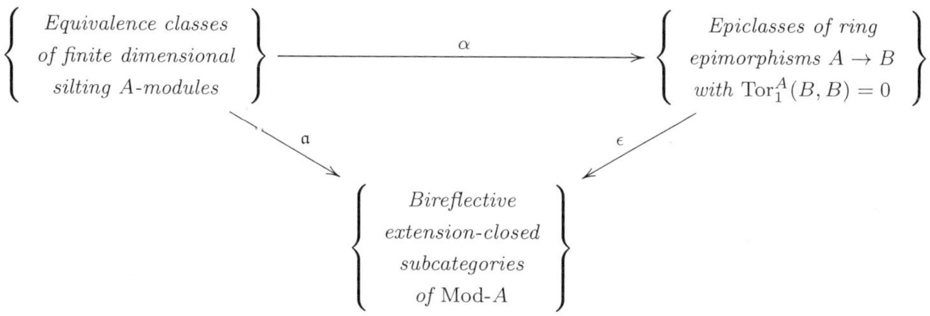

The maps are defined as in Corollary 3.8. *Moreover, it follows that all ring epimorphisms $A \longrightarrow B$ with $\operatorname{Tor}_1^A(B,B) = 0$ are universal localisations with $\dim B < \infty$.*

PROOF. By Theorem 4.2 and Lemma 4.6, every torsion class in Mod-A is generated by a finite dimensional silting A-module. The diagram of bijections then follows from Corollary 3.8. For the last statement we refer back to Theorem 2.5. □

THEOREM 4.8. *The following statements are equivalent for a finite dimensional algebra A.*

(1) *A is τ-tilting finite.*
(2) *Every torsion class $\mathcal{T}$ in* Mod-A *has the form $\mathcal{T} = \operatorname{Gen}(T)$ for a finite dimensional silting module T.*
(3) *Every silting A-module is finite dimensional up to equivalence.*
(4) *There are only finitely many epiclasses of ring epimorphisms $A \longrightarrow B$ with $\operatorname{Tor}_1^A(B,B) = 0$.*

PROOF. (1) $\Rightarrow$ (2) is Lemma 4.6, and (2) $\Rightarrow$ (3) is clear. Moreover, (1) $\Rightarrow$ (4) is a direct consequence of Theorem 4.7 and (4) $\Rightarrow$ (1) follows from Corollary 3.8.

(3) $\Rightarrow$ (1): If A is not τ-tilting finite, there must be a torsion class in mod-A that is not functorially finite. As argued in the proof of [**DIJ**, Theorem 3.8], it follows that there is an infinite sequence of finite dimensional silting modules $T_1, T_2, \ldots$ inducing a strictly decreasing chain of torsion classes

$$\text{mod-}A \supset \operatorname{gen}(T_1) \supset \operatorname{gen}(T_2) \supset \ldots$$

Passing to Mod-A, we note that the equality $\operatorname{Gen}(T_i) = \operatorname{Gen}(T_{i+1})$ would imply $\operatorname{add}(T_i) = \operatorname{add}(T_{i+1})$. Hence, there is also a strictly decreasing chain of torsion classes in Mod-A

$$\text{Mod-}A \supset \operatorname{Gen}(T_1) \supset \operatorname{Gen}(T_2) \supset \ldots$$

and its intersection gives rise to a definable torsion class $\bigcap \operatorname{Gen}(T_i)$ which is a silting class as a consequence of [**AH**, Corollary 3.8]. By assumption, there is a finite dimensional silting module T such that $\bigcap \operatorname{Gen}(T_i) = \operatorname{Gen}(T)$. It follows that $\bigcap \operatorname{gen}(T_i) = \operatorname{Gen}(T) \cap \text{mod-}A = \operatorname{gen}(T)$, contradicting [**DIJ**, Lemma 3.10 (b)]. □

Following Theorem 4.8, we provide a further class of examples of τ-tilting finite algebras. This class was discussed in [**GdP**].

EXAMPLE 4.9. Suppose that the algebra A fulfills the following two conditions:

- there are elements c_1 and c_2 in the center of A such that the algebra A/c_iA is representation-finite for $i = 1, 2$;
- the sum of the annihilators of c_1 and c_2 is A.

As argued in [**GdP**, Subsection 2.4], then there are only finitely many epiclasses of ring epimorphisms $A \longrightarrow B$. In particular, by Theorem 4.8, A is τ-tilting finite. An explicit example of a representation-infinite algebra fulfilling our assumptions, as found in [**GdP**, Subsection 2.4], is the quotient of the path algebra of

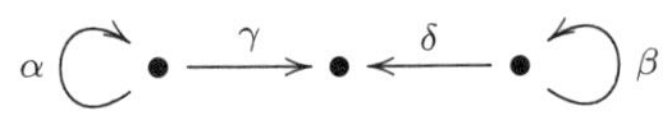

by the ideal generated by the relations $\alpha^2 = \gamma\alpha = \beta^3 = \delta\beta^2 = 0$. The elements c_1 and c_2 above can be chosen to be, respectively, the elements α and β^2.

REMARK 4.10. If A is τ-tilting finite, it follows from Theorem 4.7 that whenever $A \longrightarrow B$ is a ring epimorphism fulfilling $\mathrm{Tor}_1^A(B,B) = 0$, then also the algebra B is finite dimensional (and τ-tilting finite). We do not know about the converse. Is it true that A does not admit an infinite dimensional ring epimorphism $A \longrightarrow B$ with $\mathrm{Tor}_1^A(B,B) = 0$ if and only if it is τ-tilting finite?

References

[AIR] T. Adachi, O. Iyama, and I. Reiten, *τ-tilting theory*, Compos. Math. **150** (2014), no. 3, 415–452, DOI 10.1112/S0010437X13007422. MR3187626

[AH] L. Angeleri Hügel and M. Hrbek, *Silting modules over commutative rings*, Int. Math. Res. Not. IMRN **13** (2017), 4131–4151, DOI 10.1093/imrn/rnw147. MR3671512

[AMV1] L. Angeleri Hügel, F. Marks, and J. Vitória, *Silting modules*, Int. Math. Res. Not. IMRN **4** (2016), 1251–1284, DOI 10.1093/imrn/rnv191. MR3493448

[AMV2] L. Angeleri Hügel, F. Marks, and J. Vitória, *Silting modules and ring epimorphisms*, Adv. Math. **303** (2016), 1044–1076, DOI 10.1016/j.aim.2016.08.035. MR3552543

[AS] L. Angeleri Hügel and J. Sánchez, *Tilting modules over tame hereditary algebras*, J. Reine Angew. Math. **682** (2013), 1–48. MR3181497

[Asai] S. Asai, *Semibricks*, Int. Math. Res. Not. IMRN, DOI:10.1093/imrn/rny150.

[CB1] W. Crawley-Boevey, *Locally finitely presented additive categories*, Comm. Algebra **22** (1994), no. 5, 1641–1674, DOI 10.1080/00927879408824927. MR1264733

[CB2] W. Crawley-Boevey, *Infinite-dimensional modules in the representation theory of finite-dimensional algebras*, Algebras and modules, I (Trondheim, 1996), CMS Conf. Proc., vol. 23, Amer. Math. Soc., Providence, RI, 1998, pp. 29–54. MR1648602

[DIJ] L. Demonet, O. Iyama and G. Jasso, *τ-tilting finite algebras, bricks and g-vectors*, Int. Math. Res. Not. IMRN, 2019, no. 3, 852–892. DOI:10.1093/imrn/rnx135.

[GdP] P. Gabriel and J. A. de la Peña, *Quotients of representation-finite algebras*, Comm. Algebra **15** (1987), no. 1-2, 279–307, DOI 10.1080/00927878708823421. MR876981

[GL] W. Geigle and H. Lenzing, *Perpendicular categories with applications to representations and sheaves*, J. Algebra **144** (1991), no. 2, 273–343, DOI 10.1016/0021-8693(91)90107-J. MR1140607

[IT] C. Ingalls and H. Thomas, *Noncrossing partitions and representations of quivers*, Compos. Math. **145** (2009), no. 6, 1533–1562, DOI 10.1112/S0010437X09004023. MR2575093

[I] O. Iyama, *Rejective subcategories of artin algebras and orders*, preprint (2003), arxiv: math/0311281.

[MS1] F. Marks and J. Šťovíček, *Torsion classes, wide subcategories and localisations*, Bull. Lond. Math. Soc. **49** (2017), no. 3, 405–416, DOI 10.1112/blms.12033. MR3723626

[MS2] F. Marks and J. Šťovíček, *Universal localisations via silting*, Proc. Roy. Soc. Edinburgh Sect. A. DOI:10.1017/prm.2018.37.

[Mi] Y. Mizuno, *Classifying τ-tilting modules over preprojective algebras of Dynkin type*, Math. Z. **277** (2014), no. 3-4, 665–690, DOI 10.1007/s00209-013-1271-5. MR3229959

[RR] I. Reiten and C. M. Ringel, *Infinite dimensional representations of canonical algebras*, Canad. J. Math. **58** (2006), no. 1, 180–224, DOI 10.4153/CJM-2006-008-1. MR2195596

[R] C. M. Ringel, *Wild Algebras: Two Examples*, preprint (2013), arxiv.org/abs//1307.6509.

[Sch] A. H. Schofield, *Representation of rings over skew fields*, London Mathematical Society Lecture Note Series, vol. 92, Cambridge University Press, Cambridge, 1985. MR800853

[Wei] J. Wei, *Semi-tilting complexes*, Israel J. Math. **194** (2013), no. 2, 871–893, DOI 10.1007/s11856-012-0093-1. MR3047094

Dipartimento di Informatica - Settore di Matematica, Università degli Studi di Verona, Strada le Grazie 15 - Ca' Vignal, I-37134 Verona, Italy
Email address: lidia.angeleri@univr.it

Institut für Algebra und Zahlentheorie, Universität Stuttgart, Pfaffenwaldring 57, 70569 Stuttgart, Germany
Email address: marks@mathematik.uni-stuttgart.de

Department of Mathematics, City, University of London, Northampton Square, London EC1V 0HB, United Kingdom
Email address: jorge.vitoria@city.ac.uk

Contemporary Mathematics
Volume **730**, 2019
https://doi.org/10.1090/conm/730/14712

Describing models of Th($\mathbb{Z}$) in adelic terms

Angus Macintyre

ABSTRACT. We consider abelian groups Γ elementarily equivalent to $\mathbb{Z}$, the additive group of integers. We give an exact classification of the isomorphism types of those Γ which admit a pure embedding of $\mathbb{Z}$, by using homological algebra and adelic ideas to be found in Tate's thesis (and given an elementary presentation in an expository paper of Keith Conrad).

1. Introduction

We work in the usual first-order language L_{ab} for abelian groups, based on $+, -, 0$.

[**Prest**] gives an excellent account of the state of the model theory of abelian groups.

Let T ($=$ Th($\mathbb{Z}$)) be the first-order theory of $\mathbb{Z}$ in L_{ab} . A complete set of axioms for models G of T is well-known [Prest] namely

(1) G is nontrivial and torsion-free ;
(2) For each positive integer n, G/nG is cyclic of order n.

Because $x \to -x$ is an automorphism of any abelian group, the element 1 of $\mathbb{Z}$ is not L_{ab} - definable.

$\mathbb{Z}$ is embeddable in any model G of T, but not uniquely and not necessarily purely. T has no minimal model, since any infinite subgroup of $\mathbb{Z}$ is also a model of T.

Let Γ be any nonstandard (i.e not cyclic) model of T. Consider an extension (exact diagram)

$$0 \to \mathbb{Z} \to \Gamma \to \Gamma/\mathbb{Z} \to 0. \tag{1}$$

Suppose $\mathbb{Z} \to \Gamma$ is pure . Suppose n is an integer bigger than 1. Let γ be the image of 1 in Γ. Then $\gamma + n\Gamma$ generates $\Gamma/\mathbb{Z}$, cyclic of order n. It follows that $\Gamma/\mathbb{Z}$ is divisible. But also $\Gamma/\mathbb{Z}$ is torsion-free, by purity. So

$$\Gamma/\mathbb{Z} \cong \bigoplus_{\alpha} \mathbb{Q}$$

for a unique cardinal α.

Conversely, if Γ is any abelian group in an extension

2010 *Mathematics Subject Classification.* Primary 03C60, 13C05, 18Gxx.

$$0 \to \mathbb{Z} \to \Gamma \to \bigoplus_{\alpha} \mathbb{Q} \to 0$$

then $\mathbb{Z} \to \Gamma$ is pure (since $\bigoplus_\alpha \mathbb{Q}$) is torsion-free). But also Γ is torsion-free, for if $\gamma \in \Gamma$ and $n\gamma = 0$ then γ is in kernel of the map from Γ to $\bigoplus_\alpha \mathbb{Q}$, so $\gamma \in \mathbb{Z}$ so $\gamma = 0$.

Finally, we show that for any positive integer n $\Gamma/n\Gamma$ is cyclic of order n. Let $\gamma \in \Gamma$. Then $\gamma = n\delta$ for some δ, by exactness, and so, again by exactness $\gamma - n\delta \in \mathbb{Z}$, so we conclude using the purity already proved.

Thus Γ is a model of T.

The preceding analysis thus classifies (up to isomorphism) those models Γ of $Th(\mathbb{Z})$, which admit a pure embedding of $\mathbb{Z}$ into Γ, by cardinals α and elements of

$$Ext^1(\bigoplus_{\alpha} \mathbb{Q})$$

(and indeed by $PExt^1(\bigoplus_\alpha \mathbb{Q})$, the corresponding pure extensions). Note that there are models Γ of $Th(\mathbb{Z})$, for example $\bigoplus_p \mathbb{Z}_p$, which admit no pure embeddings of $\mathbb{Z}$.

The purpose of this paper is to make the classification much more illuminating, via an adelic interpetation.

2. Applying the long exact sequence

Now we consider the exact

$$0 \to \mathbb{Z} \to \mathbb{Q} \to \mathbb{Q}/\mathbb{Z} \to 0 \tag{2}$$

(where $\mathbb{Q}/\mathbb{Z}$ is divisible and torsion), and apply the long exact sequence of cohomology [**Rotman**] to it, against

$$\bigoplus_{\alpha} \mathbb{Q},$$

to get

$$\begin{aligned} &Hom(\bigoplus_{\alpha} \mathbb{Q}, \mathbb{Z}) \to Hom(\bigoplus_{\alpha} \mathbb{Q}, \mathbb{Q}) \to Hom(\bigoplus_{\alpha} \mathbb{Q}, \mathbb{Q}/\mathbb{Z}) \to \\ &\to Ext(\bigoplus_{\alpha} \mathbb{Q}, \mathbb{Z}) \to Ext(\bigoplus_{\alpha} \mathbb{Q}, \mathbb{Q}) \to Ext(\bigoplus_{\alpha} \mathbb{Q}, \mathbb{Q}/\mathbb{Z}) \end{aligned} \tag{3}$$

Now clearly

$$Hom(\bigoplus_{\alpha} \mathbb{Q}, \mathbb{Z}) = 0$$

and

$$Ext^1(\bigoplus_{\alpha} \mathbb{Q}, \mathbb{Q}) = 0.$$

This gives exact

$$0 \to Hom(\bigoplus_{\alpha} \mathbb{Q}, \mathbb{Q}) \to Hom(\bigoplus_{\alpha} \mathbb{Q}, \mathbb{Q}/\mathbb{Z}) \to Ext(\bigoplus_{\alpha} \mathbb{Q}, \mathbb{Z}) \to 0 \tag{4}$$

Now $Hom(\bigoplus_\alpha \mathbb{Q}, \mathbb{Q})$ is naturally isomorphic to $\prod_\alpha Hom(\mathbb{Q}, \mathbb{Q})$, i.e to $\prod_\alpha \mathbb{Q}$. Similarly, $Hom(\bigoplus_\alpha \mathbb{Q}, \mathbb{Q}/\mathbb{Z})$ is naturally isomorphic to $\prod_\alpha Hom(\mathbb{Q}, \mathbb{Q}/\mathbb{Z})$.

It follows from (3) that

$$Ext^1(\bigoplus_\alpha \mathbb{Q}, \mathbb{Z}) \cong \prod_\alpha \Big((Hom(\mathbb{Q}, \mathbb{Q}/\mathbb{Z}))/\mathbb{Q}\Big) \tag{5}$$

This still leaves us far short of a useful explicit isomorphism, and of any useful description of $Hom(\mathbb{Q}, \mathbb{Q}/\mathbb{Z}))/\mathbb{Q}$.

Before investigating the possibilities here, we note that if we put $\alpha = 1$ in (5) we get

$$Ext^1(\mathbb{Q}, \mathbb{Z}) \cong (Hom(\mathbb{Q}, \mathbb{Q}/\mathbb{Z}))/\mathbb{Q}.$$

2.1. It is standard that for A and B abelian the elements of $Ext^1(B, A)$ are classified (up to the usual coboundary conditions) by factor sets

$$[,] : B \times B \to A$$

satisfying cocycle conditions [**Rotman**], for $f, g, h \in B$

$$[g, 0] = [0, g] = 0$$

and

$$[g, h] - [f + g, h] + [f, g + h] - [f, g] = 0.$$

Note that if θ is in $Hom(C, B)$ then

$$(g, h) \to [\theta(f), \theta(g)]$$

is a factor set from C to A, for any factor set [.] from $B \times B$ to A, and the process respects coboundaries.

Let us <u>fix</u> a factor set [,] for (2). (We will specify one shortly). Then the map

$$Hom(\bigoplus_\alpha \mathbb{Q}, \mathbb{Q}/\mathbb{Z}) \to Ext(\bigoplus_\alpha \mathbb{Q}, \mathbb{Z})$$

in the middle of (3) is given by the recipe of the previous paragraph, with $A = \mathbb{Z}$, $B = \mathbb{Q}/\mathbb{Z}$, $C = \bigoplus_\alpha \mathbb{Q}$, and θ maps to the factor set

$$(g, h) \to [\theta(f), \theta(g)].$$

To see this, one simply examines the standard argument for the long exact sequence.

Which factor set to choose?

2.2. Factor set for (2). For $x \in \mathbb{R}$, $\{x\}_\mathbb{R}$ (the real fractional part of x) is uniquely defined by

$$\{x\}_\mathbb{R} \in [0, 1)$$

and

$$x - \{x\}_\mathbb{R} \in \mathbb{Z}.$$

$\{x\}_\mathbb{R}$ is not additive, but

$$\{x + y\}_\mathbb{R} - (\{x\}_\mathbb{R} + \{y\}_\mathbb{R}) \in \mathbb{Z}.$$

Now

$$x \to e^{2\pi i \{x\}_\mathbb{R}}$$

is an epimorphism from $(\mathbb{R}, +)$ to the unit circle T (with multiplicative structure) in $\mathbb{C}$, and

$$\mathbb{R}/\mathbb{Z} \cong \mathbb{T}.$$

$\mathbb{T}$ is of course compact in the usual Euclidean topology. Note that $\mathbb{Q}/\mathbb{Z}$ is the torsion subgroup of $\mathbb{R}/\mathbb{Z}$, thus naturally identified with the subgroup of $\mathbb{T}$ consisting of the complex roots of unity.

Now we define the promised factor set, thus (for $r, s \in \mathbb{Q}$).

$$[r + \mathbb{Z}, s + \mathbb{Z}] = \{r + s\}_{\mathbb{R}} - (\{r\}_{\mathbb{R}} + \{s\}_{\mathbb{R}})$$

taking values in $\mathbb{Z}$.

This comes directly from the standard process for going from an extension to a factor set.

2.3. The remaining big problem is to get an explicit description of $Hom(\bigoplus_\alpha \mathbb{Q}, \mathbb{Q}/\mathbb{Z})$.

Recall that naturally

$$Hom(\bigoplus_\alpha \mathbb{Q}, \mathbb{Q}/\mathbb{Z}) \cong \prod_\alpha Hom(\mathbb{Q}, \mathbb{Q}/\mathbb{Z})$$

so we are reduced to getting an explicit description of $Hom(\mathbb{Q}, \mathbb{Q}/\mathbb{Z})$.

We could not find a down-to-earth treatment of this in a standard textbook. Note that $Hom(\mathbb{Q}, \mathbb{Q}/\mathbb{Z})$ is a subgroup of $Hom(\mathbb{Q}, \mathbb{R}/\mathbb{Z})$, the character group of $\mathbb{Q}$, and one can find the basic ideas in Tate's great thesis [**CF**], in the setting of harmonic analysis. But we need less than that, and we had the good fortune to find [**Conrad**], a very lucid account by Keith Conrad. This is the basis of the remainder of our analysis.

The main ingredients are p-adic fractional parts, and $\mathbb{A}_{\mathbb{Q}}$, the ring of adeles over $\mathbb{Q}$.

2.4. p-adic fractional parts. Fix a prime p. We define, for $x \in \mathbb{Q}_p$ a "p-adic fractional part" in $\mathbb{Q}$.

Take the standard representation of x as a sum of terms $c_n.p^n$ where $0 \leq c_n < p$, where all but finitely many terms with negative n have $a_n = 0$, and define x_p as the sum of the $c_n.p^n$ for n negative. Then $x_p \in \mathbb{Q}$.

Though this fractional part is not additive, one has (a little calculation needed [**Conrad**]) that for $r, s \in \mathbb{Q}$

$$\{r + s\}_p - (\{r\}_p + \{s\}_p) \in \mathbb{Z}$$

To pursue the analogy with.$_{\mathrm{R}}$ one considers the map

$$x \to e^{2\pi i\{x\}_p}$$

and notes that for all our fractional part maps the corresponding exponential is the standard exponential of a complex number (since the p-adic fractional part takes only rational values).

2.5. Characters of $\mathbb{Q}$. We follow Conrad's notation in defining

$$\Psi_\infty(x) = e^{-2\pi ix}$$

and

$$\Psi_p(x) = e^{-2\pi i\{x\}_p}$$

for a prime p. The first Ψ is defined on R and the second on $\mathbb{Q}_p$.

For any $a \in$ R the map

$$x \to \Psi_\infty(ax)$$

is in $Hom(\mathrm{R}, \mathbb{T})$, and for $a \in \mathbb{Q}_p$ the map

$$x \to \Psi_p(ax)$$

is in $Hom(\mathbb{Q}_p, \mathbb{T})$.

These maps give us a large supply of characters of $\mathbb{Q}$, since each is defined on $\mathbb{Q}$.

2.6. Adeles. $\mathbb{A}_{\mathbb{Q}}$, the adele ring of $\mathbb{Q}$ is beautifully explained at various places in [**CF**].

The diagonal embedding of $\mathbb{Q}$ in the adele ring is of basic importance below. We bring together the materials of 1.9 in an adelic construction.

Fix an adele a with real component a_∞ and p-adic components a_p (so $a_p \in \mathbb{Z}_p$ for all but finitely many p.

Now define Ψ_a in $Hom(\mathbb{Q}, \mathbb{T})$ by

$$\Psi_a(r) = \Psi_\infty(a_\infty r) \cdot \prod_p \Psi_p(a_p \cdot r).$$

This is well-defined , since $a_p \cdot r \in \mathbb{Z}_p$ for all but finitely many p, and in those cases $\Psi_p(a_p) = 1$.

Ψ_a is a homomorphism, and

$$\Psi_{a+b} = \Psi_a \cdot \Psi_b.$$

Ψ_0 is the trivial character.

Because of the judicious choice of the minus sign in the definition of Ψ_∞ , one shows [Conrad] that for any $s \in \mathbb{Q}$ Ψ_s is trivial, and that this gives exactly the a for which Ψ_a is trivial. Moreover the Ψ_a give all the characters of $\mathbb{Q}$.

Thus $\widehat{\mathbb{Q}}$, the Pontrjagin dual of $\mathbb{Q}$, is identified as $\mathbb{A}_{\mathbb{Q}}/\mathbb{Q}$.

We want a good description of $Hom(\mathbb{Q}, \mathbb{Q}/\mathbb{Z})$, a subgroup of the character group of $\mathbb{Q}$.

So suppose $\Psi_a(r)$ takes only torsion values for $r \in \mathbb{Q}$. The $\Psi_\infty(r)$ takes only torsion values, so $a_\infty \in \mathbb{Q}$, and conversely.

Now consider $a' = a - a_\infty$. $\Psi_a = \Psi_{a'}$ by the preceding. Now a' has real coordinate 0. Such adeles are called finite adeles, and they form an ideal $(\mathbb{A}_{\mathbb{Q}})^f$ in the ring of adeles. This is in fact the restricted product of the $\mathbb{Q}_p$ with respect to the $\mathbb{Z}_p$.

We use the natural identification of the finite adeles inside the adeles.We define $(\Psi_b)^f$ for a finite adele b, construed as living in the restricted product of the $\mathbb{Q}_p$ with respect to the $\mathbb{Z}_p$, as $\Psi_{b'}$ where b' is got from b by making the real component 0 and leaving the others unchanged. So we have proved that any element of $Hom(\mathbb{Q}, \mathbb{Q}/\mathbb{Z})$ is of the form $(\Psi_b)^f$ for a finite adele b. Moreover (and in contrast to the case of $\widehat{\mathbb{Q}}$, b is now unique. For if some c also works, then $\Psi_b = \Psi_c$, so b and c differ by a rational, and that rational must be 0 by considering the real coordinates. so

THEOREM 2.1. *$Hom(\mathbb{Q}, \mathbb{Q}/\mathbb{Z})$ is naturally isomorphic to $(\mathbb{A}_{\mathbb{Q}})^f$, via $b \to (\Psi_b)^f$.*

Proof. Done.

Note Of course we claim no originality for this.

2.7. Now finally we describe $Ext^1(\mathbb{Q}, \mathbb{Q}/\mathbb{Z})$. The image of $Hom(\mathbb{Q}, \mathbb{Q})$ in $Hom(\mathbb{Q}, \mathbb{Q}/\mathbb{Z})$ is clearly the group of all Ψ_r, for $r \in \mathbb{Q}$, and thus

THEOREM 2.2. $Ext^1(\mathbb{Q}, \mathbb{Q}/\mathbb{Z}) \cong \mathbb{A}^f_{\mathbb{Q}}/\mathbb{Q}$

Proof. Done, with help from [**Conrad**].

2.8. The recipe. It remains to make the above isomorphism explicit, by assigning to each finite adele b (modulo $\mathbb{Q}$) a cocycle, so that all elements of $Ext^1(\mathbb{Q}, \mathbb{Z})$ are obtained in this way. Here is the recipe, where $r, s \in \mathbb{Q}$ and b is a finite adele:

$$[r, s]_b = \{\sum_p \{b_p r\}_p\}_{\mathrm{R}} + \{\sum_p \{b_p s\}_p\}_{\mathrm{R}} - \{\sum_p \{b_p(r+s)\}_p\}_{\mathrm{R}}.$$

3. Conclusions

Time constraints force us to conclude the paper here. A sequel will consider the topological connections between the finite adeles and the spaces of 1-types realized in the models attached to the factor sets just displayed.

There is more to be done in considering models of $Th(\mathbb{Z})$ which do not realize the type of 1. We are grateful to the anonymous referee for some illuminating comments on this point.

On page 45 of [**Prest**] Mike Prest alludes to the absence of a structure theorem for countable abelian groups. This paper does provide a structure theorem in adelic terms, for models of $Th(\mathbb{Z})$ that realize the type of 1, but this appears to have little possibility of generalization.

References

[Conrad] K. Conrad. The character group of $\mathbb{Q}$. *An expository paper* at `https://kconrad.math.uconn.edu/`.

[Prest] M. Prest, *Model theory and modules*, London Mathematical Society Lecture Note Series, vol. 130, Cambridge University Press, Cambridge, 1988. MR933092

[CF] J. W. S. Cassels and A. Fröhlich (eds.), *Algebraic number theory*, Proceedings of an instructional conference organized by the London Mathematical Society (a NATO Advanced Study Institute) with the support of the International Mathematical Union, Academic Press, London; Thompson Book Co., Inc., Washington, D.C., 1967. MR0215665

[Rotman] J. J. Rotman, *An introduction to homological algebra*, 2nd ed., Universitext, Springer, New York, 2009. MR2455920

SCHOOL OF MATHEMATICS, QUEEN MARY UNIVERSITY & SCHOOL OF MATHEMATICS, UNIVERSITY OF EDINBURGH

Email address: `a.macintyre@qmul.ac.uk`

Contemporary Mathematics
Volume **730**, 2019
https://doi.org/10.1090/conm/730/14713

Valued modules on skew polynomial rings and Bézout domains

Françoise Point

Abstract. We review results obtained on the model theory of valued modules over skew polynomial rings and Bézout domains in the definablility and decidability point of view.

1. Introduction

This survey paper is based on joint works with Luc Bélair [**2**], [**3**] and with Sonia L'Innocente [**25**], on the model theory of certain valued modules. The valued modules that we analysed were on one hand the additive part of a valued field (possibly endowed with extra structure, such as an endomorphism or a derivation) and on the other hand the additive part of a commutative Bézout domain, endowed with its valuation map into its group of divisibility. The two main classes of examples that have been guiding us are:

(1) the valued fields of Witt vectors $W(F)$ endowed with the Witt Frobenius, and of Laurent series $F((t))$ endowed with the Frobenius map $x \to x^p$, with F a field of characteristic p,
(2) the ring of algebraic integers and the ring of holomorphic functions over $\mathbb{C}$.

First we try to capture in the formalism of valued modules certain properties of these algebraic structures. Then we add divisibility conditions to the corresponding theory of valued modules in order to get quantifier elimination in a reasonable language. Finally we translate back these model-theoretic properties into algebraic properties of these valued modules.

Even when the full field (respectively ring) theory is well-understood, it is still informative to identify the additive part of the theory. When K is a field of characteristic p, the polynomials of the form $\sum_i x^{p^i}$ are called *additive polynomials* (or p-polynomials); the field K is said to be p-closed if the associated polynomial maps are surjective. O. Ore studied the algebraic properties of the ring formed by these additive polynomials under addition and composition. More generally one can form the skew polynomial ring $K[t; \sigma]$, where σ is an endomorphism of K, with the commutation rule given by $r.t = t.r^\sigma$ for any $r \in K$ and investigate the properties of $(K, +, 0, v, \sigma)$ as a $K[t; \sigma]$-valued module.

2010 *Mathematics Subject Classification.* Primary 03C60, 16D, 16S36, 13A18, 06F15.
Research Director at the "Fonds de la Recherche Scientifique (F.R.S.-F.N.R.S.)".

More generally, we take a valued field (K, v, σ, ∂) endowed with an endomorphism σ and a σ-derivation ∂ (namely $\partial(a.b) = \partial(a).\sigma(b) + a.\partial(b)$) and we consider the class of $K[t;\sigma,\partial]$-modules. Modules over $K[t;\sigma,\partial]$ can be reduced to modules either over $K[t;\sigma]$ or over $K[t;\partial]$ where ∂ is a (usual) derivation [**6**, Theorem 2.1.3 (i)]. In our analysis, we distinguish whether the endomorphism σ is an isometry ($v(\sigma(x)) = v(x)$) and ∂ is contractive ($v(\partial(x)) \geq v(x)$), or σ is contractive ($v(\sigma(x)) \geq v(x)$). In that last case we will always assume that σ is not the identity.

In [**2**], [**3**], we consider the class of $K[t;\sigma,\partial]$-modules endowed with a value map going to a totally ordered set endowed with an action of the value group of K, as abelian structures and in [**2**, section 6] as two-sorted structures.

In [**2**], we consider the case when σ is an isometry and ∂ is contractive. Under a certain divisibility property, which in the case of modules over $W(F)[t;\sigma]$, reflects that F is p-closed and Hensel's lemma holds in $W(F)$ for certain additive polynomials, we prove quantifier elimination results in the one-sorted and two-sorted settings [**2**, sections 4, 6] (see Theorem 4.1). Under certain conditions on the theory of the ordered set (which implied that it is NIP), we transfer the NIP property to the corresponding theory of valued modules [**2**, section 7], using the above quantifier elimination result.

Then in [**3**], we consider the case where σ is a contractive endomorphism (and so we will assume that $\partial = 0$), assuming that the field K is separably closed of finite imperfection degree. We enrich the language adding λ-functions in order to control how far from being an isometry σ is. We also endow the value group with the induced action by σ (and get what K. Pal [**34**] called a multiplicative ordered difference abelian group). In that enriched language, we show a quantifier elimination result for certain theories of valued modules over $K[t,\sigma]$ considered as abelian structures [**3**, Proposition 6.3] (see Theorem 4.2). As a consequence, we identified various completions [**3**, Corollaries 8.3, 8.4] and get decidability results.

When F is algebraically closed, the theory of the valued difference field $W(F)$ with the Witt Frobenius, is well understood [**4**]. In contrast, there is the still open problem of the decidability of the field of Laurent series over $\mathbb{F}_p$. However the theory of valued modules over $\mathbb{F}_p((t))[t;\sigma]$, with σ the Frobenius, was described by T. Rowher [**41**] and later by G. Onay [**33**]. The methods used by Rowher do not apply to the case when K is the separable closure of $\tilde{\mathbb{F}}_p((t))$ with $\tilde{\mathbb{F}}_p$ the algebraic closure of $\mathbb{F}_p$. (Indeed in that case, K does not have a weak σ-valuation basis [**3**, Lemma 4.16], an hypothesis used by Rohwer.)

Then with Sonia L'Innocente, we study valued modules over commutative Bézout domains B. Recall that Bézout domains can be endowed with a valuation map with values in their groups of divisibility. In contrast with the previous cases, these groups are lattice-ordered groups (so in general no longer totally ordered groups). One of the motivations was to study the ring of holomorphic functions over the complexes as a valued module over itself. Unlike the case of ring of algebraic integers where we know the decidability of the theory as a ring, the case of the positive existential theory of the ring of holomorphic functions over $\mathbb{C}$ is still open (see section 5.5). As a consequence of our work, we obtain a partial answer to this last decidability question (see subsection 5.6.4).

The relationship with the preceding work is that we reduce ourselves to consider modules over the localizations of B at each maximal ideal (see Proposition 5.3). When B is a valuation domain, the image of the value map is now totally ordered, so we can use the same techniques as in [**2**] for analysing valued modules (see Proposition 5.4). Our main result is Theorem 5.5.

Related works are [**26**] and [**27**], the difference with the present account is that we focus on the definability point of view in these classes of valued modules, even though for decidability applications we also consider the Ziegler spectrum [**48**] associated with the indecomposable pure-injective modules over these rings (and use our previous analysis to get a manageable description).

The paper is organized as follows. In the first section, we review certain notions such as ℓ-valuation maps on a ring (subsection 2.1) and various properties of the maximal spectrum of a ring (subsection 2.2). In section 3, we point out to other works on valued modules (and in particular on valued groups) and we detail our precise set-up. Even though we mainly consider abelian structures in order to be able to use the Baur-Monk-Fisher pp-elimination result, we also consider two-sorted structures; one sort being the module sort and the other sort an ordered set acted upon by the value group. In the last two sections, we try to give a flavour of the results we obtained and the kind of techniques we used, for valued modules over skew polynomial rings on one hand and over Bézout domains on the other hand.

2. Preliminaries

All rings R will be associative rings, with 1 and the only non-commutative rings that we will consider, are skew polynomial rings. Denote by $U(R)$ the set of invertible elements of R, by $R^\star := R \setminus \{0\}$ and more generally for any set of elements E containing 0, we denote by $E^\star := E \setminus \{0\}$.

Let D be a commutative domain endowed with an endomorphism σ and a σ-derivation ∂ and form the skew polynomial ring $R := D[t; \sigma, \partial]$ with the commutation law given by: $d.t = t.d^\sigma + \partial(d)$ with $d \in D$ [**6**, chapter 2, section 1]. Then R has no zero-divisors provided σ is injective. If D is a field, then R is a principal right ideal domain [**6**, Proposition 2.1.1].

A (commutative) Bézout ring R is a ring where any finitely generated ideal is principal, equivalently for any $a, b \in R^\star$, $\gcd(a, b)$ exists and belongs to the ideal generated by a, b ($\gcd(a, b)$ is well defined up to invertible elements).

Examples of Bézout rings which are not principal ideal rings (in particular non-noetherian rings) are:

(1) the ring of algebraic integers: $\tilde{\mathbb{Z}}$, or $\tilde{\mathbb{Z}} \cap \mathbb{R}$, $\tilde{\mathbb{Z}} \cap \mathbb{Q}_p$ [**9**], [**38**],
(2) the ring of holomorphic functions over $\mathbb{C}$: $\mathfrak{H}(\mathbb{C})$ or the integral closure $\tilde{\mathfrak{H}}(\mathbb{C})$ of $\mathfrak{H}(\mathbb{C})$, or more generally the ring of holomorphic functions $\mathfrak{H}(\Omega)$, where Ω is an open subset of $\mathbb{C}$ [**42**, Theorem 15.15],
(3) von Neumann regular rings [**36**, section 16.2].

Note that both the theory of modules over von Neumann regular rings [**36**, Proposition 16.17], as well as the ring theory of algebraically closed commutative von Neumann regular rings [**29**] are well-understood. The theory of the ring of algebraic integers is decidable [**9**] and more generally, of good Rumely domains [**10**, Corollary 4.14], [**37**].

Decidability questions for rings of holomorphic functions have been considered usually in the language of rings with an extra constant. It is still open whether the positive existential theory of $\mathfrak{H}(\mathbb{C})$ is decidable (see section 5.5).

2.1. ℓ-valuations. Let $(\Gamma, ., 1, \wedge)$ be an abelian lattice-ordered group and $\bar{\Gamma} := \Gamma \cup \{+\infty\}$. Let B be a (commutative) integral domain, the map $v: B \to \bar{\Gamma}$ satisfying (1) up to (3) below, is called an ℓ-valuation on B [**44**, Definition 1]:

(1) $v(a+b) \geq v(a) \wedge v(b)$,
(2) $v(a.b) = v(a).v(b)$,
(3) $v(1) = 1$ and $v(0) = \infty$.

One can show that every ℓ-valuation on B has a unique extension to its field of fractions $Q(B)$ [**44**, Corollary to Proposition 2].

More generally, let $(\Gamma, ., 1, \leq)$ be a partially ordered abelian group (with the group law written multiplicatively). Let K be a field, the map $v: K \to \bar{\Gamma}$, is a semi-valuation [**47**, section 2], if it satisfies: for all $a, b \in K$,

(1) $v(a-b) \geq \inf\{v(a), v(b)\}$,
(2) $v(a.b) = v(a).v(b)$,
(3) $v(1) = 1$ and $v(0) = \infty$.

D. Zelinsky introduced these semi-valuations on his way to characterise topological fields whose topology is induced by a valuation.

Note that the set $\{a \in K^\star : v(a) \geq 1\}$ forms a subring of K. Conversely given a subring R of K and defining the positive elements of K to be those belonging to R, it induces a (partial) pre-order on K. Consider the multiplicative group $K^\star/U(R)$ endowed with the partial order defined by: $a.U(R) \leq b.U(R)$ if $(b.a^{-1} \in R)$. Denote by $\Gamma := (K^\star/U(R), ., \leq)$ and by Γ^+ the positive elements.

Define the map $v: K \to K^\star/U(R)$ sending $a \in K^\star$ to $a.U(R)$. Then this map v is a semi-valuation, setting $v(0) = +\infty$ [**32**, section 1, c].

If B is a commutative Bézout domain, then the group $\Gamma(B) := (Q(B)^\times/U, 1.U, .)$, with $U = U(B)$, is a lattice-ordered group with the lattice operations on $Q(B)$ defined by: $a.U \wedge b.U := \gcd(a,b).U$ and $a.U \vee b.U = \operatorname{lcm}(a,b).U$. It is called the group of divisibility of B and note that $(\Gamma(B)^+, ., \wedge, \vee)$ is interpretable in the ring $(B, +, ., 1)$.

By the Jaffard-Ohm's theorem, we know that any lattice ordered group Γ can be realised as the group of divisibility of a Bézout domain B [**14**, Chapter 3, Theorem 5.3]. (One defines B inside the group algebra $k[\Gamma]$ with k any field.)

This group $\Gamma(B)$ also reflects (some of) the model-theoretic properties of the class of B-modules. For instance, G. Puninski and C. Toffalori showed that the lattice $L(B)$ of positive primitive formulas (pp-formulas) over B has no width if and only if $\Gamma(B)$ contains a densely ordered subchain [**40**].

When B, B' are two valuation domains (and so $\Gamma(B)$, $\Gamma(B')$ are totally ordered abelian groups, also called the value groups), L. Gregory observed that the Ziegler spectrum Zg_B, Zg'_B are homeomorphic if and only if the value groups $\Gamma(B)$, $\Gamma(B')$ are isomorphic [**17**].

2.2. Maximal spectrum. Denote by $MSpec(B)$ the space of maximal ideals of B with basic closed sets $V(a) := \{\mathcal{M} \in MSpec(B) : a \in \mathcal{M}\}$, $a \in B$. A constructible subset is a boolean combination of basic closed subsets. Since B is Bézout, we get: $V(a) \cap V(b) = V(\gcd(a,b))$, with $a, b \in B^\star$ and we denote by $(a:b) := \frac{a}{\gcd(a,b)}$ (it is well-defined up to an element of U; $V(a) \setminus V(b) \subseteq V((a:b))$.) One says that B has *good factorization* [**10**, 2.10]: if for a, $b \in B^\star$ there exist a_1, b_1 such that $a = a_1.b_1$, $\gcd(a_1, b) = 1$ and b belongs to $\bigcap_{\mathcal{M} \in V(b_1)} \mathcal{M}$ so $V(a) \setminus V(b) = V(a_1)$. Therefore if B has good factorization, every constructible subset of $MSpec(B)$ is either a basic closed set or its complement (a basic open set) [**10**, Lemma 2.12].

(1) The space $MSpec(B)$ is homeomorphic to the space of minimal prime ideals of $\Gamma(B)$ [**43**, Proposition 8].
(2) If B has a good factorization, then $\Gamma(B)$ is projectable.

Note that if B has good factorisation, then B is adequate and so B is an elementary divisor domain and any prime ideal is contained in a unique maximal ideal [**14**, Exercice 6.4, page 118].

Examples of such Bézout domains with good factorization are the rings of holomorphic functions and the ring of algebraic integers. (For more examples see: [**10**, section 5], for the ring of holomorphic functions, see [**25**, Lemma 6.5] and [**27**].)

3. Valued modules

Let us revert to the case of a non necessarily commutative ring R. Before making precise the notion of valued R-modules that we will use, we quickly review previous works on valued modules. We certainly do not claim to be exhaustive.

Early references on valued modules may be found in the works of P. Conrad [**7**, chapter 1] where the values are taken in a partially ordered set Δ, of I. Fleischer [**13**] who assumes that Δ is totally ordered with a minimum; later M. Lazard [**24**, section 2.1] used that notion in his work on p-adic analytic groups (there $\Delta = \mathbb{R}_+$).

In the model theory point of view, van den Dries proved a quantifier elimination result for valued vector spaces [**8**]. If we restrict ourselves to abelian groups, P. Schmitt showed the class of all p-valuated (torsion-free) abelian groups is hereditarily undecidable, using an undecidability result for pairs of abelian groups due to W. Baur) [**45**]. Recall that a valued group is p-valuated (p a fixed prime number) if for $g \neq 0$, $v(p.g) > v(g)$. Schmitt then identified a decidable subclass, namely the *tamely p-valuated groups* [**46**]. These p-valuated abelian groups also occur as multiplicative subgroups of certain valued fields, for instance in $(\mathbb{Q}_p, +, \cdot, (1+p)^{\mathbb{Z}})$, $p \neq 2$ [**31**, section 3]. N. Mariaule [**31**, section 2] and F. Guignot [**18**] axiomatized the theory of the group of integers $(\mathbb{Z}, +, v_p)$ with the p-adic valuation v_p, proving a quantifier elimination result.

From a different point of view, valued abelian groups appeared as groups endowed with a C-relation (compatible with the group law) in the work of D. Macpherson and C. Steinhorn [**30**]. A classification of C-minimal valued abelian groups was recently obtained by F. Delon and P. Simonetta [**11**]. R. Aldama studied more particularly the NIP property in theories of valued abelian groups [**1**]. In the thesis of F. Guignot, one can find an example of an abelian valued group with an IP-theory [**18**].

Let $\mathcal{L}_R := \{+, -, 0, \cdot r; r \in R\}$ be the usual language of R-modules where $\cdot r$ denotes the scalar multiplication by $r \in R$. Denote by T_R the theory of right R-modules and let M be a R-module. In order to define a structure of an R-valued module on M, we introduce a map w from M to Δ, a distributive lattice on which a lattice-ordered group Γ acts.

DEFINITION 3.1. *Let $(\Delta \cup \{+\infty\}, \wedge, \vee, 1_\Delta, +\infty)$ be a distributive lattice with two constants: $+\infty$ interpreted by the maximum and a constant 1_Δ interpreted by a distinguished element of Δ. Set $\bar{\Delta} := \Delta \cup \{+\infty\}$. Let $(\Gamma, ., \wedge, \vee, 1_\Gamma)$ be an abelian* *lattice-ordered group (written multiplicatively) and let* $\Gamma^+ := \{\gamma \in \Gamma : 1_\Gamma \leq \gamma\}$. *We will assume that for each* $\gamma \in \Gamma$, *we have a right action denoted by* $\cdot\gamma$ *on* $\bar{\Delta}$ *satisfying the following axioms:*

(1) *for all* $\delta \in \Delta$, $\delta \cdot 1_\Gamma = \delta$,
(2) *for all* $\delta \in \Delta$, *for all* $\gamma_1, \gamma_2 \in \Gamma$, $\delta \cdot (\gamma_1 \cdot \gamma_2) = (\delta \cdot \gamma_1) \cdot \gamma_2$,
(3) *for all* $\delta_1, \delta_2 \in \Delta$, *for all* $\gamma \in \Gamma$, $(\delta_1 \cdot \gamma) \wedge (\delta_2 \cdot \gamma) = (\delta_1 \wedge \delta_2) \cdot \gamma$,
(4) *for all* $\delta \in \Delta$, *for all* $\gamma_1, \gamma_2 \in \Gamma$, $\delta \cdot (\gamma_1 \wedge \gamma_2) = (\delta \cdot \gamma_1) \wedge (\delta \cdot \gamma_2)$,
(5) *for all* $\delta_1, \delta_2 \in \Delta$, *there exists* $\gamma \in \Gamma$ $(\delta_1 \cdot \gamma \geq \delta_2)$,
(6) *for all* $\delta \in \Delta$, $+\infty > \delta$ *and for all* $\gamma \in \Gamma$, $+\infty \cdot \gamma = +\infty$.

The above definition generalises the one given in [**2**, section 3], [**3**, Definition 3.1], *where* Δ *is a totally ordered set and* Γ *a totally ordered group.*

Let T_L *be the theory of such two-sorted structures:*

$$((\bar{\Delta}, \wedge, \vee, 1_\Delta, +\infty), (\Gamma, 1_\Gamma, ., \wedge, \vee), \cdot\gamma; \gamma \in \Gamma);$$

we will use the abbreviation (Δ, Γ). *(When the context is clear, we will suppress the indices of the elements* 1_Δ, 1_Γ*).*

Now we are ready to define the notion of a valued R-module that we will be working with.

DEFINITION 3.2. *Let* R_0 *be a commutative subring of* R *with no zero-divisors endowed with an* ℓ*-valuation* v *such that* $v(R_0) = \Gamma$.

A valued R-module *is a two-sorted structure* $(M, \bar{\Delta}, w)$, *where* M *is a* R*-module,* (Δ, Γ) *satisfies* T_L, *and* $w : M \to \bar{\Delta}$ *is such that*

(1) *for all* $m_1, m_2 \in M$, $w(m_1 + m_2) \geq w(m_1) \wedge w(m_2)$, *and* $w(0) = +\infty$;
(2) *for all* $m \in M^\star$, $w(m \cdot a) = w(m) \cdot v(a)$, *for each* $a \in R_0^\star$.

Examples of such valued modules are:

(1) Let (K, v) be a valued field endowed with an endomorphism σ. We consider K as a module over the skew polynomial ring $R := K[t; \sigma]$ and we let $w = v$, $R_0 = K$ and $\Delta = \Gamma = v(K^\star)$.
(2) Let B be a (commutative) Bézout domain with its ℓ-valuation map v to its group of divisibility $\Gamma(B)$. Then we consider B as a module over itself and take $w = v$ and $R_0 = R = B$.

3.1. Abelian structures. With the same set-up as in the previous subsection, in particular we will always assume that $(\Delta, \Gamma) \models T_L$, we consider the structure $(M, \bar{\Delta}, w)$ in the weaker formalism of abelian structures.

Recall that an abelian structure M is an $\mathcal{L}$-structure, where $\mathcal{L}$ contains the language of abelian groups, whose domain is an abelian group, the functions of $\mathcal{L}$ are homomorphisms and the relations are abelian subgroups [**36**, chapter 3, section 3A].

Let $\mathcal{L}_V$ be the language $\mathcal{L}_R \cup \{V_\delta,\ \delta \in \Delta\}$, where V_δ are unary predicates. To $(M, \bar{\Delta}, w)$, we associate the abelian structure $M_V := (M, (M_\delta)_{\delta \in \bar{\Delta}})$, where $M_\delta := \{m \in M : w(m) \geq \delta\} = V_\delta(M)$.

Note that for δ_1, $\delta_2 \in \Delta$, we have: $M_{\delta_1} \cap M_{\delta_2} \subseteq M_{\delta_1 \vee \delta_2}$ and $M_{\delta_1} + M_{\delta_2} \subseteq M_{\delta_1 \wedge \delta_2}$.

DEFINITION 3.3. *Let* $T_{R,V}$ *be the* $\mathcal{L}_V$*-theory, consisting of the* $\mathcal{L}_R$*-theory of* R*-modules together with:*

$(1)_V$ $V_\delta(0)$, *for each* $\delta \in \Delta$,

$(2)_V$ $\forall m_1\, \forall m_2\, ((V_{\delta_1}(m_1)\,\&\, V_{\delta_2}(m_2)) \to V_{\delta_1\wedge\delta_2}(m_1+m_2))$, *for every* $\delta_1,\ \delta_2 \in \Delta$,
$(3)_V$ $\forall m\, (V_\delta(m) \to V_{\delta\wedge\mu}(m))$, *for any* $\delta,\ \mu \in \Delta$,
$(4)_V$ $\forall m\, (V_\delta(m) \to V_{\delta\cdot v(a)}(m\cdot a))$, *for each* $a \in R_0^\star$, $\delta \in \Delta$.

When the context is clear, we will simply use the notation T_V (instead of $T_{R,V}$).

E. Fisher developed the model theory of abelian structures and most of the results known for modules transfer to that larger setting [**36**]. One can also, as it was observed by T. Kucera and M. Prest, view abelian structures as modules over a certain path algebra associated with the lattice of submodules [**22**, section 1.9].

The advantage of that weaker formalism is that it allows us to use the Baur-Monk pp-elimination theorem (see below).

A positive primitive (pp) $\mathcal{L}_V$-formula $\psi(\bar{y})$ is a positive existential formula of the form: $\exists \bar{x}\, \theta(\bar{x},\bar{y})$, where $\theta(\bar{x},\bar{y})$ is a conjunction of atomic formulas. In particular $\psi(M)$ is a subgroup of M^m, m =length($\bar{y}$).

An index sentence is a sentence of the form $(\phi/\psi) > n$, where $\psi \to \phi$ (in T_V) and $\phi(x), \psi(x)$ are pp-formulas (with one variable and without parameters), $n \in \mathbb{N}$.

THEOREM 3.4. (Baur-Monk-Fisher pp-elimination Theorem) [**21**, Appendix A.1], [**36**, chapter 3, 3.A]) *Any $\mathcal{L}_V$-formula is equivalent to a boolean combination of pp $\mathcal{L}_V$-formulas and index sentences, modulo the theory T_V.*

4. Valued Ore modules

In this subsection we consider the special case where $R := K[t;\sigma,\partial]$ is the skew polynomial ring over the valued field (K, v, σ, ∂) endowed with an endomorphism σ and a contractive σ-derivation ∂. In that case the subring R_0 will be equal to K, the group Γ the value group of K and Δ will be totally ordered. The main algebraic property of the ring R we use, is that R is right euclidean [**6**, chapter 2].

We will distinguish two cases either σ is an isometry of K and ∂ is a contractive σ-derivation, or σ is a contractive endomorphism of K (and $\partial = 0$). In that last case, we introduce an additional action on Δ which reflects the action of t (on the module).

4.1. σ is an isometry and ∂ is a contractive σ-derivation. Under these hypotheses, we are able to deal with the corresponding valued modules viewed as 2-sorted structures. Even in that relatively easy setting, our proof of the quantifier elimination result was quite involved. We haven't pushed it further in the contractive case. Below we recall the main step of that result (Theorem 4.1) and give two examples where one obtains full quantifier elimination.

Recall that a σ-derivation ∂ satisfies: $\partial(a.b) = \partial(a).\sigma(b) + a.\partial(b)$. We will assume that σ is an isometry, namely $v(x) = v(\sigma(x))$ and that ∂ is contractive, namely $v(\partial(x)) \geq v(x)$. In that case, one can extend v to a valuation on the ring $K[t;\sigma,\partial]$ with the same value group by simply defining the value of a polynomial as the minimum of the values of its coefficients (the Gaussian valuation).

We define a valued R-module M [**2**, Definition 3.1] as an R-module endowed with a surjective map w on $\bar{\Delta}$ satisfying, in addition to (1), (2) of Definition 3.2,

(3) for all $m \in M$ $w(m \cdot t) \geq w(m)$.

We denote that theory by T_w and by $T_{w,d}$ the theory T_w plus two axiom schemes, a divisibility axiom scheme: for each $r \in R$ with $v(r) = 1$.

$$\forall x \neq 0\, \exists y\, (x = y \cdot r\ \&\ w(x) = w(y)),$$

and for each $r_1, \cdots, r_n \in R$ with $v(r_i) = 1$, for each $\delta \in \Delta$, the following scheme (called *no residual identities*):

$$\forall x_1 \cdots \forall x_n \, \exists y \, (\bigwedge_i w(x_i) \geq \delta \rightarrow \bigwedge_i w(y \cdot r_i + x_i) = \delta).$$

Let $M_\delta + := \{m \in M : w(m) > \delta\}$. This last axiom implies in particular that $M_\delta / M_\delta +$ is infinite.

THEOREM 4.1. [**2**, Proposition 6.2] *Let (M, w, Δ) be a valued R-module satisfying the theory $T_{w,d}$. Then any existential formula in the module sort is equivalent to a quantifier-free formula in the module sort, an existential formula in the Δ-sort plus some residual index formulas and sentences.*

The residual index formulas and sentences are existential formulas of the form [**2**, Definition 6.4]:

(1) A residual index formula $Ind_{n,r}(\delta)$ with $r \in R$ and $v(r) \geq 1$, $\delta \in \Delta$, is of the form: $\exists x_1 \cdots \exists x_n \in M$

$$(\bigwedge_{1 \leq i < j \leq n} w(x_i - x_j) = \delta \,\&\, \bigwedge_{i=1}^{n} (w(x_i \cdot r) > \delta \,\&\, w(x_i) = \delta)).$$

(2) A residual index sentence is of the form $\exists \delta \, Ind_{n,r}(\delta)$.

The residual index sentences generalize module index sentences. We deduce a quantifier elimination result in two cases: when $(\Delta, \leq)$ is a densely ordered set and when Δ and Γ are discretely ordered with a compatible action of Γ on Δ [**2**, Corollary 6.4]. In both cases, using this quantifier elimination result, we deduce that the theory of such valued R-modules M is NIP [**2**, Proposition 7.1].

In the discrete case, denote the corresponding theory by $T_{w,d,discrete}$ and let us give two examples of models of $T_{w,d,discrete}$.

Let $K := (W(F), +, v)$ where $W(F)$ is the field of Witt vectors, F is a p-closed field of characteristic p and v the usual valuation map. As before we consider that field K as a valued module over the skew polynomial ring $R := W(F)[t; \sigma]$ where σ the Witt Frobenius [**2**, Example 3].

Let F be a differential field of characteristic 0 with an algebraically closed subfield of constants C_F; let $\tilde{F}$ be an extension of F such that $\tilde{F}$ has no proper Picard-Vessiot extensions and such that any finite subset of $\tilde{F}$ is included in a finite sequence of successive Picard-Vessiot extensions, starting with F. Consider the field of Laurent series $K := (\tilde{F}((x^{-1})), \partial, v)$ as a module over the skew polynomial ring $R := K[t; \partial]$, where v is the x^{-1}-adic valuation and ∂ the derivative with respect to x [**2**, Example 4].

Then both fields K viewed as valued R-modules admit quantifier elimination in the two-sorted language, as models of $T_{w,d,discrete}$.

4.2. σ is a contractive endomorphism. In this case, we will assume that σ is not an isometry and so we may take $\partial = 0$ [**6**, Theorem 2.1.3(i)]. We will also only deal with the corresponding abelian structures. The action of σ on K will induce an action σ_v on Γ defined by $\sigma_v(v(a)) = v(\sigma(a))$, where $a \in K^\star$. On a valued $K[t; \sigma]$-module, the action of t will induce an action on Δ, compatible with σ_v. Moreover, we assume that we have on $(\Delta, <)$ an endomorphism τ (respecting $<$) and we require in addition that

(1) $\tau(1_\Delta) = 1_\Delta$,

(2) $\forall\delta > 1_\Delta\ \tau(\delta) > \delta$ and $\forall\delta < 1_\Delta\ \tau(\delta) < \delta$,
(3) $\forall\delta\ (\tau(\delta\cdot\gamma) = \tau(\delta)\cdot\gamma^{\sigma_v})$, for all $\gamma \in \Gamma$,
(4) τ is surjective on Δ.

Let $r \in R$ be of the form $\sum_{i=0}^{d} t^i.a_i$, with $a_i \in K$ and let $\delta \in \Delta$. Under the above assumptions on Δ, there exists $\mu \in \Delta$ such that $\delta = min\{\tau^i(\mu) + v(a_i) : 0 \le i \le d\}$ [**3**, Lemma 4.5]. We denote this μ by $\Upsilon(r,\delta)$.

A valued R-module M_V is an R-module satisfying $(1)_V$ up to $(4)_V$ of Definition 3.3 together with

$(5)_V$ $\forall m\ V_\delta(m) \leftrightarrow V_{\tau(\delta)}(m\cdot t)$.

Let us denote this set of axioms by $T_{V,\sigma}$.

Now we express in the theory of R-modules that σ is non surjective on K. We assume that K is a finite-dimensional K^σ-vector space and we fix a basis $1 = c_0, \cdots, c_{n-1}$ of K over K^σ. The class of R-modules M that we consider, will always have a direct sum decomposition $M = \oplus_{i=0}^{n-1} M\cdot tc_i$. In order to express that property we expand the module language $\mathcal{L}_R$ with n unary functions symbols λ_i, $i \in n$, $n := \{0, \cdots, n-1\}$.

The theory T_σ consists of the theory of R-modules together with:

(1) $\forall x\ (x = \sum_{i\in n} \lambda_i(x)\cdot tc_i)$,
(2) $\forall x \forall (x_i)_{i\in n}\ (x = \sum_{i\in n} x_i\cdot tc_i \to \bigwedge_{i\in n} x_i = \lambda_i(x))$.

Let T_V^{sep} be the theory $T_\sigma \cup T_{V,\sigma}$ together with for all $\delta \in \Delta$, for all $r \in R$ of the form $\sum_{i=0}^{d} t^i.a_i$ with $v(a_i) \ge 1$, $0 \le i \le d$ and $a_0 \ne 0$:

(1) $\forall n\ (V_\delta(n) \to \exists m(V_{\Upsilon(r,\delta)}(m)\ \&\ m\cdot r = n))$,
(2) $\forall m\ \exists n\ V_\delta(m - n\cdot t)$.

The second axiom scheme reflects in the module theory that the separable closure of a valued field of characteristic p is dense in its algebraic closure (thinking of the action of t as the Frobenius map).

If K is a separably closed field of characteristic p, of finite imperfection degree n and σ the Frobenius map, then K as a $K[t;\sigma]$-valued module is a model of T_V^{sep}.

THEOREM 4.2. [**3**, Corollary 6.5] *In T_V^{sep}, any $\mathcal{L}_V$-formula is equivalent to a quantifier-free $\mathcal{L}_V$-formula up to index sentences.*

This last result allows us to identify the various completions of T_V^{sep}, analysing the torsion [**3**, sections 7, 8].

5. Valued modules over a Bézout domain

Now we consider the case where B is a (commutative) Bézout domain with v its ℓ-valuation to its group of divisibility $\Gamma(B)$. In this case we set $R_0 = B = R$ and for sake of simplicity, we assume that $\Gamma = \Gamma(B)$, $\Delta = \Gamma(B)^+$ and $1_\Delta = 1_\Gamma$. We denote the corresponding theory by $T_{B,V}$ (see Definition 3.3). Given a valued B-module $(M, \bar{\Gamma}(B), w)$, we associate as before an abelian structure M_V.

We will consider a subclass of the class of abelian structures satisfying $T_{B,V}$, namely those which satisfy in addition

$$(5) \qquad \forall m\ V_{1_\Gamma}(m),$$

and the following divisibility scheme:

$$(6)_{V,div} \qquad \forall m\ \exists n\ (V_{v(a)}(m) \to n\cdot a = m), \text{for each } a \in B^\star.$$

DEFINITION 5.1. *We will denote by* $T_{B,V,div}$ *the theory* $T_{B,V}$ *together with the axiom* (5) *and the divisibility axioms scheme* $(6)_{V,div}$.

It is easy to see that the theory $T_{B,V,div}$ is consistent. Indeed, taking $M_V := (B, (B_{v(a)})_{a \in B^\star})$, we have that M_V is a model of $T_{B,V,div}$. Indeed, by definition of the valuation on B, we have that for a, $b \in B^\star$, $v(b) \geq v(a)$ iff $b.a^{-1} \in B$ and so B (viewed as a B-module) satisfies axiom (5) and axiom scheme $(6)_{V,div}$.

REMARK 5.2. [**25**, Remark 4.5]

(1) In fact, $T_{B,V,div}$ is what is called *an expansion by definitions* of the theory of B-modules. Explicitly, it means that given any B-module M, we can expand it to a model of $T_{B,V,div}$ and in any model of $T_{B,V,div}$, the new predicates V_γ are definable in the language of B-modules.

Indeed, given any B-module M, we set $V_\delta(M) := \{m \in M : \exists n\, m = n.a\}$ with $v(a) = \delta$, $a \in B^\star$. To check that this expansion satisfies all the axioms $(1)_V$ up to $(6)_{V,div}$ is rather straightforward. Let us check for instance $(2)_V$. Let $\delta_1 = v(a_1)$ and $\delta_2 = v(a_2)$ and write $a_1 = a'.\gcd(a_1, a_2)$, $a_2 = a''.\gcd(a_1, a_2)$. Assume $V_{\delta_1}(m_1) \,\&\, V_{\delta_2}(m_2)$ holds. So for some n_1, n_2 we have $m_1 = n_1.a_1$ and $m_2 = n_2.a_2$ and $m_1 + m_2 = (n_1.a' + n_2.a'').\gcd(a_1, a_2)$. Therefore $V_{\delta_1 \wedge \delta_2}(m_1 + m_2)$.

By definition of the ℓ-valuation on B, we have that for a, $b \in B^\star$, $v(b) \geq v(a)$ iff $b.a^{-1} \in B$. Using this, one can easily deduce axioms $(3)_V$, $(5)_V$ and $(6)_{V,div}$.

Conversely, if $M_V \models T_{B,V,div}$, then the subgroup $V_{v(a)}(M)$, $a \in B^\star$ is definable by the $\mathcal{L}_B$-formula: $\exists x\; y = x.a$. It follows from $(4)_V$, $(5)_V$ and $(6)_{V,div}$.

(2) Let $M_V \models T_{B,V,div}$ and assume M is a torsion-free B-module. Let $m \in M$ and suppose that $V_{\delta.v(b)}(m.b)$ holds with $b \in B^\star$, $\delta \in \Gamma^+$. Then let us show that $V_\delta(m)$ holds. Let $a \in B^\star$ be such that $v(a) = \delta$. By axiom $(6)_{V,div}$, for some $n \in M$, $m.b = n.a.b$. So $(m - n.a).b = 0$. Since M is torsion-free, $m = n.a$ and so $V_\delta(m)$ holds.

5.1. Localizations. Let $\mathcal{M} \in MSpec(B)$ and $B_\mathcal{M}$ be the localization of B at $\mathcal{M}$. Since B is Bézout, $B_\mathcal{M}$ is a valuation domain and $\Gamma(B_\mathcal{M})$ is a totally ordered abelian group. Denote by $v_\mathcal{M}$ the ℓ-valuation induced on $B_\mathcal{M}$ to its group of divisibility $\Gamma(B_\mathcal{M})$, namely the map sending $a.s^{-1}$ to $a.s^{-1}.U_\mathcal{M}$, with $a \in B$, $s \in B \setminus \mathcal{M}$ and $U_\mathcal{M}$ the subgroup of invertible elements of $(B^\star_\mathcal{M}, \cdot, 1)$. The map $f : (\Gamma^+(B), ., \wedge, 1) \to \prod_{\mathcal{M} \in MSpec(B)} (\Gamma^+(B_\mathcal{M}), ., \wedge, 1) : a.U \to (a.U_\mathcal{M})_{\mathcal{M} \in MSpec(B)}$ is an embedding of lattice-ordered monoids.

Denote by $f_\mathcal{M} : \Gamma(B)^+ \to \Gamma(B_\mathcal{M})^+ : a.U \to a.U_\mathcal{M}$, $a \in B^\star$.

Let (M, w) be a valued B-module. and let $ann_{B \setminus \mathcal{M}} M := \{m \in M : \exists r \in B \setminus \mathcal{M}\; m.r = 0\}$. One can define in an explicit way the localization $M_\mathcal{M}$ of M at $\mathcal{M}$ by $M_\mathcal{M} := \{m + ann_{B \setminus \mathcal{M}} M : \; m \in M\}$. Since we have started with a valued B-module M, we get that $(M_\mathcal{M}, w_\mathcal{M})$ is a valued $B_\mathcal{M}$-module, by defining $w_\mathcal{M}(m_\mathcal{M}) := f_\mathcal{M}(w(m))$.

We can also start with M_V the associated abelian structure and each $M_\mathcal{M}$ can be endowed with a structure of valued module as follows: $V_\delta(m + ann_{B \setminus \mathcal{M}} M)$ iff there exists $n \in m + ann_{B \setminus \mathcal{M}} M$ such that $V_\delta(n)$. In this case, we can extend a result due to Garavaglia [**15**] on the relationship between M_V and the direct product of its localizations $M_\mathcal{M}$, $\mathcal{M} \in MSpec(B)$.

PROPOSITION 5.3. [**25**, Proposition 5.2] *Let M_V be a model of $T_{B,V}$. Then there is an elementary embedding of M_V into the direct product $\prod_{\mathcal{M}\in MSpec(B)}(M_{\mathcal{M}})_V$, where the localised modules $M_{\mathcal{M}}$ are viewed as models of $T_{B_{\mathcal{M}},V}$.*

The proof is a straightforward adaptation of Garavaglia's proof. It will be used in the proof of the main result of this section (Theorem 5.5).

5.2. Valuation domains. Let A be a valuation domain. We have the following quantifier elimination result, whose proof is inspired by the one of [**2**, Proposition 4.1].

PROPOSITION 5.4. [**25**, Theorem 3.2] *The theory $T_{A,V,div}$ admits positive quantifier elimination, namely any pp $\mathcal{L}_V$-formula is equivalent to a conjunction of atomic $\mathcal{L}_V$-formulas. In particular, any pp $\mathcal{L}_V$-formula $\phi(x)$ (in one free variable) is equivalent to a formula of the form*

$$x\cdot a=0\ \&\ \bigwedge_{i=1}^{n} V_{\delta_i}(x.b_i),$$

with $a,\ b_i\in A$, $\delta_i\in\Gamma^+$, $1\le i\le n$, $\delta_1.v(b_1)^{-1}>\cdots>\delta_n.v(b_n)^{-1}$ and $v(b_1)>\cdots>v(b_n)$.

This above result was essentially known but our method of proof gives additional informations in the case when we start with a Bézout domain B and we consider the class of valuation domains obtained by localizing B at maximal ideals $\mathcal{M}$. Given a pp $\mathcal{L}_V$-formula $\phi(\mathbf{x})$ with coefficients in B, considering it in the localised rings $B_{\mathcal{M}}$, we can trace how the corresponding equivalent quantifier-free (positive) formula depends on divisibility conditions among the coefficients of the original formula $\phi(\mathbf{x})$. We will use that additional information in Theorem 5.5.

Finally note that for $a,\ b\in B$, we have that

$$B_{\mathcal{M}}\models a|b \text{ iff } \Gamma(B_{\mathcal{M}})\models v_{\mathcal{M}}(a)\le v_{\mathcal{M}}(b) \text{ iff } (a:b)\notin\mathcal{M} \text{ iff } \mathcal{M}\notin V((a:b))$$

[**9**, section 1].

5.3. Feferman-Vaught. Recall that the classical Feferman-Vaught theorem relates satisfaction in a product of structures to: on one hand satisfaction in the factors of that product and on the other hand satisfaction in a Boolean algebra of the subsets of the index set. In Proposition 5.3 we saw that any B-module M elementarily embeds in the direct product of the localizations of M by the maximal ideals of B. In Proposition 5.4, we proved a quantifier elimination result for modules over valuation domains and we observed that the procedure was uniform on certain constructible subsets of $MSpec(B)$. Putting those two ingredients together, we obtain:

THEOREM 5.5. [**25**, Theorem 5.6] *Let M_V be a model of $T_{B,V,div}$ and let $\phi(\mathbf{y})$ be a pp $\mathcal{L}_V$-formula. Then there are finitely many quantifier-free positive $\mathcal{L}_V$-formulas $\theta_i(\mathbf{y})$, $i\in I$, (obtained explicitly from ϕ) and a finite covering of $MSpec(B)$ into constructible subsets O_i such that for any $\mathbf{u}\in M$, we have:*

$$M_V\models\phi(\mathbf{u})\leftrightarrow(\bigwedge_{i\in I}\text{ for all }\mathcal{M}\in \mathrm{O_i}\ (\mathcal{M}_{\mathcal{M}})_{\mathrm{V}}\models\theta_{\mathrm{i}}(\mathbf{u}_{\mathcal{M}})),$$

where $\mathbf{u}_{\mathcal{M}}$ denotes the image of the tuple $\mathbf{u}$ in $M_{\mathcal{M}}$.

We will use the above theorem together with the following remark to deduce decidability results. We will translate back in the ring the divisibility conditions that occur in the localisations, using the following observation due to van den Dries.

Let A be a commutative ring, and $f,\ g \in A$. The radical Jacobson relation is defined by: $g \in rad(f)$ if all maximal ideals containing f, do contain g and the prime radical relation by: $g \in Rad(f)$ if all prime ideals containing f, do contain g.

REMARK 5.6. [**9**, Lemma 1.3] $(B_{\mathcal{M}} \models a|b) \leftrightarrow (a:b) \notin \mathcal{M}$.
Further, for $a_i, b_i, c_j, d_j \in B$, I, J finite, the following equivalences hold:

$$\forall \mathcal{M} \in MSpec(B)\ B_{\mathcal{M}} \models (\bigvee_{i \in I} a_i \not| b_i \vee \bigvee_{j \in J} c_j | d_j),$$

$$\forall \mathcal{M} \in MSpec(B)\ (\bigwedge_{j \in J} (c_j : d_j) \in \mathcal{M} \to \prod_{i \in I} (a_i : b_i) \in \mathcal{M}),$$

$$B \models \prod_{i \in I} (a_i : b_i) \in rad(\gcd((c_j : d_j)_{j \in J})),$$

where $\gcd((c_j : d_j)_{j \in J})$ denotes a generator of the ideal generated by the elements $(c_j : d_j)$, $j \in J$.

In case B has good factorization, $MSpec(B)$ is homeomorphic to the space of prime ideals of $\Gamma(B)$, and so the subsets O_i are either basic open or basic closed sets of $MSpec(B)$ (see section 2.2). So under that hypothesis, we may assume that the covering of $MSpec(B)$ by constructible subsets occurring in Theorem 5.5 is simply a partition by basic open or basic closed subsets.

5.4. Effectively given rings and decidability. Let B be a Bézout domain, we will call assumption (EF) on B the following: B is a countable ring, it can be enumerated as $(r_n :\ n \in \omega)$ in such a way, there are algorithms to perform the following operations: given $a,\ b \in B$, produce $a+b$, $-a$, $a.b$, decide whether $a = b$ or not and the relation $\{(a,b) \in R^2 :\ a|b\}$ is recursive.

Note that if B satisfies (EF), then the relations $c = \gcd(a,b)$ and $c = (a:b)$ are recursive.

FACT 5.7. [**17**, Lemma 3.2] Suppose any prime ideal is an intersection of maximal ideals and that T_B is decidable, then the relation $a \in rad(b)$ is recursive. One shows that:

$$a \in rad(b) \Leftrightarrow T_B \models (\exists x \neq 0\ x \cdot b = 0 \to \exists x \neq 0\ x \cdot a = 0).$$

FACT 5.8. [**9**, Fact 2] Suppose the ring R satisfies hypothesis (EF) and that the Jacobson radical relation is equal to the prime radical relation, then the Jacobson radical relation rad on R is recursive.

PROPOSITION 5.9. [**25**, Proposition 5.8] *Suppose that B is a countable Bézout domain satisfying hypothesis (EF) and that the Jacobson radical relation rad is recursive. Assume also that for each $\mathcal{M} \in MSpec(B)$, the quotient $B/\mathcal{M}$ is infinite. Then T_B is decidable.*

Recall that $T_{B,V,div}$ is an expansion by definition of the theory T_B of all B-modules. In the proof of the proposition, we use Theorem 5.5 and the above remark.

This applies in particular to $B := \widetilde{\mathbb{Z}}$, $\widetilde{\mathbb{Z}} \cap \mathbb{R}$, $\widetilde{\mathbb{Z}} \cap \mathbb{Q}_p$, namely for these rings B, we have the theory T_B is decidable [**25**, Corollaries 6.3, 6.4].

The decidability of the theory of modules of the ring of algebraic integers $\widetilde{\mathbb{Z}}$ has also been obtained by S. L'Innocente, G. Puninski and C. Toffalori, using the description of the Ziegler spectrum attached to that ring [**26**].

Of course none of the above can be applied to the ring $\mathfrak{H}(\mathbb{C})$ of holomorphic functions. First for the obvious reason that this ring is uncountable. Then, as I. Kaplansky pointed out that there are prime, non maximal ideals in $\mathfrak{H}(\mathbb{C})$ and so the prime radical relation does not coincide with the maximal radical relation. However, the divisibility relation in the localisations of $\mathfrak{H}(\mathbb{C})$ by maximal ideals is well understood as well as the lattice of the constructible subsets of $MSpec(\mathfrak{H}(\mathbb{C}))$. In the remainder of the article, we will revisit Theorem 5.5 for $B = \mathfrak{H}(\mathbb{C})$.

We first review what is known on the decidability question for rings of holomorphic functions (in expansions of the ring language).

5.5. On decidability results for the rings of holomorphic functions. Let $\mathcal{L}_{rings} := \{+, -, \cdot, 0, 1\}$ and $\mathcal{L}_c := \mathcal{L}_{rings} \cup \{c\}$ where c is a new constant. Let $(R, |.|)$ be an integral domain with an absolute value $|.|$.

We will consider two cases: first assume $|.|$ is archimedean, then let $\mathfrak{H}(R) := \{\sum_{n\geq 0} a_n.z^n : \lim_{m\to+\infty} \sum |a_m|.r^m = 0, \text{ for all } r \in \mathbb{R}^+\}$. Second assume that $|.|$ is non-archimedean, then let $\mathfrak{H}(R) := \{\sum_{n\geq 0} a_n.z^n : \lim_{m\to+\infty} |a_m|.r^m = 0, \text{ for all } r \in \mathbb{R}^+\}$.

Concerning the case of the archimedean absolute value, we have the following undecidability result due to L. Lipshitz and T. Pheidas.

THEOREM 5.10. [**28**] *The $\mathcal{L}_c$-theory of $\mathfrak{H}(\mathbb{C})$ is undecidable, where c is interpreted by the element z.*

However, it is still unknown whether the positive existential $\mathcal{L}_c$-theory of $\mathfrak{H}(\mathbb{C})$ is undecidable. In [**35**], T. Pheidas and K. Zahidi point out a gap in an earlier argument of the first author for proving the undecidability of the positive existential theory of $\mathfrak{H}(\mathbb{C})$, in: *The Diophantine theory of a ring of analytic functions. J. Reine Angew. Math. 463 (1995), 153-167.*

Recently some advances have been made by N. Garcia-Fritz and H. Pasten in the non-archimedean case, for instance in the case of $\mathfrak{H}(\mathbb{C}_p)$.

THEOREM 5.11. [**16**] *Let R be an integral domain of positive characteristic with a non-archimedean absolute value $|.|$. Then the positive existential theory of $\mathfrak{H}(R)$ is undecidable. (Moreover one has a uniform (positive existential) interpretability of the ring $\mathbb{Z}$ in such class of integral domains.)*

In the following subsection, we apply our previous results considering the theory $T_{\mathfrak{H}(\mathbb{C}),V}$ of certain valued modules over the ring of holomorphic functions over $\mathbb{C}$. As pointed before, $\mathfrak{H}(\mathbb{C})$ viewed as a valued module over itself is a model of the theory $T_{\mathfrak{H}(\mathbb{C}),V}$. This will give us a partial answer to the decidability question of the positive existential theory of $\mathfrak{H}(\mathbb{C})$. Given an $\mathcal{L}_V$-sentence we will reduce its satisfaction in $\mathfrak{H}(\mathbb{C})$ to whether two zero-sets are included in one another (see subsection 5.6.4). Part of the discussion below may be found in [**27**].

5.6. Valued modules over $\mathfrak{H}(\Omega)$.

5.6.1. *Preliminaries on rings of holomorphic functions.* Let Ω be a nonempty connected open subset of $\mathbb{C}$ and denote by $\mathfrak{H}(\Omega)$ the ring of holomorphic functions in Ω and by $\mathcal{F}(\Omega)$ the field of meromorphic functions of Ω. Let $f \in \mathfrak{H}(\Omega)^\star$ and denote by $Z(f)$ the set of zeroes of f. We will use two basic results on the ring $\mathfrak{H}(\Omega)$: Weierstrass factorization theorem [**42**, Theorem 15.11] and Mittag-Leffler theorem [**42**, Theorem 13.10]. It implies that $\mathfrak{H}(\Omega)$ is a Bézout domain [**19**, Theorem 1], [**42**, Theorem 15.15] and a good factorization domain [**25**, Lemma 6.5]. In particular, the group of divisibility $\Gamma(\mathfrak{H}(\Omega))$ is a projectable ℓ-group.

Also as pointed out before, the fact that $\mathfrak{H}(\Omega)$ has good factorization implies that any prime ideal of $\mathfrak{H}(\Omega)$ is contained in a unique maximal ideal. We denote by $\mathcal{P}_{\mathcal{M}}$ the set of prime ideals included in a given maximal ideal $\mathcal{M}$. This set is totally ordered by inclusion [**20**, Corollary of Theorem 4].

Let $f \in \mathfrak{H}(\Omega)$, we denote by $\mu_f(z)$ the multiplicity of f at z if $f(z) = 0$ and if $f(z) \neq 0$, we set $\mu_f(z) := 0$. For $h \in \mathcal{F}(\Omega)$ of the form $\frac{f}{g}$ with $f,\ g \in \mathfrak{H}(\Omega)$, then $\mu_h(z) := \mu_f(z) - \mu_g(z)$. For $f \in \mathfrak{H}(\Omega)$, it is convenient to denote the supremum of the function μ_f by $m(f)$: it is equal either to $+\infty$ or to an element of $\mathbb{N}$. For $\mu : \Omega \to \mathbb{N}$ denote by $supp(\mu) := \{z \in \Omega : \mu(z) \neq 0\}$. Denote by $\urcorner^+ := \{\mu \in \mathbb{N}^\Omega$ with $supp(\mu)$ a countable discrete subset of Ω without limit points$\}$. Using the elementary facts on $\mathfrak{H}(\Omega)$ recalled above, one can show that for $\mu \in \urcorner^+$, there exists $f \in \mathfrak{H}(\Omega)$ such that $\mu = \mu_f$. We endow $\urcorner^+$ with the induced operations of $(\mathbb{N}, +, \leq)$. The induced partial order induces a lattice operation $\wedge$; we define $\mu \wedge \nu$ by: $(\mu \wedge \nu)(z) := \inf\{\mu(z), \nu(z)\}$. We have that $(\urcorner^+, +, \wedge)$ is a partially ordered commutative monoid and we will denote by $\urcorner$ the group generated by this monoid, extending the lattice operation $\wedge$ to the whole group $\urcorner$. The map $v : \mathfrak{H}(\Omega) \to \urcorner :\ f \to \mu_f$ is an ℓ-valuation (with the group operation on $\urcorner$ denoted by $+$). Indeed, we have: $\mu_{f.g} = \mu_f + \mu_g$, $f|g$ in $\mathfrak{H}(\Omega)$ iff $\mu_f \leq \mu_g$ $(\star)$ (in particular f is invertible iff $\mu_f = 0$) and for $f,\ g \in \mathfrak{H}(\Omega)$, $\mu_{\gcd(f,g)} = \mu_f \wedge \mu_g$. The lattice-ordered group $\urcorner$ is isomorphic to the group of divisibility $\Gamma(\mathfrak{H}(\Omega))$.

It is easily seen that the lattice of zero-sets of elements of $\mathfrak{H}(\Omega)$ (which coincides with the lattice of supports of the elements $\mu_f \in \urcorner^+$) is not only distributive but relatively complemented. It is is also atomic (namely every non-zero element contains an atom). So its theory is complete by a result of Y. Ershov [**12**] (he showed that the theory of distributive, relatively complemented, atomic lattices with a smallest element and without greatest element is complete). Moreover the theory is model-complete if one adds a unary predicate for the set of atoms, and it admits quantifier elimination in the language $\{\wedge, \vee,\ ^c, 0, 1, <_n;\ n \in \omega^\star\}$, where $a <_1 b$ if ($a \wedge b = a$ and $a \neq b$) and $a <_n b$ if there are $a_0, a_1, \cdots a_n$ with $a = a_0 <_1 a_1 \cdots <_1 a_n = b$ [**5**].

5.6.2. *Maximal ideals and ultrafilters.* Let us now recall how to associate with a basic subset $V(a)$, $a \in \mathfrak{H}(\Omega)$, of $MSpec(\mathfrak{H}(\Omega))$ the set of all ultrafilters on $Z(a)$. In case $Z(a)$ is infinite, we identify it with ω and for $\mathcal{M} \in V(a)$, we denote by $\mathfrak{U}_{\mathcal{M}}$ the corresponding ultrafilter. More generally, given any proper ideal I of $\mathfrak{H}(\Omega)$, we associate $Z_I := \bigcap_{f \in I} Z(f)$. Note that any finite sub-intersection is non-empty since it corresponds to the zero set of the greatest common divisor of the corresponding elements [**19**, Theorem 1]. Moreover if Z_I is not empty, then it is finite (since there is no infinite decreasing chain of infinite subsets of ω with nonempty intersection) and so there exists an element f of I such that $Z_I = Z(f)$. If Z_I is empty, one says that I is *free*, otherwise I is *fixed*.

Recall that an ideal I of a (commutative) Bézout domain B is weakly prime if for any $f_1,\ f_2 \in B$, whenever $\mathrm{lcm}(f_1, f_2) \in I$ implies that either $f_1 \in I$ or $f_2 \in I$.

In each free maximal ideal $\mathcal{M}$ of $\mathfrak{H}(\Omega)$, one can find an element $f \in \mathcal{M}$ with $\mu_f = 1$. We denote by $f_{\mathcal{M}}$ such element with μ_f minimal (of course f is by no means unique). If I is a weakly prime ideal, in case there is an element $f \in I$ with μ_f taking only finitely many values, there is an element $f \in I$ with μ_f constant. If I is a prime, non maximal ideal, then for any $f \in I$, $m(f) = \infty$.

LEMMA 5.12. *Let $a \in \mathfrak{H}(\Omega)$. Then there is a map from the set of weakly prime ideals I that contains a to the set of ultrafilters $\mathfrak{U}_I$ on $Z(a)$. Moreover there is a bijection between the ultrafilters on $Z(a)$ and maximal ideals in $V(a)$.* □

REMARK 5.13. In order to distinguish between weakly prime ideals I (or prime ideals) and maximals ideals we have to look at the following subset of $\mathbb{N}^\omega/\mathfrak{U}$: $\{[\mu_f]_{\mathfrak{U}} : f \in I\}$, where $\mathfrak{U}$ is the ultrafilter associated with I. This subset is always an upward cut of $\mathbb{N}^\omega/\mathfrak{U}$. If I is prime, it is closed under the archimedean equivalence relation $\sim$ (namely if $f \in I$ and if some $n \in \mathbb{N}$, $n.[\mu_g]_{\mathfrak{U}} \geq [\mu_f]_{\mathfrak{U}}$, then $g \in I$). In particular, the elements of $\mathcal{P}_{\mathcal{M}}$ are in one to one correspondance with upward cuts in $\mathbb{N}^\omega_{\mathfrak{U}_{\mathcal{M}}}/\sim$.

LEMMA 5.14. *Let $\mathcal{M}$ be a maximal ideal of $\mathfrak{H}(\Omega)$ and let $f,\ g \in \mathfrak{H}(\Omega)$. Then $f|g$ in $\mathfrak{H}(\Omega)_{\mathcal{M}}$ iff $(\mathbb{N}^\omega_{\mathfrak{U}_{\mathcal{M}}}, \leq) \models [\mu_f]_{\mathfrak{U}_{\mathcal{M}}} \leq [\mu_g]_{\mathfrak{U}_{\mathcal{M}}}$.* □

5.6.3. *Localizations.* Let $\mathcal{M}$ be a maximal fixed ideal of $\mathfrak{H}(\Omega)$ with $Z_{\mathcal{M}} = \{z_0\}$, then the localization $\mathfrak{H}(\Omega)_{\mathcal{M}}$ of $\mathfrak{H}(\Omega)$ at $\mathcal{M}$ is a valuation ring which is isomorphic to the ring of germs of $\mathfrak{H}(\Omega)$ at z_0. Its field of fractions $\mathcal{F}_{z_0}$ is the valued field of germs of meromorphic functions at z_0. Its theory (as a valued field) is decidable and it is an elementary substructure of the field of formal power series $\mathbb{C}((t))$ [**23**, Theorem 3].

LEMMA 5.15. *Let $\mathcal{M}$ be a maximal ideal of $\mathfrak{H}(\Omega)$, then the residue field of the localization $\mathfrak{H}(\Omega)_{\mathcal{M}}$ is isomorphic to $\mathbb{C}^\omega_{\mathfrak{U}_{\mathcal{M}}}$.*

PROOF. If $\mathcal{M}$ is a fixed maximal ideal, the residue field is isomorphic to $\mathbb{C}$.

Assume that $\mathcal{M}$ is a maximal free ideal. We enumerate the subset $Z(f_{\mathcal{M}}) \subset \Omega$ as $(z_n)_{n \in \omega}$. Define a map from $\mathfrak{H}(\Omega)_{\mathcal{M}}/\mathcal{M}$ to $\mathbb{C}^\omega_{\mathfrak{U}_{\mathcal{M}}}$ by $g + \mathcal{M} \to [g(z_n)]_{\mathcal{M}}$ with $g \in \mathfrak{H}(\Omega)_{\mathcal{M}}$. Let $[c_n]_{\mathfrak{U}_{\mathcal{M}}} \in \mathbb{C}^\omega_{\mathfrak{U}_{\mathcal{M}}}$. By the interpolation theorem [**42**, Theorem 15.13], there is $h \in \mathfrak{H}(\Omega)$ such that $h(z_n) = c_n$ for each $n \in \omega$ and so $h + \mathcal{M} \to [c_n]_{\mathfrak{U}_{\mathcal{M}}}$. □

5.6.4. Finally let us read off Theorem 5.5, for $B = \mathfrak{H}(\Omega)$, using Lemma 5.14. Let $\mathcal{L}_m := \{+, -, \wedge\}$. This will allow us to partially answer the decidability question for the theory of modules over $\mathfrak{H}(\Omega)$.

Let M be a module over $\mathfrak{H}(\Omega)$ and consider its expansion M_V to a model of $T_{\mathfrak{H}(\Omega),V,div}$ (see Remark 5.2 (1)). Likewise, given any maximal ideal $\mathcal{M}$, we consider the expansion of the localization $(M_{\mathcal{M}})_V$.

Let $\phi(\mathbf{y})$ be a pp $\mathcal{L}_V$-formula, let $\bar{f} := (f_1, \cdots, f_n) \in \mathfrak{H}(\Omega)$ be the coefficients occurring in that formula, let $\bar{\delta} := (\delta_1, \cdots, \delta_m)$ be the elements of Γ^+ occurring as indices of the predicates V_δ's and let $\bar{h} := (h_1, \cdots, h_m) \in \mathfrak{H}(\Omega)$ be such that $\mu_{h_j} = \delta_j$, $1 \leq j \leq m$.

Then there are finitely many quantifier-free positive $\mathcal{L}_V$-formulas $\theta_i(\mathbf{y})$, finitely many quantifier-free positive $\mathcal{L}_m$-formulas $\chi_i(\bar{x})$, $i \in I$, (all obtained explicitly from

ϕ), such that for any tuple of elements $\mathbf{u} \in M$, we have:

$$M_V \models \phi(\mathbf{u}) \leftrightarrow \text{for all maximal ideals } \mathcal{M}$$

$$\bigwedge_{i\in I}(\mathbb{N}^\omega/\mathfrak{U}_\mathcal{M} \models \chi_i(\overline{[\mu_f]}_{\mathfrak{U}_\mathcal{M}}, \overline{[\mu_h]}_{\mathfrak{U}_\mathcal{M}}) \to (\mathcal{M}_\mathcal{M})_V \models \theta_i(\mathbf{u}_\mathcal{M})),$$

where $\mathbf{u}_\mathcal{M}$ denotes the image of the tuple $\mathbf{u}$ in $M_\mathcal{M}$ and $\mathbb{N}^\omega/\mathfrak{U}_\mathcal{M} \models \bigvee_{i\in I} \chi_i(\overline{[\mu_f]}_{\mathfrak{U}_\mathcal{M}}, \overline{[\mu_h]}_{\mathfrak{U}_\mathcal{M}})$. (Note that the quantifier-free positive formulas θ_i are in fact conjunctions of atomic formulas.)

By Theorem 3.4, any $\mathcal{L}_V$-sentence τ is equivalent, modulo the theory $T_{\mathfrak{H}(\Omega),V}$, to a boolean combination of index sentences. Since for every maximal ideal $\mathcal{M}$, $\mathfrak{H}(\Omega)/\mathcal{M}$ is infinite, we may only consider index sentences of the form $(\phi/\psi) > 1$, where ϕ, ψ are pp $\mathcal{L}_V$-formulas. Given such index sentence, let us denote by $[\phi/\psi]$ the corresponding basic open subset in $Zg_{\mathfrak{H}(\Omega)}$. Recall that a point in the Ziegler spectrum is (the isomorphism class of) an indecomposable pure-injective $\mathfrak{H}(\Omega)$-module and so a $\mathfrak{H}(\Omega)_\mathcal{M}$-module for some maximal ideal $\mathcal{M}$ of $\mathfrak{H}(\Omega)$ [**48**, Theorem 5.4]. By standard reductions [**39**, proof of Theorem 6.2], a sentence τ holds in some $\mathfrak{H}(\Omega)$-module iff a basic open subset in $Zg_{\mathfrak{H}(\Omega)}$ is included in a finite union of other basic open sets, namely $[\phi_0/\psi_0] \subset \bigcup_{j=1}^n [\phi_j/\psi_j]$, where ϕ_j, ψ_j are pp $\mathcal{L}_V$-formulas obtained from τ, and $\psi_j \to \phi_j$, $0 \leq j \leq n$ $(\star)$.

Given the above finite set of pp $\mathcal{L}_V$-formulas ϕ_j, ψ_j, $0 \leq j \leq n$, we apply to each of these formulas the quantifier elimination procedure described in Theorem 5.5. (Recall that $\mathfrak{H}(\Omega)$ has good factorisation.) So we obtain (in an effective way) a finite partition of $MSpec(\mathfrak{H}(\Omega))$ into basic open or closed subsets: $C_\ell \subset MSpec(\mathfrak{H}(\Omega))$, $\ell \in L$, over which each of these pp $\mathcal{L}_V$-formulas is equivalent to a formula of the form: $x.a_\ell = 0 \ \& \ \bigwedge_{i\in I} V_{\delta_{\ell_i}}(x.b_{\ell_i})$, a_ℓ, $b_{\ell_i} \in \mathfrak{H}(\Omega)_\mathcal{M}, \delta_{\ell_i} \in \Gamma^+_\mathcal{M}$. Using the duality functor [**36**, Chapter 8], we may only consider open sets of the form $[x.b = 0 \,\&\, V_{\delta_1}(x)/V_{\delta_2}(x.c)]$, $b, c \in \mathfrak{H}(\Omega)_\mathcal{M}$, $\delta_1, \delta_2 \in \Gamma^+_\mathcal{M}$. Then one can show that whether $(\star)$ holds, is equivalent on each element C_ℓ, $\ell \in L$, of the partition of $MSpec(\mathfrak{H}(\Omega))$, to order relations in the localizations between the valuation of elements of $\mathfrak{H}(\Omega)$ and between the δ's that occur in the above formulas.

Finally, using Remark 5.6 and replacing the elements δ of the value group by the corresponding $v(h)$, for $h \in \mathfrak{H}(\Omega)$, this can be expressed using the Jacobson radical relation in $\mathfrak{H}(\Omega)$. So in order to check whether the sentence τ holds in some $\mathfrak{H}(\Omega)$-module, we have to answer statements which are finite conjunctions of the following form: $\prod_{i\in I'}(r'_i : s'_i) \in rad(\gcd(a'_j : c'_j)_{j\in J'})$, where the (finite) index sets I', J' and the elements a'_j, c'_j, r'_i, s'_i can be effectively determined from the formulas ϕ_j, ψ_j, $0 \leq j \leq n$. Then we observe that $\prod_{i\in I'}(r'_i : s'_i) \in rad(\gcd(a'_j : c'_j)_{j\in J'})$ iff $Z(\prod_{i\in I'}(r'_i : s'_i)) \subset Z(\gcd(a'_j : c'_j)_{j\in J'})$.

References

[1] De Aldama Sanchez R., *Chaînes et dépendance*, Thèse Lyon 1, septembre 2009.

[2] L. Bélair and F. Point, *Quantifier elimination in valued Ore modules*, J. Symb. Logic **75** (2010) 1007-1034. MR2723780. Corrigendum to: *Quantifier to elimation in valued Ore modules*, J. Symb. Logic **77** (2012), no. 2, 727-728. MR2963033

[3] L. Bélair and F. Point, *Separably closed fields and contractive Ore modules*, J. Symb. Log. **80** (2015), no. 4, 1315–1338, DOI 10.1017/jsl.2015.42. MR3436370

[4] L. Bélair, A. Macintyre, and T. Scanlon, *Model theory of the Frobenius on the Witt vectors*, Amer. J. Math. **129** (2007), no. 3, 665–721, DOI 10.1353/ajm.2007.0018. MR2325101

[5] Z. Chatzidakis, *La représentation en termes de faisceaux des modèles de la théorie élémentaire de la multiplication des entiers naturels* (French), Model theory and arithmetic (Paris, 1979), Lecture Notes in Math., vol. 890, Springer, Berlin-New York, 1981, pp. 90–110. MR644997

[6] P. M. Cohn, *Skew fields: Theory of general division rings*, Encyclopedia of Mathematics and its Applications, vol. 57, Cambridge University Press, Cambridge, 1995. MR1349108

[7] P. F. Conrad, *Embedding theorems for abelian groups with valuations*, Amer. J. Math. **75** (1953), 1–29, DOI 10.2307/2372611. MR0053933

[8] L. van den Dries, *Quantifier elimination for linear formulas over ordered and valued fields*, Proceedings of the Model Theory Meeting (Univ. Brussels, Brussels/Univ. Mons, Mons, 1980), Bull. Soc. Math. Belg. Sér. B **33** (1981), no. 1, 19–31. MR620959

[9] L. van den Dries, *Elimination theory for the ring of algebraic integers*, J. Reine Angew. Math. **388** (1988), 189–205, DOI 10.1515/crll.1988.388.189. MR944190

[10] L. van den Dries and A. Macintyre, *The logic of Rumely's local-global principle*, J. Reine Angew. Math. **407** (1990), 33–56. MR1048527

[11] F. Delon and P. Simonetta, *Abelian C-minimal valued groups*, Ann. Pure Appl. Logic **168** (2017), no. 9, 1729–1782, DOI 10.1016/j.apal.2017.03.008. MR3659410

[12] Ju. L. Eršov, *Decidability of the elementary theory of relatively complemented lattices and of the theory of filters* (Russian), Algebra i Logika Sem. **3** (1964), no. 3, 17–38. MR0180490

[13] I. Fleischer, *Maximality and ultracompleteness in normed modules*, Proc. Amer. Math. Soc. **9** (1958), 151–157, DOI 10.2307/2033414. MR0093693

[14] L. Fuchs and L. Salce, *Modules over non-Noetherian domains*, Mathematical Surveys and Monographs, vol. 84, American Mathematical Society, Providence, RI, 2001. MR1794715

[15] S. Garavaglia, *Direct product decomposition of theories of modules*, J. Symbolic Logic **44** (1979), no. 1, 77–88, DOI 10.2307/2273705. MR523490

[16] N. Garcia-Fritz and H. Pasten, *Uniform positive existential interpretation of the integers in rings of entire functions of positive characteristic*, J. Number Theory **156** (2015), 368–393, DOI 10.1016/j.jnt.2015.04.018. MR3360345

[17] L. Gregory, *Decidability for theories of modules over valuation domains*, J. Symb. Log. **80** (2015), no. 2, 684–711, DOI 10.1017/jsl.2014.1. MR3377362

[18] Guignot F., *Théorie des modèles des groupes abéliens valués*, Thèse Paris 7, novembre 2016.

[19] M. Henriksen, *On the ideal structure of the ring of entire functions*, Pacific J. Math. **2** (1952), 179–184. MR0047928

[20] M. Henriksen, *On the prime ideals of the ring of entire functions*, Pacific J. Math. **3** (1953), 711–720. MR0059479

[21] W. Hodges, *Model theory*, Encyclopedia of Mathematics and its Applications, vol. 42, Cambridge University Press, Cambridge, 1993. MR1221741

[22] T. G. Kucera and M. Prest, *Imaginary modules*, J. Symbolic Logic **57** (1992), no. 2, 698–723, DOI 10.2307/2275302. MR1169204

[23] S. Kochen, *The model theory of local fields*, ISILC Logic Conference (Proc. Internat. Summer Inst. and Logic Colloq., Kiel, 1974), Lecture Notes in Math., vol. 499, Springer, Berlin, 1975, pp. 384–425. MR0568318

[24] M. Lazard, *Groupes analytiques p-adiques* (French), Inst. Hautes Études Sci. Publ. Math. **26** (1965), 389–603. MR0209286

[25] L'Innocente S., Point F., *Bézout domains and lattice-valued modules*, preprint, arXiv:1604.05922.

[26] S. L'Innocente, C. Toffalori, and G. Puninski, *On the decidability of the theory of modules over the ring of algebraic integers*, Ann. Pure Appl. Logic **168** (2017), no. 8, 1507–1516, DOI 10.1016/j.apal.2017.02.003. MR3650351

[27] S. L'Innocente, F. Point, G. Puninski, and C. Toffalori, *The Ziegler spectrum of the ring of entire complex valued functions*, J. Symb. Logic **84** (2019), no. 1, 160–177.

[28] L. Lipshitz and T. Pheidas, *An analogue of Hilbert's tenth problem for p-adic entire functions*, J. Symbolic Logic **60** (1995), no. 4, 1301–1309, DOI 10.2307/2275889. MR1367211

[29] L. Lipshitz and D. Saracino, *The model companion of the theory of commutative rings without nilpotent elements*, Proc. Amer. Math. Soc. **38** (1973), 381–387, DOI 10.2307/2039295. MR0439624

[30] D. Macpherson and C. Steinhorn, *On variants of o-minimality*, Ann. Pure Appl. Logic **79** (1996), no. 2, 165–209, DOI 10.1016/0168-0072(95)00037-2. MR1396850

[31] N. Mariaule, *The field of p-adic numbers with a predicate for the powers of an integer*, J. Symb. Log. **82** (2017), no. 1, 166–182, DOI 10.1017/jsl.2016.22. MR3631281

[32] J. Ohm, *Semi-valuations and groups of divisibility*, Canad. J. Math. **21** (1969), 576–591, DOI 10.4153/CJM-1969-065-9. MR0242819

[33] Onay G., *Modules valués*, Thèse Paris 7, 2011.

[34] K. Pal, *Multiplicative valued difference fields*, J. Symbolic Logic **77** (2012), no. 2, 545–579, DOI 10.2178/jsl/1333566637. MR2963021

[35] T. Pheidas and K. Zahidi, *Undecidability of existential theories of rings and fields: a survey*, Hilbert's tenth problem: relations with arithmetic and algebraic geometry (Ghent, 1999), Contemp. Math., vol. 270, Amer. Math. Soc., Providence, RI, 2000, pp. 49–105, DOI 10.1090/conm/270/04369. MR1802009

[36] M. Prest, *Model theory and modules*, London Mathematical Society Lecture Note Series, vol. 130, Cambridge University Press, Cambridge, 1988. MR933092

[37] A. Prestel and J. Schmid, *Existentially closed domains with radical relations: an axiomatization of the ring of algebraic integers*, J. Reine Angew. Math. **407** (1990), 178–201, DOI 10.1515/crll.1990.407.178. MR1048534

[38] A. Prestel and J. Schmid, *Decidability of the rings of real algebraic and p-adic algebraic integers*, J. Reine Angew. Math. **414** (1991), 141–148, DOI 10.1515/crll.1991.414.141. MR1092628

[39] G. Puninski, V. Puninskaya, and C. Toffalori, *Decidability of the theory of modules over commutative valuation domains*, Ann. Pure Appl. Logic **145** (2007), no. 3, 258–275, DOI 10.1016/j.apal.2006.09.002. MR2286415

[40] G. Puninski and C. Toffalori, *Some model theory of modules over Bézout domains. The width*, J. Pure Appl. Algebra **219** (2015), no. 4, 807–829, DOI 10.1016/j.jpaa.2014.04.031. MR3282111

[41] T. Rohwer, *Valued difference fields as modules over twisted polynomial rings*, ProQuest LLC, Ann Arbor, MI, 2003. Thesis (Ph.D.)–University of Illinois at Urbana-Champaign. MR2705045

[42] W. Rudin, *Analyse réelle et complexe* (French), Masson et Cie, Éditeurs, Paris, 1975. Traduit de l'anglais par N. Dhombres et F. Hoffman. MR0389461

[43] W. Rump, *Abelian lattice-ordered groups and a characterization of the maximal spectrum of a Prüfer domain*, J. Pure Appl. Algebra **218** (2014), no. 12, 2204–2217, DOI 10.1016/j.jpaa.2014.03.011. MR3227300

[44] W. Rump and Y. C. Yang, *Jaffard-Ohm correspondence and Hochster duality*, Bull. Lond. Math. Soc. **40** (2008), no. 2, 263–273, DOI 10.1112/blms/bdn006. MR2414785

[45] P. H. Schmitt, *Undecidable theories of valuated abelian groups*, Mém. Soc. Math. France (N.S.) **16** (1984), 67–76. Logic (Paris, 1983). MR792494

[46] P. H. Schmitt, *Decidable theories of valuated abelian groups*, Logic colloquium '84 (Manchester, 1984), Stud. Logic Found. Math., vol. 120, North-Holland, Amsterdam, 1986, pp. 245–276, DOI 10.1016/S0049-237X(08)70466-X. MR861428

[47] D. Zelinsky, *Topological characterization of fields with valuations*, Bull. Amer. Math. Soc. **54** (1948), 1145–1150, DOI 10.1090/S0002-9904-1948-09141-8. MR0028303

[48] M. Ziegler, *Model theory of modules*, Ann. Pure Appl. Logic **26** (1984), no. 2, 149–213, DOI 10.1016/0168-0072(84)90014-9. MR739577

Département de Mathématique, Université de Mons, Le Pentagone, 20, Place du Parc, B-7000 Mons, Belgium

Email address: `point@math.univ-paris-diderot.fr`

Contemporary Mathematics
Volume **730**, 2019
https://doi.org/10.1090/conm/730/14714

Multisorted modules and their model theory

Mike Prest

ABSTRACT. Multisorted modules, equivalently representations of quivers, equivalently additive functors on preadditive categories, encompass a wide variety of additive structures. In addition, every module has a natural and useful multisorted extension by imaginaries. The model theory of multisorted modules works just as for the usual, 1-sorted modules. A number of examples are presented, some in considerable detail.

1. Introduction

Multisorted modules encompass a wide variety of additive structures; we illustrate this and point out that every module, even a 1-sorted one, has a natural and useful multisorted extension. At the same time we show how to set up and use the model theory of multisorted modules.

Throughout, the emphasis is on examples and the running theme is *everything works*, meaning that if something is meaningful/true for ordinary modules then ditto for multisorted modules.

Throughout K will denote a commutative (1-sorted) ring with 1. It could be a field but it doesn't have to be.

2. Multisorted modules, quiver representations and additive functors

In this section we will see that the types of object listed in the title are really the same: they are just different ways of presenting the same information.

A normal, 1-sorted, module is a set with some specified algebraic structure on that set. A multisorted module is a disjoint collection of sets with algebraic structure on each of those sets and also between those sets. To be more precise, a module in the usual sense can be specified as an abelian group M together with a set of linear actions on that group; a multisorted module is an indexed collection $\{M_i\}_i$ of abelian groups together with a set of linear actions, each being a linear map from some M_i to some M_j. The model-theoretic way to make this precise starts by specifying a multisorted language but we'll introduce that later and, rather, start with the structures themselves, specifying them as representations of quivers.

A **quiver** Q is a directed graph: a set Q_0 of vertices and a set Q_1 of arrows between them, multiple directed edges and loops being allowed. We will use the

2010 *Mathematics Subject Classification.* Primary 03C60, 16G20, 16D90, 18E05.

quiver A_3

$$1 \xrightarrow{\alpha} 2 \xrightarrow{\beta} 3$$

as a running example.

We allow quivers to have infinitely many vertices, for instance A_∞^∞:

$$\cdots \to \bullet \to \bullet \to \bullet \to \dots$$

We will see an example of a much larger quiver in the section on Nori motives.

A K-**representation** V of a quiver Q is given by associating, to each vertex i, a K-module V_i, and, to each arrow $i \xrightarrow{\alpha} j$, a K-linear map $T_\alpha : V_i \to V_j$.

Thus, a K-representation of the one-vertex quiver $\bullet$ with no arrow is just a K-module. A K-representation of the quiver $\bullet \to \bullet$ is an ordered pair of K-modules with a homomorphism from the first to the second. A K-representation of the quiver $1 \underset{\beta}{\overset{\alpha}{\rightrightarrows}} 2$ is given by two K-modules and a pair of K-linear maps from the first to the second. A representation of the quiver with just one vertex and a loop is a K-module together with a specified endomorphism, equivalently a module over the polynomial ring $K[T]$.

A **morphism** f from the K-representation V of Q to the K-representation V' of the same quiver is (given by) a set $f_i : V_i \to V_i'$ $(i \in Q_0)$ of K-linear maps such that, for each arrow $\alpha \in Q_1$, the natural diagram (below) commutes.

$$\begin{array}{ccc} V_i & \xrightarrow{T_\alpha} & V_j \\ {\scriptstyle f_i}\downarrow & & \downarrow{\scriptstyle f_j} \\ V_i' & \xrightarrow[T_\alpha']{} & V_j' \end{array}$$

Thus we obtain the category $\mathrm{Rep}_K Q$ of K-representations of Q. Like the category of modules over any ring, this is an abelian Grothendieck category, with the additive group structure on the set (V, V') of morphisms from V to V' being given by the natural abelian group structures on the (V_i, V_i') for $i \in Q_0$

EXAMPLE 2.1. Take Q to be A_3 and suppose that K is a field. Then a K-representation of A_3 is given by three K-vectorspaces V_1, V_2, V_3 and two K-linear maps $T_\alpha : V_1 \to V_2$ and $T_\beta : V_2 \to V_3$. If $f = (f_1, f_2, f_3)$ and $g = (g_1, g_2, g_3)$ are morphisms from one representation to another, then the sum of these morphisms is given by $f + g = (f_1 + g_1, f_2 + g_2, f_3 + g_3)$.

The abelian category $\mathrm{Rep}_K A_3$ is particularly simple because the quiver A_3 is of **finite representation type** meaning that there are, up to isomorphism, just finitely many indecomposable[1] representations and every representation is a direct sum of copies of these indecomposables. It is easy, using elementary linear algebra, to identify the indecomposable representations, which are as follows; we use the notation 1_K to denote any isomorphism of a 1-dimensional vector space to another which thereby identifies them as copies of the same object.

[1]meaning direct-sum-indecomposable: a representation is such if it is not the direct sum of two non-zero representations

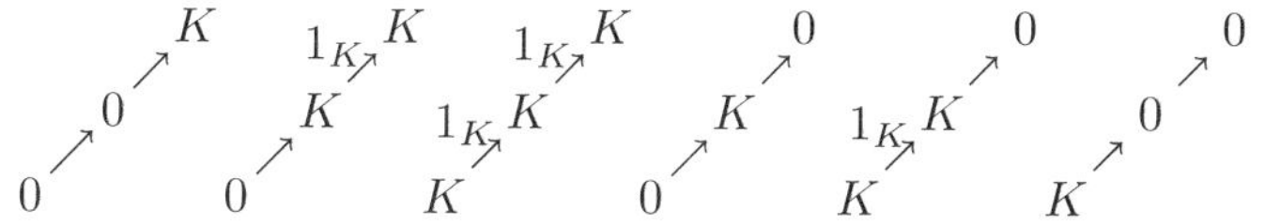

There is a nice way of displaying these indecomposables and the morphisms between them. It is the **Auslander-Reiten quiver**[2] for the category of representations. This quiver has a vertex for each finite-dimensional indecomposable representation and an arrow for each member of an independent[3] set of of irreducible morphisms between indecomposables (a morphism between indecomposables is **irreducible** it it has no non-trivial factorisation). In the case of a quiver, like A_3, of finite representation type, every morphism between indecomposables is a linear combination of compositions of irreducible morphisms so, in such a case, this really does give a complete picture of the category of representations. In the diagram we have expanded each vertex to show which representation is there, at the same time simplifying our notation by giving just the dimension of the vector space at each vertex of a representation.

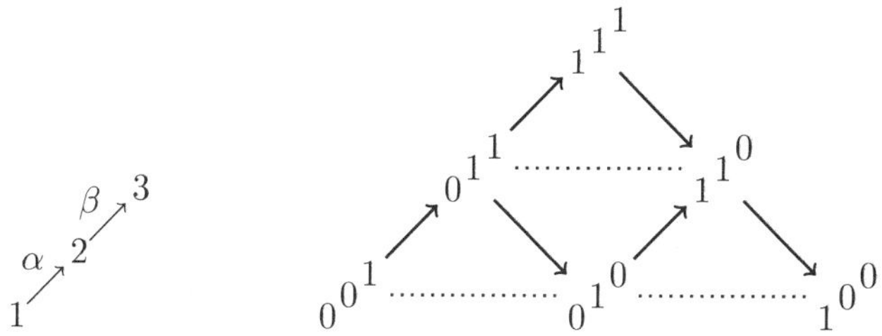

For example, the leftmost two maps are the obvious inclusions of ${}_0 0^1$ in ${}_0 1^1$ and of that in ${}_1 1^1$. The inclusion of the first in the third is not irreducible since it is the composition of the other two maps.

The dotted lines in the diagram above indicate almost split sequences (also called Auslander-Reiten sequences): these are the "minimal" exact sequences in the category. In the diagram we can see three:

$$0 \to {}_0 0^1 \to {}_0 1^1 \to {}_0 1^0 \to 0$$

$$0 \to {}_0 1^1 \to {}_1 1^1 \oplus {}_0 1^0 \to {}_1 1^0 \to 0$$

$$0 \to {}_0 1^0 \to {}_1 1^0 \to {}_1 0^0 \to 0.$$

Of course we mean essentially three - for example, the extensions of ${}_0 1^0$ by ${}_0 0^1$ form a vector space but it is 1-dimensional, so there is, up to scalar multiple, just one non-split extension.

[2]For Auslander-Reiten theory see, for instance, [**5**] or [**6**].

[3]In the sense of being linearly independent modulo the space of morphisms which have non-trivial factorisations.

R-Mod, of left R-modules, is isomorphic to the category $(R, \mathbf{Ab})$ of additive functors from R, regarded as a preadditive category, to **Ab**. Therefore, if $\mathcal{R}$ is any skeletally small preadditive category, then we refer to additive functors from $\mathcal{R}$ to **Ab** as left $\mathcal{R}$-modules, writing $\mathcal{R}$-Mod for the category they form.

For instance, if $\mathcal{R} = R$-mod is the category of finitely presented modules over some ring (or preadditive category) R then the functor category $(R\text{-mod}, \mathbf{Ab})$ may be regarded as the category of R-mod-modules. Thus the model theory of multisorted modules includes the model theory of additive functors on skeletally small preadditive categories.

Right R-modules are contravariant functors from R, regarded as a 1-object category, to **Ab**, equivalently as left modules over the opposite ring R^{op}, and we extend the terminology and notation to multisorted modules in the obvious ways.

EXAMPLE 2.3. What is a finitely presented multisorted module? A module M over a 1-sorted ring R is finitely presented if it has a projective presentation of the form ${}_RR^m \to {}_RR^n \to M \to 0$: an exact sequence where ${}_RR$ denotes the ring regarded as a left module over itself. This module is the image of the object $*_R$ under the (contravariant) Yoneda embedding of R, regarded as above as a 1-object category, to R-Mod, that is, it is the representable functor $(*_R, -)$. In the general case there are many objects in $\mathcal{R}$, so the finitely generated free modules are replaced by any finite direct sums of objects of the form $(p_i, -)$ with the p_i objects of $\mathcal{R}$, giving the definition of finitely presented $\mathcal{R}$-module.

Equivalently, an $\mathcal{R}$-module M is **finitely presented** if there is an exact sequence $Q \to P \to M \to 0$ where P and Q are finitely generated projective $\mathcal{R}$-modules (these being the direct summands of direct sums of representable functors $(p, -)$).

Finally in this section, we illustrate how to go directly from representations of quivers to functors on preadditive categories.

EXAMPLE 2.4. The quiver A_3 is not a category but there is an obvious category, namely the free one, that we can build from it, as from any quiver Q. The objects of that category, which we denote $\overrightarrow{Q}$, are the vertices of Q and the morphisms from i to j are the **paths** from i to j (compatible concatenations of arrows where the head of one is at the tail of the next, working from right to left), including a "lazy path" at each vertex i (the identity at i). Composition of morphisms is given by concatenation where compatible and 0 where not.

In the case of A_3 this adds the identity map at each vertex and the path $\beta\alpha$ from 1 to 3.

The free preadditive category $\mathbb{Z}\overrightarrow{Q}$ on Q is obtained by replacing each set $\overrightarrow{Q}(i,j)$ of morphisms by the free abelian group on that set, and extending the composition of morphisms by requiring that it be bilinear.

In our example, we see that $\mathbb{Z}\overrightarrow{A_3}$ is such that (i, j) is $\mathbb{Z}$ if $i \le j$ and is 0 otherwise.

Of course we can replace $\mathbb{Z}$ by any commutative ring K and then we have the free K-Mod-enriched preadditive category $K\overrightarrow{Q}$ on Q. Note that this is not identical to the K-path algebra KQ: the latter, as a preadditive category, has just one object, whereas $K\overrightarrow{A_3}$ has three. But they are almost the same as categories, in that they have the same idempotent-splitting additive completion. The free additive completion of KQ is obtained by formally adding finite direct sums of copies of the objects and the obvious morphisms; then the idempotent-splitting completion adds kernels and cokernels of idempotent maps to that. (Considering the Yoneda embedding which takes each object to the corresponding representable functor, that is left module, one sees that the additive completion is the opposite of the category of finitely generated free KQ-modules and the idempotent-splitting completion of that is the opposite of the category of finitely generated projective KQ-modules.) One can easily see that starting with $K\overrightarrow{Q}$ leads to the same result. In particular, the categories have the same representation theory: $KQ\text{-Mod} \simeq K\overrightarrow{Q}\text{-Mod}$.

3. Setting up linear algebra in multisorted modules

Linear algebra over a field (more generally, over a von Neumann regular ring) is about systems of linear equations. But over other rings, projections of solution sets of systems of linear equations are in general not solution sets of systems of linear equations, and existential quantifiers are needed to define them. Consider, for example, the relation of divisibility over $R = \mathbb{Z}$: the condition $3|x$ ("3 divides x") is obtained from the linear equation $x - 3y = 0$ by projecting out the second coordinate: $\exists y\,(x - 3y = 0)$. There is no way of avoiding that quantifier so we must accept that linear algebra over general rings necessarily involves existentially quantified systems of linear equations.

Fortunately, such expresssions - existentially quantified systems of linear equations - are familiar from model theory where they are treated as mathematical objects, referred to as positive primitive (pp for short) formulas[5]. Model theory has the tools to handle these and their solution sets.

There are many accounts of this, so here I recall only what I need. For background on model theory of modules one may consult, say, [**35**] or [**36**] (the former deals more with the model theory *per se*). For general model theory, see [**30**] for instance. Multisorted languages are usually introduced rather briefly, if at all, in texts on model theory (though accounts of categorical model theory will often say more). The summary I give in [**39**, §9] is fairly typical, though there is a more leisurely account in [**29**], which also describes the imaginaries construction that we will see in Section 7.

A **pp formula** is an existentially quantified system of linear equations. A system of linear equations for left R-modules has the form $H\overline{x} = 0$ where H is a rectangular matrix with entries from R and $\overline{x}$ is a (column) vector of variables. So a pp formula has the form $\exists \overline{y}\; G(\overline{x}\,\overline{y}) = 0$ for some matrix G over R. We can denote such a formula by ϕ or, showing its free=unquantified variables, $\phi(\overline{x})$. The **solution set**, $\phi(M)$, of ϕ in a right R-module M is $\{\overline{a} \in M^n : \exists \overline{b} \in M^m \text{ such that } G(\overline{a}\,\overline{b}) = 0\}$ where n, m are the lengths of the tuples $\overline{x}$, $\overline{y}$ respectively. This is a **pp-definable subgroup** of M (more precisely, a subgroup of M^n pp-definable in M). The intersection and then sum of two such pp-definable subgroups are again pp-definable.

Such solution sets are preserved by morphisms.

PROPOSITION 3.1. *(see [**36**, 1.1.7]) Let $f : M \to N$ be a morphism of R-modules. Then $f\phi(M) \leq \phi(N)$.*

In particular every subgroup of M^n pp-definable in M is an End(M)*-submodule of M^n, where* End(M) *acts diagonally on M^n.*

[5]or "regular" formulas in categorical model theory

Therefore each pp formula ϕ defines an additive functor from Mod-R to **Ab** (equivalently to K-Mod if R is a K-algebra), which we denote by F_ϕ, given on objects by $F_\phi(M) = \phi(M)$.

PROPOSITION 3.2. *(see [**36**, 1.2.31]) Suppose that* $M = \varinjlim_\lambda M_\lambda$ *is a direct limit (= directed colimit) of modules and let* ϕ *be a pp formula. Then* $\phi(M) = \varinjlim_\lambda \phi(M_\lambda)$.

In particular, each functor of the form $M \mapsto \phi(M)$ *with* ϕ *pp is determined by its action on finitely presented modules.*

The second statement follows since every R-module is a direct limit of finitely presented R-modules. We will use this in Section 8.

4. Modules in any abelian category

The underlying structure of a module, when first met, is usually an abelian group, or perhaps a vector space over a field. But, in the definition of an $\mathcal{R}$-module as an additive functor from $\mathcal{R}$ to **Ab**, where $\mathcal{R}$ is any skeletally small abelian category, we may replace the category **Ab** by any abelian category $\mathcal{A}$ (normally we would ask for more - that $\mathcal{A}$ is Grothendieck - in order to allow infinitary constructions on modules, such as forming direct limits). So an additive functor from $\mathcal{R}$ to an abelian category $\mathcal{A}$ could be described as an $\mathcal{R}$-module with values in $\mathcal{A}$. If, for instance, $\mathcal{R}$ were a ring R in the usual sense and $\mathcal{A}$ were the category Mod-S of modules over some ring S, then functoriality would force the actions of R and S to commute on "S-module-valued-R-modules". In this more general context, the solution sets $\phi(M)$ of pp formulas on $\mathcal{A}$-valued $\mathcal{R}$-modules M will be objects of $\mathcal{A}$. For instance, a formula such as $rx = 0$ would define the kernel of the action of r (which will be an object of $\mathcal{A}$ since that category has kernels). So the terminology "pp-definable subgroup" really refers to an object of $\mathcal{A}$ not necessarily, despite the name, to an object of **Ab**. In fact, in most of our examples, it will be a vector space over some field.

5. Multisorted modules as structures

We set up the model theory of multisorted modules in the obvious way. If we're specifying the module using a quiver Q, then we introduce a sort σ_i for each vertex $i \in Q_0$ and a function symbol f_α from sort σ_i to σ_j for each arrow $\alpha : i \to j$ in Q. We want our structures to be additive, so we equip each sort with a symbol $+$ (that is, $+_i$) for the addition and a constant symbol 0 (that is 0_i) for the zero element of that sort. If we want to consider K-representations then we should add, for each i and each $\lambda \in K$, a 1-ary function symbol to express multiplication by λ on sort σ_i.

With that language we can then write down the requirements that each sort σ_i have the structure of a K-module, that each function f_α be K-linear and, if we started with a quiver with relations, that all the imposed relations are satisfied. In this way the category of K-representations of Q (with any relations) becomes the category of models of a set of equations in the language we just set up. This gives us the view of multisorted modules as structures for some language, where the **sorts** of a multisorted module, that is, a K-representation M of Q, are the K-modules M_i for $i \in Q_0$.

If the quiver has just one vertex then we get essentially[6] the usual language for modules over a K-algebra (that K-algebra being determined by the arrows of Q and any imposed relations). But this multisorted set-up allows us to encompass many more examples, some of which are described in the next, and later, sections. The main methodological point is that the model theory of multisorted modules works exactly as does that for 1-sorted modules. Of course, one has to say some things slightly differently because of there being more than one sort and one has to bear in mind that any single formula refers to only finitely many sorts of the structure (variables are sorted, as are the function and constant symbols). So a pp-definable subgroup of a multisorted module will be a subgroup of a product $M_{i_1} \times \cdots \times M_{i_n}$ of finitely many sorts, in particular, the whole module is not a definable set if there are infinitely many vertices in Q.

What changes would be needed if we were to use a small preadditive category (a multisorted ring in another view), instead of a quiver, to specify multisorted modules? That is, if we were to interpret "multisorted module" as meaning a K-linear functor from such a category to K-Mod. Noting that a category can be seen as a quiver, what is extra here is the addition in each hom-set, so we will have more axioms to write down, expressing relations between arrows of that quiver which involve addition as well as composition. Otherwise everything is as before.

The construction above gives the "covariant" language but sometimes, see 6.2 below, it is more natural to consider right modules = contravariant functors, in which case, for each arrow $\alpha : i \to j$ of the quiver (or category) we would introduce a function symbol from sort σ_j to σ_i.

What about categories, such as R-mod, that are skeletally small but not small? Then we just choose some full subcategory which contains at least one copy of each object and apply the above.

6. Examples of multisorted modules

EXAMPLE 6.1. **Chain complexes of modules:** Suppose that R is any ring. A **chain complex** of R-modules is a sequence $(M_i)_{i\in\mathbb{Z}}$ of R-modules and morphisms $(d_i : M_i \to M_{i-1})_i$ such that $d_{i-1}d_i = 0$ for every i. The usual way of writing this is

$$\cdots \to M_2 \xrightarrow{d_2} M_1 \xrightarrow{d_1} M_0 \xrightarrow{d_0} M_{-1} \xrightarrow{d_{-1}} M_{-2} \to \dots,$$

which makes it clear that this is naturally an R-representation of the quiver $A_\infty^\infty = \cdots \to \cdot \to \cdot \to \cdot \to \dots$ (the notion of K-representation of a quiver doesn't require the ring K to be commutative).

Thus complexes are multisorted modules and the language for them has sorts indexed by $\mathbb{Z}$. The language, as well as being able to express the R-module structure in each sort, has a function symbol for each i to express the differential d_i. The conditions $d_{i-1}d_i = 0$ give axioms $\forall x_i\, d_{i-1}d_ix_i = 0_{i-2}$, where subscripts on the variable and the constant symbol indicate their sorts. Each of these axioms says that a certain quotient of pp-definable groups is 0, in this case, the quotient $(x_i = x_i)/(d_{i-1}d_ix_i = 0_{i-2})$. These infinitely many pp conditions cut out the chain complexes as an axiomatisable (indeed, "definable" in the sense we discuss in Section 10) subclass of the category of R-representations of A_∞^∞.

[6]The language based on a quiver does not have function symbols for K-polynomial combinations of function symbols but these can be added, just as in forming the path algebra of a quiver.

Similarly, the exact sequences are obtained as those on which all the pp-pairs $(d_i x_i = 0)/(\exists x_{i+1}\,(x_i = d_{i+1}x_{i+1}))$ are "closed" (that is, 0). Such pp-pairs, that is the various homologies of such a complex, will appear as new sorts in an enriched language which we define in the next section. An example of a subcategory that cannot be defined by only finitely many formulas (saying that certain pp-pairs are "closed") is that consisting of the complexes concentrated at a given vertex i (that, which are 0 on every other vertex).

EXAMPLE 6.2. **Finitely accessible categories:** A category $\mathcal{C}$ is **finitely accessible** if it has direct limits, if its finitely presented objects form a set up to isomorphism and if every object of $\mathcal{C}$ is a direct limit of finitely presented objects. An object $X \in \mathcal{C}$ is **finitely presented** if the covariant representable functor $(X, -) : \mathcal{C} \to \mathbf{Set}$ commutes with direct limits. This does coincide with the more familiar definition - as being finitely generated and finitely related - when the latter makes sense. The category of modules over a ring (or over a small preadditive category) is an example as is, for instance, the category of groups. A standard reference is [**1**].

There is a natural language for a finitely accessible (additive) category $\mathcal{C}$, based on its full, skeletally small, subcategory $\mathcal{C}^{\mathrm{fp}}$ of finitely presented objects. Namely the language which was described in the previous section, based on a skeletally small category, with a sort for each finitely presented object (in some small version of $\mathcal{C}^{\mathrm{fp}}$), *but* the contravariant version. Contravariance is natural because each object C of $\mathcal{C}$ is naturally a structure for the contravariant language. Namely, the set of objects of C of sort $X \in \mathcal{C}^{\mathrm{fp}}$ is defined to be the set, (X, C), of morphisms from X to C. Then a morphism $\alpha : X \to Y$ induces, by composition, a map from (Y, C) to (X, C), hence the contravariance. Of course, here we are interested in finitely accessible categories which are additive. It also turns out that, in order to have model theory work essentially as in modules, we should require $\mathcal{C}$ to have products, but that is enough - every finitely accessible category with products is definable in the sense of Section 10.

Note that, for $\mathcal{C} = R$-Mod the language we obtain this way is richer than the usual 1-sorted language since it has at least one sort for every isomorphism type of finitely presented module A. Recall that $({}_RR, M)$ is naturally isomorphic to M for $M \in R$-Mod (associate to $f :_R R \to M$ the element $f1$), so this language contains the usual 1-sorted language. But the language has many more sorts though, as we will see next, they are all definable in terms of that "home sort".

For instance $({}_RR^n, M) \simeq M^n$, so the n-tuples from a module are the elements of one of its sorts. In fact, each of the new sorts $(A, -)$ is (non-canonically) isomorphic, given a choice of a finite sequence $\overline{a}$ of generators for A, to a definable subset of the sort $(-)^n$ where n is the length of $\overline{a}$. The identification takes a morphism $f : A \to M$ to the image $f\overline{a}$. The formula defining the set of possible images is quantifier-free, consisting of the conjunction of any finitely many generators for the R-linear relations between the entries of $\overline{a}$. So the enrichment of a module to this language is part of the enrichment obtained by adding all imaginary sorts (in the sense that we will discuss in Section 10). Indeed it consists of the enrichment by pp-pairs of the form $\theta(\overline{x})/(\overline{x} = \overline{0})$ where θ is a quantifierfree pp formula.

EXAMPLE 6.3. **Comodules over a coalgebra:** Let C be a K-coalgebra; then the structure map of a C-comodule M has the form $\rho : M \to M \otimes_K C$. At the outset, it was not at all clear how, or whether, such structures could be treated model-theoretically. However, if K is a field, then the category C-Comod of C-comodules is finitely accessible, indeed is locally finitely presented, which implies finitely accessible. More generally, this is true if K is noetherian and C is a projective K-module, in which case C-Comod is equivalent to the category of "rational" modules over the dual algebra and it is the case that every rational module over that dual algebra is a direct limit of finitely presented rational modules (see [**41**]).

We conclude then, from the previous example, that we can develop a model theory of comodules as multisorted modules - see [**19**], [**41**], which also have references for the relevant background - with sorts corresponding to the finitely presented rational modules over the dual algebra.

EXAMPLE 6.4. **Presheaves and sheaves:** If $(X, \mathcal{O}_X)$ is any ringed space then the category of presheaves over this space is locally finitely presented (see, e.g., [**4**, §I.3]), hence finitely accessible, so has a natural model theory based on the finitely presented presheaves. The finitely presented presheaves are the extensions by 0 of the restrictions of the structure sheaf $\mathcal{O}_X$ to the various open subsets. If the space is noetherian then the category of sheaves is a nice Gabriel localisation of the category of presheaves and it follows that it also is locally finitely presented (for a more general result see [**16**]). However, over general ringed spaces, the category of sheaves need not be finitely accessible (see, e.g., [**36**, §16.3.4]).

For some model theory of sheaves based on this language, see [**40**].

EXAMPLE 6.5. **Locally coherent sheaves:** If $(X, \mathcal{O}_X)$ is a nice enough scheme (for instance, if it is compact and quasi-separated [**23**, 6.9.12]), then the category of quasicoherent sheaves over it is locally finitely presented, hence these sheaves may be treated as multisorted modules, with the coherent sheaves, these being the finitely presented quasicoherent sheaves, labelling the sorts.

7. Adding new sorts

The imaginaries construction in model theory adds definable or, more generally, interpretable sets as new sorts to a language. In the additive context, all sorts should have an induced abelian-group structure, and that forces us (see [**18**, 2.1]) to restrict to adding sorts which are defined by pp formulas.

Given a language $\mathcal{L}$ for, possibly already multisorted, modules, we add, as new sorts, the pairs, denoted ϕ/ψ and referred to as **pp-pairs**, of pp formulas where ϕ, ψ are in the same free variables and, ψ implies ϕ in the sense that $\psi(M) \leq \phi(M)$ for every module M. We add symbols $+, 0$ in each new sort to express the induced abelian-group structure. We also add, for each pair, ϕ/ψ and ϕ'/ψ' of sorts, a function symbol for every **pp-definable function** from ϕ/ψ to ϕ'/ψ'. By that we mean a function which is given by a pp-definable relation from ϕ to ϕ' which well-defines a functional and total relation from $\phi(M)/\psi(M)$ to $\phi'(M)/\psi'(M)$ for every module M.[7] We denote this new language by $\mathcal{L}^{\mathrm{eq}+}$. Every module M is naturally enriched to an $\mathcal{L}^{\mathrm{eq}+}$-structure, denoted $M^{\mathrm{eq}+}$, by taking the collection of all groups $\phi(M)/\psi(M)$ and all pp-definable maps between them - this is naturally an $\mathcal{L}^{\mathrm{eq}+}$-structure.

[7] It is not necessary to check in every module M: the condition that one pp formula imply another is equivalent to a simple algebraic condition on the matrices used to express the formulas, see [**36**, 1.1.13].

Note that an additive $\mathcal{L}^{\mathrm{eq}+}$-structure is nothing more than an additive functor from the category, which we denote $\mathbb{L}^{\mathrm{eq}+}$, of pp-sorts and pp-definable functions between them. We refer to this as the **category of pp-pairs** or **pp-imaginaries category** and, in the case that we started with a language for left $\mathcal{R}$-modules, we denote it ${}_{\mathcal{R}}\mathbb{L}^{\mathrm{eq}+}_{\mathcal{R}}$. This category is, in fact, abelian ([**24**, §1]), and we will look at the significance of that later. We will also compute some examples of these categories in Sections 9, 11 and 14.

It is easy to check that the class of $\mathcal{L}^{\mathrm{eq}+}$-structures which arise as $M^{\mathrm{eq}+}$ for some module M is axiomatisable (indeed definable in the sense considered in Section 10) and that $M \mapsto M^{\mathrm{eq}+}$ is an equivalence between the original module category and the full subcategory, of the category of all $\mathbb{L}^{\mathrm{eq}+}$-modules, on these structures. So, in some sense, in replacing a module M by $M^{\mathrm{eq}+}$, nothing has changed: we still have the same objects and morphisms but each has been replaced by a highly-enriched version.

Just to be clear, there will in general be many $\mathbb{L}^{\mathrm{eq}+}$-modules which are not of the form $M^{\mathrm{eq}+}$; indeed the latter are those which, regarded as functors on $\mathbb{L}^{\mathrm{eq}+}$, are **exact** (take exact sequences to exact sequences), see 10.4 below.

We saw already in Example 6.2 how the modules over a ring become structures for a richer language which has a sort for each finitely presented module. But that enriched language is just a part of the full pp-imaginaries language. In fact, those sorts - the pairs $\theta/0$ where θ is a system of linear equations, are exactly the projective objects in the category $\mathbb{L}^{\mathrm{eq}+}$ (and, more generally, the pp formulas ϕ, regarded as pp-pairs $\phi\overline{x}/\overline{x} = \overline{0}$, are the objects of projective dimension ≤ 1, see [**36**, 10.2.14]).

This operation of forming the pp-imaginaries category is idempotent in the sense that $(M^{\mathrm{eq}+})^{\mathrm{eq}+}$ is naturally isomorphic to $M^{\mathrm{eq}+}$ (model-theoretically, this is clear; an algebraic reason is remarked in [**38**, §5].

8. Three categories

Associated to any ring or skeletally small preadditive category R are the following three skeletally small abelian categories:

- the category ${}_R\mathbb{L}^{\mathrm{eq}+}$ of pp imaginaries for left R-modules;
- the category $(R\text{-mod}, \mathbf{Ab})^{\mathrm{fp}}$ of finitely presented additive functors on finitely presented R-modules - this is equivalent to $(R\text{-mod}, K\text{-Mod})^{\mathrm{fp}}$ if R is a K-algebra with each element of K acting centrally;
- the **free abelian category** on R - Freyd [**20**], see also [**2**], showed that there is an embedding $R \to \mathrm{Ab}(R)$ of R into an abelian category which has the following universal property: for every additive functor $M : R \to \mathcal{A}$, where $\mathcal{A}$ is an abelian category, there is a unique-to-natural-equivalence extension of M to an exact functor $\widetilde{M} : \mathrm{Ab}(R) \to \mathcal{A}$ making the following diagram commute.

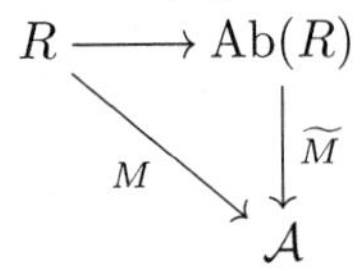

It follows that the free abelian category $\mathrm{Ab}(R)$ on R is unique up to natural equivalence. In fact, these categories all are equivalent.

THEOREM 8.1. *(see [**36**, 10.2.30, 10.2.37]) For any ring or small preadditive category R, there are natural equivalences ${}_R\mathbb{L}^{\mathrm{eq}+} \simeq (R\text{-mod}, \mathbf{Ab})^{\mathrm{fp}} \simeq \mathrm{Ab}(R)$. Furthermore, with reference to the diagram above, $\widetilde{M} = M^{\mathrm{eq}+}$.*

The second statement is explained by thinking of $M : R \to \mathcal{A}$ as an R-module taking values in $\mathcal{A}$ and the domain of $\widetilde{M}$ as being the category ${}_R\mathbb{L}^{\mathrm{eq}+}$ of pp-imaginaries for R-modules. Then $\widetilde{M}$ is equivalent to the functor which assigns to each pp-sort ϕ/ψ the object $\phi(M)/\psi(M) \in \mathcal{A}$ and similarly for pp-definable maps (for both are exact extensions of M).

If we think of the domain of $\widetilde{M}$ as the functor category $(R\ \mathrm{mod}, \mathbf{Ab})^{\mathrm{fp}}$ then $\widetilde{M}$ is $\mathrm{ev_M}$ - evaluation-at-M - which takes a functor $F \in (R\text{-mod}, \mathbf{Ab})^{\mathrm{fp}}$ to its value FM (at least if $\mathcal{A} = \mathbf{Ab}$; more generally the value of F on M where evaluation takes place in $\mathcal{A}$ rather than $\mathbf{Ab}$). At least, if M is finitely presented then that makes sense since the functors are functors on finitely presented modules. For general M, we use that each functor F on finitely presented modules has an essentially unique extension which commutes with direct limits, to a functor $\overrightarrow{F}$ on all modules. That functor is (well-)defined on a module M by representing M as a direct limit of finitely presented modules, applying the given functor F to these and taking the direct limit of the resulting diagram of abelian groups (or objects of the Ind-completion of $\mathcal{A}$ in the general case). So we should write $\mathrm{ev}_M : F \to \overrightarrow{F}(M)$: in this way we may regard the functors in $(R\text{-mod}, \mathbf{Ab})^{\mathrm{fp}}$ as acting on the whole of R-Mod.

These are two complementary views of what $M^{\mathrm{eq}+}$ is: as the functor just described, alternatively as the "image" of that functor - literally as the collection of images of the objects of the category and all the image maps between these - that is precisely $M^{\mathrm{eq}+}$ regarded as an ${}_R\mathbb{L}^{\mathrm{eq}+}$-structure.

9. An example: A_3

We illustrate the ideas and results above by computing the free abelian category on the path algebra $R = KA_3$ where K is a field. We will compute the objects of that category as functors on finitely presented R-modules and as pp-pairs.

In fact, this example is particularly simple, first in that R has finite representation type. That implies that the functor category $(R\text{-mod}, \mathbf{Ab})^{\mathrm{fp}}$ is actually equivalent to the category of modules over the **Auslander algebra** of R. That is $S = \mathrm{End}({}_RM_0)$ where M_0 is the direct sum of one copy of each indecomposable R-module. If we regard S as acting on the left of R-modules, then the functor category is equivalent to the category of left S-modules. We will prove a somewhat more general result in Section 11 which will be useful in computing localisations of this functor category.

The further simplification in this example is that the Auslander algebra will itself prove to be of finite representation type (that need not be the case in general, even if the ring R is of finite representation type). It is a fairly short exercise to compute all the indecomposable functors and the morphisms between them. Since S is an **artin algebra** (finitely generated as a module over an artinian centre), Auslander-Reiten theory can be used to compute its modules, just as for $R = KA_3$.

One may present S as a subring of the ring of 6×6 matrices over K, because there are 6 indecomposable R-modules and the morphisms between any pair of indecomposable R-modules form a ≤ 1-dimensional space over K. Alternatively,

S may be represented as the K-path algebra of the Auslander-Reiten quiver of R, including the relations which come from the three Auslander-Reiten sequences in that quiver. We will use the latter representation since then the S-modules may be given as representations of that quiver. For this particular algebra S, it is the case that each indecomposable is determined by the dimensions of the K-vector spaces at the 6 vertices (of the AR-quiver of R), so it is enough, in describing the indecomposables, just to give these dimensions. The maps within each representation are the "obvious" ones: each arrow of the Auslander-Reiten quiver of R being represented by an isomorphism if possible, the zero map otherwise and, of course, all relations in the Auslander-Reiten quiver must be satisfied. For example the picture

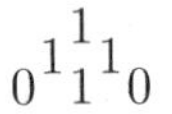

means the representation (equivalently, S-module)

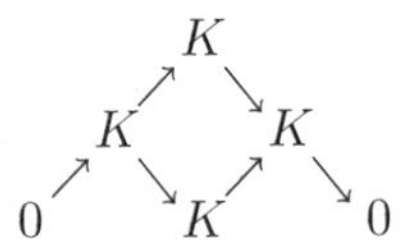

where each arrow shown is 0 if it must be and otherwise an isomorphism ("the identity") and where, necessarily, the composition of the top two arrows is the negative of the composition of the lower two (reflecting one of the three Auslander-Reiten sequences in the AR-quiver of R which was seen in Section 2).

Here, below, then, is the Auslander-Reiten quiver of S. It is reasonable to assert that this gives a complete picture of the category of all S-modules: it shows all indecomposable S-modules (and every S-module is a direct sum of these). Furthermore, every morphism between indecomposable S-modules is a K-linear combination of compositions of the irreducible maps (the maps shown on the AR-quiver). In fact, in this very simple case, every map is just a scalar multiple of compositions of irreducible maps. Then every morphism between S-modules can be given as a matrix of morphisms between its indecomposable components. So all this information is essentially in the following quiver with relations (those being given by the Auslander-Reiten sequences)

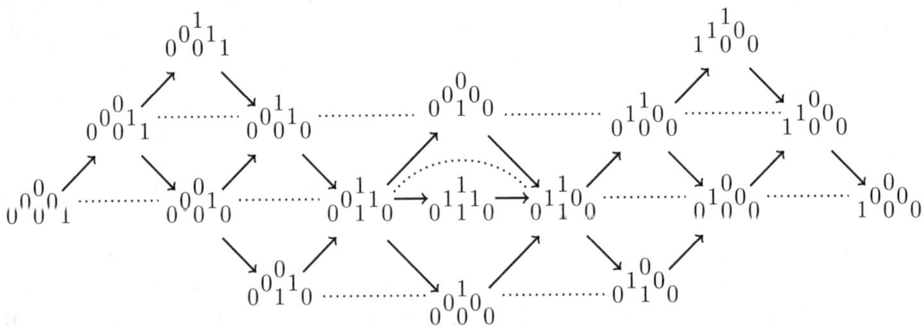

Bear in mind that, by 8.1, this is also a complete picture of the category of (finitely presented) functors on finitely presented R-modules, equivalently of the category of pp-sorts and pp-definable maps between them. So everything about the, admittedly rather simple, model theory of KA_3-modules is contained in this picture.

This picture does illustrate much that is general, so let us look in detail at some of these 17 indecomposable functors/pp-sorts. Let us label them as F_i with i as in the following diagram.

$$\begin{array}{ccccccccccccc}
 & & 3 & & & & & & & 14 & & & \\
 & 2 & & 5 & & 8 & & & 12 & & 16 & & \\
1 & & 4 & & 7 & 9 & 11 & & 15 & & 17 & & \\
 & & & 6 & & 10 & & 13 & & & & &
\end{array}$$

And let us also label the indecomposable R-modules as follows, locating each at its position in the Auslander-Reiten quiver of R.

$$\begin{array}{ccccc}
 & & P_1 & & \\
 & P_2 & & I_2 & \\
P_3 & & S_2 & & I_1
\end{array}$$

Here S_i, P_i, I_i respectively denote the simple module at vertex i, the projective cover of S_i and the injective hull of S_i. So we also have $S_3 = P_3$, $I_3 = P_1$, $S_1 = I_1$.

Each indecomposable R-module N gives us the representable functor $(N, -)$ and these are precisely the indecomposable projective functors. Consider the indecomposable functor F_9. This is both projective and injective (one can see that from the AR-quiver since it neither ends nor begins an almost split sequence) and is readily identified as $(P_2, -)$ since, as is easily seen from the AR-quiver of R, the dimensions $\dim(P_2, X)$ as X ranges over the six indecomposable R-modules, give exactly the pattern of 0s and 1s seen in F_9. As a 3-sorted module P_2 is cyclic, generated by an element of sort 2, so $(P_2, -)$ is isomorphic to the pp-pair $(x_2 = x_2)/(x_2 = 0)$. If, instead, we think of P_2 as a 1-sorted R-module, then we can use the pp-pair $(e_1 x = 0 \wedge e_3 x = 0)/(x = 0)$.

Next, let us look at the factor F_{11} of $F_9 = (P_2, -)$: clearly this is $(P_2, -)$ modulo its socle but it is perhaps easier to express this in terms of pp formulas for R-modules if we give F_{11} as the cokernel of a map between representable functors. So replace the socle of $(P_2, -)$ by its projective cover, which is clearly (from consulting the list of representable functors) $(I_2, -)$. Therefore we get the exact sequence

$$\begin{smallmatrix} & & 0 & & \\ & 0 & & 1 & \\ 0 & & 0 & & 1 \end{smallmatrix} \to \begin{smallmatrix} & & 1 & & \\ & 1 & & 1 & \\ 0 & & 1 & & 0 \end{smallmatrix} \to \begin{smallmatrix} & & 1 & & \\ & 1 & & 0 & \\ 0 & & 1 & & 0 \end{smallmatrix} \to 0$$

That is,

$$(I_2, -) \xrightarrow{(f,-)} (P_2, -) \to F_{11} \to 0$$

where $f : P_2 \to I_2$ is the obvious morphism of R-modules. Therefore, for any R-module M, $F_{11}(M) = (P_2, M)/\mathrm{im}(f, M)$ - the morphisms from P_2 to M modulo those which factor through f, that is $e_2 M/\alpha M$. Therefore, as a pp-pair, we can write F_{11} as $(x_2 = x_2)/(\exists x_1\ \alpha x_1 = x_2)$ if we are using the 3-sorted language for R-modules, or as $(x = e_2 x)/(\exists y\ \alpha y = x)$ if using the 1-sorted language.

By 8.1 every morphism in this category of pp-pairs is pp-definable. As remarked already, in this example every non-isomorphism between indecomposables is a K-linear combination of compositions of the morphisms seen as arrows in the Auslander-Reiten quiver above (this follows from finite representation type using that the irreducible morphisms generate the radical of the ring with many objects which is R-mod). So, in order to give pp-definitions of the morphisms, it is enough to do that for those arrows. That can be done explicitly and easily using the recipe from [**17**] for computing a pp formula which defines a given natural transformation between finitely presented functors (see the proof of [**36**, 10.2.30]). In fact, in this

to isomorphism, fewer objects; we will illustrate all this in the next section). This abelian category $\mathcal{A}(\mathcal{D})$ of functors is intrinsically associated to $\mathcal{D}$ - note that $\mathcal{D}$ may be found, up to equivalence, as a definable subcategory of many module categories. We say that an additive category $\mathcal{D}$ is **definable** if it is equivalent to a definable subcategory of some (possibly multisorted) module category.

THEOREM 10.4. *(see* [**38**, 12.10, 10.8]*) Suppose that $\mathcal{D}$ is a definable category. Then its category $\mathcal{A}(\mathcal{D})$ of pp-pairs is equivalent to the category $(\mathcal{D}, \mathbf{Ab})^{\Pi\to}$ of additive functors from $\mathcal{A}$ to $\mathbf{Ab}$ which commute with direct products and direct limits.*

On the other hand, evaluation-at-D for $D \in \mathcal{D}$, gives an equivalence of $\mathcal{D}$ with the category of exact functors on $\mathcal{A}(\mathcal{D})$: $\mathcal{D} \simeq \mathrm{Ex}(\mathcal{A}(\mathcal{D}), \mathbf{Ab})$.

If $\mathcal{D}$ is the definable subcategory of R-Mod generated by a module M, then we also write $\mathcal{A}(M)$ for $\mathcal{A}(\mathcal{D})$. Note that $M^{\mathrm{eq}+}$, an exact functor on $\mathrm{Ab}(R)$ factors through $\mathcal{A}(M)$, indeed if we think of $M^{\mathrm{eq}+}$ as an exact functor on $\mathcal{A}(M)$ then it is faithful and is a more intrinsic representation of M than was the functor on $\mathrm{Ab}(R)$. In the paper [**39**] I argue that the functor $M^{\mathrm{eq}+} : \mathcal{A}(M) \to \mathbf{Ab}$ is the most complete and intrinsic representation of any module; in particular this representation includes all the ways that a single module may be found (literally or up to inter-interpretability) over many different rings.

11. An example: KA_3 again

We will compute some localisations of the functor category for KA_3-modules, K a field, corresponding to some definable subcategories of KA_3-Mod. The following result tells us that each will be the category of finitely presented modules over the endomorphism ring of a suitable generator of the definable category. The result extends the appearance of the Auslander algebra (in the case where the definable category is a category of modules) and it appears (e.g.[**33**, §1.2]) in this rather more general form in Nori's work on motives. It may well appear elsewhere. I include a proof.

THEOREM 11.1. *Suppose that M is a finitely presented (right or left) R-module which is noetherian over its endomorphism ring $S = \mathrm{End}(M)$, which we let act on the left. Then $\mathcal{A}(M)$ - the functor category of the definable subcategory generated by M = the category of pp-pairs for M, is equivalent to S-mod.*

The equivalence is given, in one direction, by taking a functor F to FM.

PROOF. First note that the functor ev_M which takes each $F \in \mathcal{A}(M)$ to its value at M, is an equivalence of $\mathcal{A}(M)$ with its image (a non-full subcategory of $\mathbf{Ab}$). We will prove the equivalence of S-mod and this category, which is the category of pp-definable subgroups of M and pp-definable maps between them. We know that every subgroup of M^n pp-definable in M is an $\mathrm{End}(M) = S$-submodule of M^n and that every pp-definable map commutes with that S-module structure, so we have to show that the module ${}_SS$, with its full endomorphism ring, occurs in this way. That will be enough since the smallest abelian subcategory of S-mod containing ${}_SS$ with its full endomorphism ring is S-mod itself.

Choose a finite generating tuple $\overline{a} = (a_1, \dots, a_n)$ for M as an R-module. Then $S \to M^n$ defined by $s \in S \mapsto s(\overline{a}) = (sa_1, \dots, sa_n)$ embeds S as a left S-submodule of M^n. Since, by the finiteness conditions on M, every S-submodule of M^n is pp-definable (see [**36**, 1.2.12]), it follows that there is a pp formula $\phi(\overline{x})$ such that $\phi(M) = S\overline{a}$. We must show that the right multiplications by elements of S (the endomorphisms of ${}_SS$) and the images under ev_M of the pp-definable endomorphisms of the sort $\phi(\overline{x})/(\overline{x} = \overline{0})$ coincide.

So let $t \in S$ and consider a pp formula $\rho_t(\overline{x}, \overline{x}')$ which generates the pp-type in M of $(\overline{a}, t\overline{a})$ (this exists since M is a finitely presented R-module, see [**36**, 1.2.6]). So $M \models \rho_t(\overline{a}, t\overline{a})$. Then, for every $s \in S$, we have $M \models \rho_t(s\overline{a}, st\overline{a})$. Since every $\overline{b} \in \phi(M)$ has the form $s\overline{a}$ for some $s \in S$, the relation defined by ρ_t is therefore total on $\phi(M)$. It is also functional since if $s \in S$ is such that $s\overline{a} = \overline{0}$ then $sM = 0$ so also $st\overline{a} = (sta_1, \dots, sta_n) = \overline{0}$. Therefore right multiplication by t is a pp definable map, as required. (Note that our identification of ${}_SS$ with $\phi(M)$ pairs $s \in S$ with $s\overline{a}$, so ρ_t is acting as right multiplication by t.)

Suppose, for the converse, that ρ is a pp formula defining an endomorphism of $\phi(M)$. Say $M \models \rho(\overline{a}, \overline{b})$. Since $\overline{b} \in \phi(M) = S\overline{a}$, there is $t \in S$ such that $ta_i = b_i$ for each i (t is unique, since $\overline{a}$ generates M_R). For each $s \in S$ we have $M \models \rho(s\overline{a}, s\overline{b} = st\overline{a})$, so the action of ρ on $\phi(M)$ is right multiplication by t, as required. □

In the other direction in 11.1, the equivalence takes a finitely presented S-module A to the functor determined by taking M to A. We can be more explicit. Take a projective presentation of A: $S^l \to S^m \to A \to 0$. Consider the exact sequence of pp-sorts which is $\phi^l \xrightarrow{\eta} \phi^m \to \mathrm{coker}(\eta)$ where ϕ is as in the proof above and where the pp formula η defines the morphism between projectives which appears in the presentation of A (and where ϕ^m means the direct sum of m copies of the pp-pair $\phi(\overline{x})/(\overline{x} = \overline{0})$). Then the functor corresponding to A is given by any pp-pair which defines the functor $\mathrm{coker}(\eta)$, for example $(\bigwedge_{j=1}^m \phi(\overline{x^j}))/(\exists \overline{y^1}, \dots, \overline{y^l}\ \eta(\overline{y^1}, \dots, \overline{y^l}, \overline{x^1}, \dots, \overline{x^m}))$.

The case we are interested in - definable subcategories generated by finitely many finite-length modules over an artin algebra - will also be covered by the following variation.

COROLLARY 11.2. *Let $\mathcal{D}$ be a definable subcategory of R-*Mod *which consists of the direct sums of copies of the finitely many R-modules $N_1, \dots, N_t$.*

*Then $\mathcal{A}(\mathcal{D}) = S$-*mod *where S is the endomorphism ring* $\mathrm{End}(\bigoplus_{i=1}^t N_i)$.

If A is a finitely presented S-module, then the functor it defines on $\mathcal{D}$ is given on objects by $D \mapsto \mathrm{Hom}_S(\mathrm{Hom}_R(D, \bigoplus_{i=1}^t N_i), A)$.

PROOF. We follow the account of the Auslander algebra in [**14**, §4.9]).

Set $M = \bigoplus_{i=1}^t N_i$. The functor from $\mathcal{A}(\mathcal{D})$ to S-Mod is defined by taking $F \in \mathcal{A}(D)$ to FM, which, as a subgroup pp-definable in M, is an S-submodule (of some power of M) and taking a natural transformation, which we can regard as a pp-definable map, $\rho : F \to G$ in $\mathcal{A}(\mathcal{D})$, to $\rho(M) : F(M) \to G(M)$. As a pp-definable action, that commutes with the action of S. So this is just as in 11.1.

In the other direction, we associate a functor F_A on $\mathcal{D}$ to each finitely presented left S-module A according to the formula given. We must check that these are inverse processes. (That the functors F_A are in $\mathcal{A}(\mathcal{D})$ can be seen by using 10.4 or just by directly checking that they are pp-definable on M.)

The functor F_A applied to M gives $\mathrm{Hom}_S(\mathrm{Hom}_R(M, M), A) = \mathrm{Hom}_S(S, A) = A$, as required. In the other direction, given a functor G, we have $F_{GM} : D \to \mathrm{Hom}_S(\mathrm{Hom}_R(D, M), GM)$ so $F_{GM}(M) = GM$ and, given that every $D \in \mathcal{D}$ is a direct sum of copies of direct summands of M, two functors which agree on M are the same functor, so $F_{GM} = G$, as required (checking the actions on morphisms is left as an exercise). □

EXAMPLE 11.3. We consider the KA_3-module $T = P_1 \oplus P_2 \oplus S_2$ and set $R' = \mathrm{End}_R(T)$ to be its endomorphism ring, acting on the left. Thus we can regard T as both an R-module and an R'-module. It is the case that these actions are interdefinable. That is, the structure of T as an R'-module is already part of $({}_RT)^{\mathrm{eq}+}$ and the structure of T as an R-module is already part of $({}_{R'}T)^{\mathrm{eq}+}$. Essentially, the same module T lies over (at least) two different rings. As an enriched, ${}^{\mathrm{eq}+}$-structure it is just one structure and the choice of which ring to view it over is simply a choice of which (finite) generating set of sorts to extract as "the ring" (literally, the endomorphism ring in the functor category of the direct sum of those sorts is "the ring"). See [**39**] for a development of this theme.

Those familiar with tilting theory will recognise that T is a tilting module and that what we are illustrating here is the fact that classical tilting can be seen as making a new (but reversible) choice of generating sorts of a module.

The functor $(T, -)$ gives an equivalence between the definable subcategory of R-modules which satisfy $\mathrm{Ext}^1_R(T, -) = 0$ and the definable subcategory of R'-modules which satisfy $\mathrm{Tor}_1^{R'}({}_{R'}T, -) = 0$. (If a module A is FP_2 then both functors $\mathrm{Ext}^1(A, -)$ and $\mathrm{Tor}_1(A, -)$ are given by pp-pairs, see [**36**, §10.2.6].)

The definable subcategory of R-Mod defined by $\mathrm{Ext}^1_R(T, -) = 0$ coincides, since T is tilting, with that generated by T, hence it is the definable subcategory $\mathcal{D}$ of those R-modules which do not contain P_3 as a direct summand, that is, of modules which are direct sums of copies of the other five indecomposable R-modules. To compute the corresponding localisation $\mathcal{A}(\mathcal{D}) = \mathrm{Ab}(R)/\mathcal{S}_{\mathcal{D}}$ of the functor category, we should determine the category $\mathcal{S}_{\mathcal{D}}$ of functors which are 0 on $\mathcal{D}$. By inspection of the AR-quiver in Section 9, there is just one such functor, namely F_{17}, so $\mathcal{S}_{\mathcal{D}}$ consists of the finite direct sums of copies of F_{17}. The effect of localisation is to make this functor 0 and also, any morphism with kernel or cokernel in $\mathcal{S}_{\mathcal{D}}$ becomes an isomorphism in the quotient category. So we see that the irreducible morphisms from F_{12} to F_{14} and from F_{15} to F_{16} become isomorphisms. No other irreducible morphisms are affected by this localisation so, since every morphism is a linear combination of compositions of irreducible maps, the quotient category $\mathcal{A}(\mathcal{D})$ is as follows, where we have represented functors by showing the dimensions of their values on the five indecomposable modules in $\mathcal{D}$.

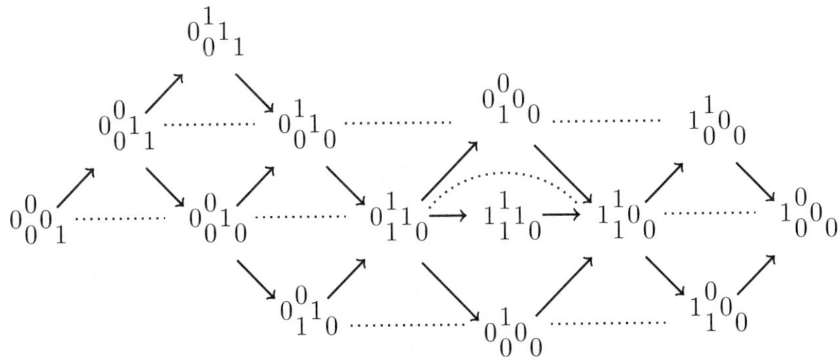

This, then, is the category of pp-pairs for the definable category $\mathcal{D}$ consisting of modules M such that $\mathrm{Ext}^1_R(T, M) = 0$. As follows from 11.2, it is the category of modules over the endomorphism ring of the direct sum of the five indecomposables in $\mathcal{D}$. Indeed that ring is just the path algebra of the quiver with relations that we obtain by deleting the vertex P_3 from the Auslander-Reiten quiver of A_3 (in Section 2).

The definable category $\mathcal{D}$ strictly contains the definable subcategory $\mathcal{E}$ generated by T. That category consists of direct sums of copies of the indecomposable direct summands P_1, P_2, S_2 of T. The category $\mathcal{A}(\mathcal{E})$ of pp-pairs for $\mathcal{E}$ can be obtained as a localisation of $\mathrm{Ab}(R)$ or, since it is a definable subcategory of $\mathcal{D}$, as a localisation of the category $\mathcal{A}(\mathcal{D})$, which we have just computed. Again, one can compute using the AR-quivers of these categories; note that, in this case, various indecomposables become isomorphic when localised and also the indecomposable $0^{\,1}_{\,1}1_0$ becomes decomposable when localised (localisation functors are exact but, outside the classical rings of fractions context, in general far from full!). One easily computes that the category $\mathcal{A}(\mathcal{E})$ has the following AR-quiver.

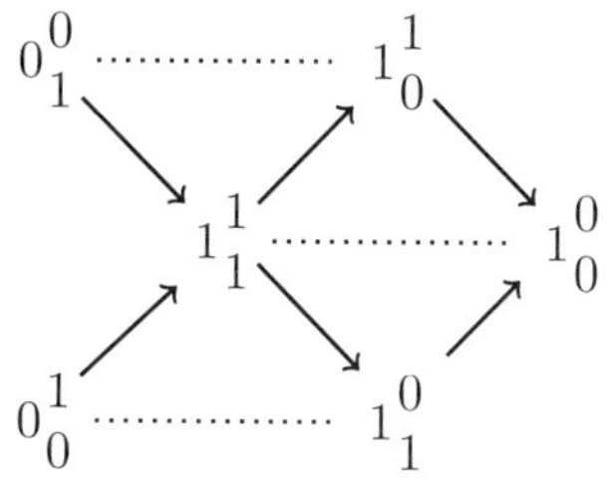

This is the path algebra of the quiver which is the subquiver of the AR-quiver of A_3 on the vertices P_2, P_3 and S_2 - that subquiver is A_3 but with a different orientation, the middle vertex pointing to each of the other two. As should be, since the effect of this kind of tilting is to change orientation of arrows in a quiver. In accordance with 11.1, the quiver, above, of the localised functor category is that of the module category over the endomorphism ring R' of T, and R' is the path algebra of A_3 with this new orientation.

12. Further examples: triangulated categories

Compactly generated triangulated categories are somewhat analogous to finitely accessible categories. Triangulated categories are seldom abelian, nor do they have direct limits, so many of the standard constructions of model theory, are not available. Nevertheless, a model theory may be set up. As with accessible categories, if finitary model theory is to be used (that is model theory using formulas which are finite, not infinite, strings of symbols) then we need some "finitary" objects to generate the category in some sense. In the triangulated context "compact" objects replace the finitely presented ones, where we say that an object C of the triangulated category $\mathcal{T}$ is **compact** if the functor $(C, -)$ commutes with infinite direct sums (we will assume that $\mathcal{T}$ has infinite coproducts - a fairly mild assumption). Then the triangulated category $\mathcal{T}$ is **compactly generated** if it has infinite direct sums, if there is, up to isomorphism, a set of compact objects and if, for every non-zero object $T \in \mathcal{T}$, there is a non-zero morphism from a compact object to T. That is, the subcategory $\mathcal{T}^c$ of compact objects generates $\mathcal{T}$ in the sense that it sees every object (but usually there will be morphisms not seen by the compact objects).

One can then set up a multisorted language based on (a small version of) the full subcategory $\mathcal{T}^c$ of compact objects and regard every object $T \in \mathcal{T}$ as a structure for this language by taking the objects of T of sort $C \in \mathcal{T}^c$ to be the morphisms in (C, T), just as for finitely accessible categories (Example 6.2). Some first observations about this language, including a quantifier-elimination result is given in [**22**, §3] (see also [**21**]).

Notice that this language essentially treats an object $T \in \mathcal{T}$ as the restricted representable functor $(-, T) : (\mathcal{T}^c)^{\mathrm{op}} \to \mathbf{Ab}$, that is, as a right $\mathcal{T}^c$-module. So it can be seen as moving $\mathcal{T}$ into the context of a (multisorted) module category. The restricted Yoneda functor, which does the moving, is neither faithful (there are morphisms not seen by the compact objects) nor full, though it is full on pure-injectives ([**28**, 1.7]). But, as is shown in the work of Krause and Beligiannis ([**28**], [**12**], [**13**]), also [**15**], in developing purity in triangulated categories, this is a very useful embedding nevertheless. In practice it gives a model theory for triangulated categories which can be used in computations in particular examples and in proving general results - see [**3**].

13. Further examples: Nori motives

The very rough idea (of Grothendieck) is that the motive of a variety is its abelian avatar: given a suitable category $\mathcal{V}$ of varieties (or schemes), there should be a functor from $\mathcal{V}$ to its category of motives. That category should be abelian and such that every homology or cohomology theory on $\mathcal{V}$ factors through the functor from $\mathcal{V}$ to its category of motives. So that functor itself should be a kind of universal (co)homology theory for $\mathcal{V}$. There are a number of accounts of the general idea, for example [**7**], [**31**]. Throughout we assume that the base field K is a subfield of the complex numbers.

In the case that $\mathcal{V}$ is the category of nonsingular projective varieties over $\mathbb{C}$, there is such a category of motives. An account of its construction can be found in, for example, [**26**]. But the question of existence for possibly singular, not-necessarily projective varieties - he conjectural category $\mathcal{MM}$ of mixed motives - is open.

In the 90s Nori described the construction of an abelian category which is a candidate for the category of mixed motives, [**33**]. His idea is to construct from a category of varieties $\mathcal{V}$ a (very large) quiver D such every (co)homology theory on $\mathcal{V}$ gives a representation of D (or D^{op}). A particular representation is used to construct this category - namely singular homology. There is more involved than this, in particular a product structure on D is needed to give a tensor product operation on the category of motives.

In fact, it turns out that Nori's category of motives is exactly the abelian category of functors/pp-pairs associated to the representation given by singular homology. In essence this first appeared in [**9**], though it is not said this way. There Caramello used the methods of categorical model theory, in particular classifying toposes for regular logic, and showed that Nori's category is the effectivisation of the regular syntactic category for the regular theory associated to Nori's diagram D. This is a much simpler construction than Nori's original one, in particular there is no need to approximate the final result through finite subdiagrams of D or to go *via* coalgebra representations.

In [**9**] additivity appears at a relatively late stage of the construction. If we build that in from the beginning then, as we show in [**11**], we are able to apply the existing model theory of additive structures and, in particular, realise Nori's category of motives as a category of pp-pairs (equivalently as a localisation of the free abelian category on the preadditive category $\mathbb{Z}\overrightarrow{D}$ generated by Nori's diagram D).

In brief, Nori's diagram is as follows. For a detailed recent account, which also describes the relations to questions about period numbers, see [**25**]

For the vertices, we take triples (X, Y, i) where $X, Y \in \mathcal{V}$, Y is a closed subvariety of X and $i \in \mathbb{Z}$,. The arrows of D are of two kinds. For each morphism $f : X \to X'$ of $\mathcal{V}$ we have a corresponding arrow $(X, Y, i) \to (X', Y', i)$ provided $fY \subseteq Y'$, for each i. Further, for each $X, Y, Z \in \mathcal{V}$ with $Y \supseteq Z$ closed subvarieties of X, we add an arrow $(Y, Z, i) \to (X, Y, i-1)$. A homology theory H on $\mathcal{V}$ then gives a representation of this quiver by sending (X, Y, i) to the relative homology $H_i(X, Y)$. Arrows of the first kind are sent to the obvious maps between relative homology objects; those of the second kind are send to the connecting maps in the long exact sequence for homology. Taking H to be singular homology, we obtain a representation of D and then the category $\mathcal{A}(\mathcal{D})$ of pp-pairs for this representation turns out to be Nori's category of motives.

As mentioned above, an important additional feature there is that there should be a tensor product structure on motives. This is needed, for example, to express the Künneth formula. In recent work [**10**] with Barbieri-Viale and Huber we extend the construction of [**11**] to include an induced tensor product. In particular we show how a tensor product on the category of R-modules induces a tensor product on the free abelian category $\text{Ab}(R)$, that is, we show how to define the induced tensor operation on pp-pairs; see the following section.

14. Extending tensor product to sorts; an example

Here I briefly report on work [**10**] with Barbieri-Viale and Huber which extends the constructions of [**9**] and [**11**] to an induced tensor product on the category of Nori motives, and I present an example, in considerable detail, illustrating a key part of that process, namely the extension of a tensor product on the category of modules to a tensor product on the category of pp-pairs. For the general construction see [**10**].

Any commutative ring R has the usual tensor product over R on R-Mod. We choose a very simple example, namely $R = K[\epsilon : \epsilon^2 = 0]$. An alternative would be to take $R = \mathbb{Z}_4$: then all the computations would be essentially the same, see [**34**, §6.6]. We compute the category of pp-pairs, that is, the free abelian category $\text{Ab}(R)$, and the induced tensor product on it. The computation of the category is, in fact, done elsewhere, [**34**, §6.8], [**38**, 4.3] but, rather than computing $\text{Ab}(R)$ as the category of modules over the Auslander algebra of R (as in those references), here we will compute it as the category $(R\text{-mod}, \mathbf{Ab})^{\text{fp}}$ of finitely presented functors on finitely presented modules. It seems very unclear how to compute it directly as a category of pairs of pp formulas, hence the usefulness of having these alternative, algebraic, descriptions of such categories.

14.1. Computing $\mathrm{Ab}(K[\epsilon])$**.** Every R-module is a direct sum of copies of the two indecomposable modules, R and the unique simple module K which is R modulo its radical and is also isomorphic to the socle of R.

There is essentially one interesting exact sequence in Mod-R, namely

$$0 \to K \xrightarrow{j} R \xrightarrow{p} K \to 0,$$

where j takes a chosen generator of K to $\epsilon \in R$ and p is projection taking $1 \in R$ to ϵ. Note that multiplication by ϵ on R is the composition jp.

Each indecomposable R-module gives an indecomposable representable functor:
(∗)
$(R,-)$ is the forgetful functor (identity functor if we think of it taking values in R-mod rather than the category of K-vector spaces);
$(K,-)$ satisfies $(K,K) = K$, $(K,R) = K$,
$(K,j) : (K,K) \to (K,R)$ is the identity 1_K,
$(K,p) : (K,R) \to (K,K)$ is 0 (any morphism $K \to R$ initially factors through j),
$(K,\epsilon) : (K,R) \to (K,R)$ is therefore 0.

Recall that the projective objects of the category of finitely presented functors are exactly the representable functors (and the injective objects are exactly those of the form $A \otimes_R -$ where A is a finitely presented *right* module), see for example [**36**, 10.1.14, 12.1.13].

The morphism p gives the exact sequence of functors

$$0 \to (K,-) \xrightarrow{(p,-)} (R,-) \to F_p \to 0;$$

so $F_pM = M/\mathrm{ann}_M(\epsilon)$ (noting that $(K,-)$ takes a module to the subspace of its elements which annihilate ϵ). The functor F_p is nonzero only on one indecomposable module, namely R, where its value is 1-dimensional, so it is a simple functor, which we will denote by S.

We also have the induced exact sequence

$$0 \to (K,-) \xrightarrow{(p,-)} (R,-) \xrightarrow{(j,-)} (K,-) \to F_j \to 0$$

(note that $(j,-)$ factors as the projection to F_p followed by the inclusion of that in $(K,-)$). The functor F_j takes a module M to $\mathrm{ann}_M(\epsilon)/\epsilon M$. Again, this is nonzero only on the one indecomposable module K and has 1-dimensional image there, so it is simple, and not isomorphic to S; we denote it T.

Thus we have the projective presentations

$$0 \to (K,-) \xrightarrow{(p,-)} (R,-) \xrightarrow{\pi_S} S \to 0$$

and

$$0 \to (K,-) \xrightarrow{(p,-)} (R,-) \xrightarrow{(j,-)} (K,-) \xrightarrow{\pi_T} T \to 0$$

of these simple functors and hence the exact sequence

$$0 \to S \to (K,-) \to T \to 0.$$

By general Auslander-Reiten theory (e.g. [**5**, IV.6.8]) we know that the simple functors are in natural bijection with the indecomposable finite-length modules; therefore we have found all the simple functors; we continue to find all the indecomposable functors.

We know that $(p,-)$ embeds $(K,-)$ as the radical of $(R,-)$, with cokernel S, that $(j,-)$ maps $(R,-)$ to $(K,-)$ with cokernel T and with kernel the copy,

$\operatorname{im}(p,-)$, of $(K,-)$ inside $(R,-)$, and that both functors $(R,-)$ and $K,-)$ are local. So we obtain S-above-T-above-S as the (unique) composition series for the forgetful functor $(R,-)$. Then we can see the following indecomposable functors (it turns out that they can be identified uniquely by giving their composition series): $(R,-)=S$-above-T-above-S (projective and injective); S-above-T (injective); $(K,-)=T$-above-S (projective); S, T. That there are no more is seen by computing the Auslander-Reiten quiver of this category (which, as we know, is the category of modules over the Auslander algebra - which is the endomorphism ring of $R\oplus K$). That AR-quiver is shown below (best drawn on a cylinder, with the dashed lines identified; dotted horizontal lines indicate Auslander-Reiten sequences). The second version shows the (unique) composition series of each indecomposable.

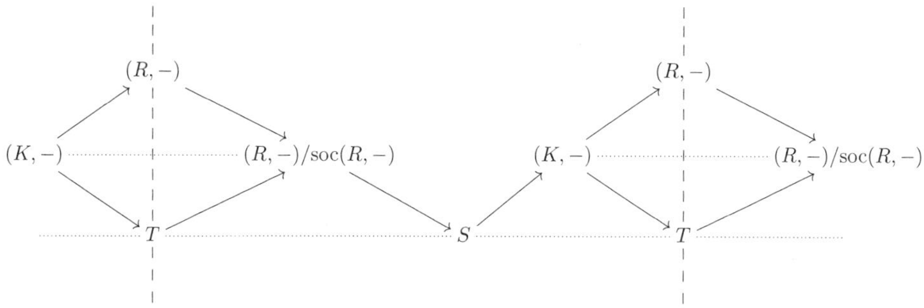

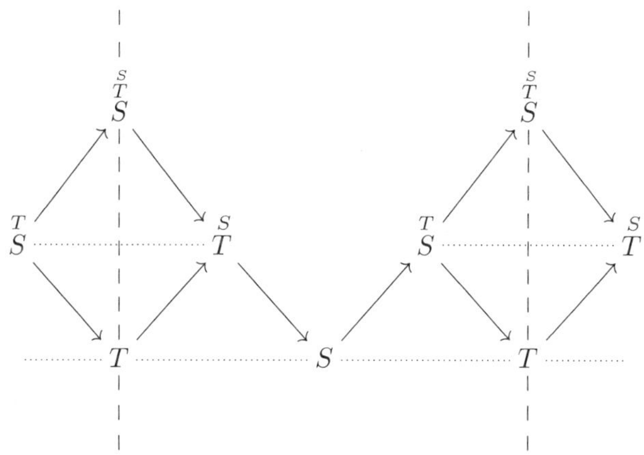

Each object of this category $(R\text{-mod},\mathbf{Ab})^{\mathrm{fp}}$ can be expressed by a pair of pp formulas and these already appeared implicitly in our computatations above:
$(R,-)$ is $(x=x)/(x=0)$, taking M to M;
$(R,-)/\operatorname{soc}(R,-)$ is $(x=x)/(\epsilon|x)$, taking M to $M/\epsilon M$;
$(K,-)$ is $(\epsilon x=0)/(x=0)$, taking M to $\operatorname{ann}_M(\epsilon)$;
S is $(\epsilon|x)/(x=0)\simeq(x=x)/(\epsilon x=0)$, taking M to $\epsilon M\simeq M/\operatorname{ann}_M(\epsilon)$;
T is $(\epsilon x=0)/(\epsilon|x)$, taking M to $\operatorname{ann}_M(\epsilon)/\epsilon M$.
As in the example in Section 11, one may compute a pp formula $\rho(x,x')$ defining any given morphism (i.e. natural transformation) between these functors. Also as in that example, all of them are rather simple, being built from $x'=x$ and $x'=0$.

14.2. A localisation. Let us compute the category $\mathcal{A}({}_RR)$ of pp-pairs for the free module ${}_RR$, equivalently for the category of projective $=$ flat R-modules, as a localisation of $\mathrm{Ab}(R)$. We need to identify the Serre subcategory $\mathcal{S}_R$ of functors F which are 0 when evaluated at R. Looking at, for example, the expression of each indecomposable functor as a pp-pair, it is immediate that only indecomposable in $\mathcal{S}_R$ is T and hence that subcategory consists just of direct sums of copies of T. What, then, is the quotient category $\mathcal{A}(R) = \mathrm{Ab}(R)/\mathcal{S}_R$? Since T becomes 0 in the quotient category we see that this category has the following shape.

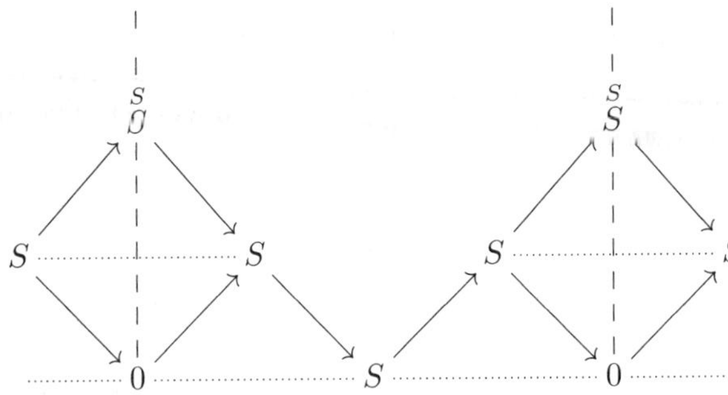

Looking at the middle two maps, an epimorphism, respectively monomorphism, in $\mathrm{Ab}(R)$, the kernel, resp. cokernel, is now 0 in the quotient category so these have become isomorphisms in $\mathcal{A}(R)$ (we are using that the functor $\mathrm{Ab}(R) \to \mathcal{A}(R)$ is exact). So we see that the quotient category is as follows.

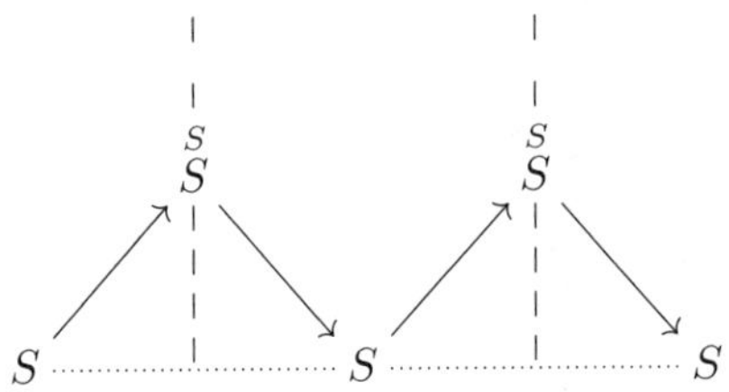

That is, $\mathcal{A}(R) \simeq R$-mod. In fact this is a general result, see [**38**, 7.1]. There it is the statement that the definable subcategory generated by the flat modules is the category of exact functors from the smallest abelian subcategory of R-Mod containing R-mod; by 10.4 this is an equivalent statement. In our case, where the ring is coherent, the statement is simpler because the flat modules already form a definable subcategory and the finitely presented modules form an abelian subcategory of the category of all modules.

14.3. Tensor product on $\mathrm{Ab}(K[\epsilon])$**.** Now we compute the tensor product on this category $\mathrm{Ab}(R)$ which is induced by the usual tensor product on modules over R.

The general definition [**10**] of the induced monoidal/tensor structure on $\mathrm{Ab}(R)$ ([**10**]) is:
given $A, B \in R$-mod, define $\otimes$ on the corresponding representable functors by $(A, -) \otimes (B, -) = (A \otimes B, -)$;
given morphisms $f : A \to A'$ and $g : B \to B'$ between finitely presented modules, define $(f, -) \otimes (g, -) = (f \otimes g, -) : (A' \otimes B', -) \to (A \otimes B, -)$.
The tensor product constructed on $\mathrm{Ab}(R)$ will be required to be right exact, so that forces the rest of the construction. Namely, a typical object of $\mathrm{Ab}(R)$ is the cokernel of a morphism between representables: it has the form F_f for some

morphism $f : A \to B$, which yields the exact sequence

$$(B,-) \xrightarrow{(f,-)} (A,-) \xrightarrow{\pi} F_f \to 0.$$

Therefore if $C \in R$-mod then the value of $(C,-) \otimes F_f$ is forced by requiring the sequence

$$(C,-) \otimes (B,-) \to (C,-) \otimes (A,-) \xrightarrow{\pi} (C,-) \otimes F_f \to 0$$

to be exact. That can then be repeated to compute the general case $F_g \otimes F_f$ (explicit computations follow).

First we need to compute $\otimes$ on R-mod. Of course, $R\otimes -$ is the identity functor on R-mod, hence $(R,-) \otimes -$ is the identity functor on $\mathrm{Ab}(R)$ - it is so on the full subcategory of projective = representable objects of $\mathrm{Ab}(R)$ therefore it is so, by right exactness of $\otimes$, on all of $\mathrm{Ab}(R)$.

As to $(K,-) \otimes -$, and noting that $K \otimes K = K$, its action is given on the subcategory of projectives by:
$(K,-) \otimes (K,-) = (K,-)$ and $(K,-) \otimes (R,-) = (K,-)$;
$(1_K,-) \otimes (p,-) = 1_{(K,-)}$ since $1_K \otimes p$ must be an epimorphism hence an isomorphism (1_K denotes the identity map on K, not a generator of the module K);
$(1_K,-) \otimes (j,-) = 0$ by direct computation (of $1_K \otimes j$ on a generator of $K \otimes K$) or just since $1_K \otimes j$ cannot be monic, since K is not flat, hence is 0.

Consider the projective presentations of S and T

$$0 \to (K,-) \xrightarrow{(p,-)} (R,-) \xrightarrow{\pi_S} S \to 0 \qquad (1)$$

$$0 \to (K,-) \xrightarrow{(p,-)} (R,-) \xrightarrow{(j,-)} (K,-) \xrightarrow{\pi_T} T \to 0 \qquad (2).$$

Applying $K \otimes -$ to (2), we obtain the right exact sequence

$$(K \otimes K,-) \xrightarrow{(1_K \otimes p,-)} (K \otimes R,-) \xrightarrow{(1_K \otimes j,-)} (K \otimes K,-) \xrightarrow{(1_K,-)\otimes \pi_T} (K,-) \otimes T \to 0$$

that is $(K,-) \xrightarrow{1} (K,-) \xrightarrow{0} (K,-) \xrightarrow{\pi_T} (K,-) \otimes T \to 0,$ from which we deduce
$(K,-) \otimes T = (K,-)$, and $(1_K,-) \otimes \pi_T = 1_{(K,-)}$ (3).

Applying $K \otimes -$ to (1), we obtain

$$(K \otimes K,-) \xrightarrow{(1_K \otimes p,-)} (K \otimes R,-) \xrightarrow{(1_K,-)\otimes \pi_S} (K,-) \otimes S \to 0,$$

that is, $(K,-) \xrightarrow{1} (K,-) \xrightarrow{(1_K,-)\otimes \pi_S} (K,-) \otimes S \to 0$, from which we deduce
$(K,-) \otimes S = 0$ and $(1_K,-) \otimes \pi_S = 0$ (4).

Applying $S \otimes -$ to (1), we obtain

$$S \otimes (K,-) \xrightarrow{1_S \otimes (p,-)} S \otimes (R,-) \xrightarrow{1_S \otimes \pi_S} S \otimes S \to 0,$$

that is, using (4), $0 \to S \xrightarrow{1_S \otimes \pi_S} S \otimes S \to 0$. So
$S \otimes S = S$, $1_S \otimes \pi_S = 1_S$ (5).

Applying $S \otimes -$ to (2), we obtain

$$S \otimes (K,-) \xrightarrow{(1_S \otimes (p,-)} S \otimes (R,-) \xrightarrow{1_S \otimes (j,-)} S \otimes (K,-) \xrightarrow{1_S \otimes \pi_T} S \otimes T \to 0,$$

that is, using (4), $0 \to S \to 0 \xrightarrow{S \otimes \pi_T} S \otimes T \to 0$, from which we deduce

$S \otimes T = 0$ and $1_S \otimes \pi_T = 0$ (6).

Applying $T \otimes -$ to (1), we obtain

$$T \otimes (K,-) \xrightarrow{1_T \otimes (p,-)} T \otimes (R,-) \xrightarrow{1_T \otimes \pi_S} T \otimes S \to 0,$$

that is, using (3) and (6), $(K,-) \xrightarrow{1_T \otimes (p,-)} T \to 0 \to 0$, from which we deduce

$1_T \otimes (p,-)$ is projection from $(K,-)$ to its top T (7).

Finally, applying $T \otimes -$ to (2), we obtain

$$T \otimes (K,-) \xrightarrow{1_T \otimes (p,-)} T \otimes (R,-) \xrightarrow{1_T \otimes (j,-)} T \otimes (K,-) \xrightarrow{1_T \otimes \pi_T} T \otimes T \to 0,$$

i.e., using (3), $(K,-) \xrightarrow{1_T \otimes (p,-)} T \xrightarrow{1_T \otimes (j,-)} (K,-) \xrightarrow{T \otimes \pi_T} T \otimes T \to 0$, from which we see $1_T \otimes (j,-) = 0$ (there being only the zero map from T to $(K,-)$) hence

$T \otimes T = (K,-)$ and $1_T \otimes \pi_T = 1_{(K,-)}$ (8).

We have, in particular, that $S \otimes S = S$, $S \otimes T = 0$, $T \otimes T = (K,-)$.

Now, S is a subobject of a representable functor, hence (Burke, see [**36**, 10.2.14]), it is a pp formula, indeed is $\epsilon|x$, that is $(\epsilon|x)/(x=0)$ as an object of this category of pp-pairs. So we have just shown that $(\epsilon|x) \otimes (\epsilon|x)$ is again the pp formula $\epsilon|x$. This is interesting, not least because it shows that the induced tensor product on pp-pairs (in particular on pp formulas) is not what one might at first guess it to be, see Section 14.4.

Lest these computations seem a little *ad hoc*, we recompute these values of $\otimes$ on $\mathrm{Ab}(R)$, directly from the definition of the extension of $\otimes$ to the functor category, using the projective presentations

$$(K,-) \xrightarrow{(p,-)} (R,-) \xrightarrow{\pi_S} S \to 0$$

and

$$(R,-) \xrightarrow{(j,-)} (K,-) \xrightarrow{\pi_T} T \to 0$$

of the simple functors S and T.

To compute $S \otimes S$:

$$\begin{array}{ccccccc}
(K \otimes K,-) & \xrightarrow{(p \otimes 1_K,-)} & (R \otimes K,-) & \xrightarrow{\pi_S \otimes (1_K,-)} & S \otimes (K,-) & \longrightarrow & 0 \\
\downarrow{\scriptstyle (1_K \otimes p,-)} & & \downarrow{\scriptstyle (1_R \otimes p,-)} & & \downarrow{\scriptstyle 1_S \otimes (p,-)} & & \\
(K \otimes R,-) & \xrightarrow{(p \otimes 1_R,-)} & (R \otimes R,-) & \xrightarrow{\pi_S \otimes (1_R,-)} & S \otimes (R,-) & \longrightarrow & 0 \\
\downarrow{\scriptstyle (1_K,-) \otimes \pi_S} & & \downarrow{\scriptstyle (1_R,-) \otimes \pi_S} & & \downarrow{\scriptstyle 1_S \otimes \pi_S} & & \\
(K,-) \otimes S & \xrightarrow{(p,-) \otimes 1_S} & (R,-) \otimes S & \xrightarrow{\pi_S \otimes 1_S} & S \otimes S & \longrightarrow & 0 \\
\downarrow & & \downarrow & & \downarrow & & \\
0 & & 0 & & 0 & &
\end{array}$$

Which simplifies just using the values of $\otimes$ on the subcategory of representables and the fact that $(R,-) \otimes -$ is the identity, together with right exactness of $\otimes$, to:

$$\begin{array}{ccccccc}
(K,-) & \xrightarrow{1} & (K,-) & \longrightarrow & 0 & \longrightarrow & 0 \\
{\scriptstyle 1}\downarrow & & \downarrow{\scriptstyle (p,-)} & & \downarrow & & \\
(K,-) & \xrightarrow{(p,-)} & (R,-) & \xrightarrow{\pi_S} & S & \longrightarrow & 0 \\
\downarrow & & \downarrow{\scriptstyle \pi_S} & & \downarrow{\scriptstyle \pi_S\otimes 1_S} & & \\
0 & \longrightarrow & S & \xrightarrow{\pi_S\otimes 1_S} & S\otimes S & \longrightarrow & 0 \\
\downarrow & & \downarrow & & \downarrow & & \\
0 & & 0 & & 0 & &
\end{array}$$

Hence $S \otimes S = S$ and $\pi_S \otimes 1_S = 1_S$.

To compute $T \otimes T$:

$$\begin{array}{ccccccc}
(R\otimes R,-) & \xrightarrow{(j\otimes 1_R,-)} & (K\otimes R,-) & \xrightarrow{\pi_T\otimes(1_R,-)} & T\otimes(R,-) & \longrightarrow & 0 \\
{\scriptstyle (1_R\otimes j,-)}\downarrow & & \downarrow{\scriptstyle (1_K\otimes j,-)} & & \downarrow{\scriptstyle 1_T\otimes(j,-)} & & \\
(R\otimes K,-) & \xrightarrow{(j\otimes 1_K,-)} & (K\otimes K,-) & \xrightarrow{\pi_T\otimes(1_K,-)} & T\otimes(K,-) & \longrightarrow & 0 \\
{\scriptstyle (1_R,-)\otimes\pi_T}\downarrow & & \downarrow{\scriptstyle (1_K,-)\otimes\pi_T} & & \downarrow{\scriptstyle 1_T\otimes\pi_T} & & \\
(R,-)\otimes T & \xrightarrow{(j,-)\otimes 1_T} & (K,-)\otimes T & \xrightarrow{\pi_T\otimes 1_T} & T\otimes T & \longrightarrow & 0 \\
\downarrow & & \downarrow & & \downarrow & & \\
0 & & 0 & & 0 & &
\end{array}$$

Which simplifies to:

$$\begin{array}{ccccccc}
(R,-) & \xrightarrow{(j,-)} & (K,-) & \xrightarrow{\pi_T} & T & \longrightarrow & 0 \\
{\scriptstyle (j,-)}\downarrow & & \downarrow{\scriptstyle 0} & & \downarrow{\scriptstyle 1_T\otimes(j,-)} & & \\
(K,-) & \xrightarrow{0} & (K,-) & \xrightarrow{1} & (K,-) & \longrightarrow & 0 \\
{\scriptstyle \pi_T}\downarrow & & \downarrow{\scriptstyle 1} & & \downarrow{\scriptstyle 1_T\otimes\pi_T} & & \\
T & \xrightarrow{(j,-)\otimes 1_T} & (K,-) & \xrightarrow{\pi_T\otimes 1_T} & T\otimes T & \longrightarrow & 0 \\
\downarrow & & \downarrow & & \downarrow & & \\
0 & & 0 & & 0 & &
\end{array}$$

From commutativity of the diagram we must have $1_T \otimes (j,-) = 0$, hence $T \otimes T \simeq (K,-)$ and $1_T \otimes \pi_T = 1_{(K,-)}$.

Finally, to compute $S \otimes T$:

$$\begin{array}{ccccccc}
(K\otimes R,-) & \xrightarrow{(p\otimes 1_R,-)} & (R\otimes R,-) & \xrightarrow{\pi_S\otimes(1_R,-)} & S\otimes(R,-) & \longrightarrow & 0\\
\downarrow{\scriptstyle (1_K\otimes j,-)} & & \downarrow{\scriptstyle (1_R\otimes j,-)} & & \downarrow{\scriptstyle 1_S\otimes(j,-)} & & \\
(K\otimes K,-) & \xrightarrow{(p\otimes 1_K,-)} & (R\otimes K,-) & \xrightarrow{\pi_S\otimes(1_K,-)} & S\otimes(K,-) & \longrightarrow & 0\\
\downarrow{\scriptstyle (1_K,-)\otimes\pi_T} & & \downarrow{\scriptstyle (1_R,-)\otimes\pi_T} & & \downarrow{\scriptstyle 1_S\otimes\pi_T} & & \\
(K,-)\otimes T & \xrightarrow{(p,-)\otimes 1_T} & (R,-)\otimes T & \xrightarrow{\pi_S\otimes 1_T} & S\otimes T & \longrightarrow & 0\\
\downarrow & & \downarrow & & \downarrow & & \\
0 & & 0 & & 0 & &
\end{array}$$

Which simplifies to:

$$\begin{array}{ccccccc}
(K,-) & \xrightarrow{(p,-)} & (R,-) & \xrightarrow{\pi_S} & S & \longrightarrow & 0\\
\downarrow{\scriptstyle 0} & & \downarrow{\scriptstyle (j,-)} & & \downarrow{\scriptstyle 1_S\otimes(j,-)} & & \\
(K,-) & \xrightarrow{1} & (K,-) & \xrightarrow{\pi_T} & S\otimes(K,-) & \longrightarrow & 0\\
\downarrow{\scriptstyle (1_K,-)\otimes\pi_T} & & \downarrow{\scriptstyle \pi_S} & & \downarrow{\scriptstyle \pi_S\otimes 1_S} & & \\
(K,-)\otimes T & \longrightarrow & T & \xrightarrow{\pi_S\otimes 1_S} & S\otimes T & \longrightarrow & 0\\
\downarrow & & \downarrow & & \downarrow & & \\
0 & & 0 & & 0 & &
\end{array}$$

From which we see that $S\otimes(K,-)=0$ and hence that $S\otimes T=0$.

Either method will give the remaining values of $\otimes$; for example as follows. Consider the exact sequence

$$0\to S\to \overset{T}{S} = (K,-)\to T\to 0$$

in $\mathrm{Ab}(R)$.

Tensor with T to get

$$T\otimes S\to T\otimes(K,-)\to T\otimes T\to 0,$$

that is, $0\to T\otimes(K,-)\to(K,-)\to 0$. So $T\otimes(K,-)=(K,-)$.

Tensor with $(K,-)$ which is $\otimes$-flat (see [**10**]), to get the exact sequence

$$0\to(K,-)\otimes S\to(K,-)\otimes(K,-)\to(K,-)\otimes T\to 0,$$

that is, $0\to(K,-)\otimes S\to(K,-)\to(K,-)\to 0$. So $(K,-)\otimes S=0$.

Next, consider the exact sequence

$$0\to T\to \overset{S}{T}\to S\to 0.$$

Tensor with $(K,-)$ to get the exact sequence

$$0\to(K,-)\otimes T\to(K,-)\otimes \overset{S}{T}\to(K,-)\otimes S\to 0,$$

that is,

$$0\to \overset{T}{S}\to(K,-)\to(K,-)\otimes \overset{S}{T}\to 0\to 0.$$

So $(K,-)\otimes \overset{S}{T}=(K,-)$.

Tensor the same sequence with S to get

$$S \otimes T \to S \otimes (K,-) \to S \otimes S \to 0,$$

that is, $0 \to S \otimes (K,-) \to S \to 0$. So $S \otimes (K,-) = S$.

Finally, consider the exact sequence

$$0 \to S \to \overset{S}{\overset{T}{S}} = (R,-) \to \overset{S}{T} \to 0.$$

Tensor with T to get

$$T \otimes S = 0 \to T \to T \otimes \overset{S}{T} \to 0.$$

So $T \otimes \overset{S}{T} = T$.

Tensor the same sequence with $\overset{S}{T}$ to get

$$\overset{S}{T} \otimes S \to \overset{S}{T} \to \overset{S}{T} \otimes \overset{S}{T} \to 0,$$

that is, $S \to \overset{S}{T} \to \overset{S}{T} \otimes \overset{S}{T} \to 0$. So $\overset{S}{T} \otimes \overset{S}{T} = \overset{S}{T}$.

In summary, here is the complete table of values of $\otimes$ on objects of $\mathrm{Ab}(R)$.

	S	T	$(K,-) = \overset{T}{S}$	$\overset{S}{T}$	$(R,-) = \overset{S}{\overset{T}{S}}$
S	S	0	0	S	S
T	0	$(K,-)$	$(K,-)$	T	T
$(K,-)$	0	$(K,-)$	$(K,-)$	$(K,-)$	$(K,-)$
$\overset{S}{T}$	S	T	$(K,-)$	$\overset{S}{T}$	$\overset{S}{T}$
$(R,-)$	S	T	$(K,-)$	$\overset{S}{T}$	$(R,-)$

14.4. Tensor product of pp-pairs. Since the objects of the functor category $\mathrm{Ab}(R)$ can be seen as pp-pairs, it follows that, in the case that there is a tensor product on R-mod, there is a tensor product on pp-pairs, computed as above using projective resolutions. (For projective resolutions of the functors defined by pp formulas and pp-pairs, see [**36**, 10.2.8, pp. 437/8].) In particular one may make sense of the tensor product of two pp formulas for any commutative ring. It is not, however, what one might at first guess it to be. Suppose throughout this subsection that we have a given monoidal operation $\otimes$ on R-mod.

One can do the following. Let ϕ and ψ be pp formulas for R-modules. Choose corresponding free realisations $(C_\phi, \overline{c}_\phi)$ and $(C_{\phi'}, \overline{c}_{\phi'})$. That is, C_ϕ is a finitely presented $\mathcal{C}$-module and c_ϕ an n-tuple of elements from C_ϕ with pp-type in C_ϕ generated by ϕ, where n is the number of free variables of ϕ. Similarly for ψ, which has, say, n' free variables. Then consider the R-module $C_\phi \otimes C_{\phi'}$ and the $n \times n'$-tuple $\overline{c_\phi} \otimes \overline{c_{\phi'}}$ in $C_\phi \otimes C_{\phi'}$. Here $\overline{c_\phi} \otimes \overline{c_{\phi'}}$ means the tuple with entries $c_i \otimes c'_j$ (using hopefully clear notation). In the case $n = n' = 1$ this means $c_\phi \otimes c_{\phi'}$. Since we are assuming that the tensor product of finitely presented modules is finitely presented, the pp-type of $\overline{c_\phi} \otimes \overline{c_{\phi'}}$ in $C_\phi \otimes C_{\phi'}$ can be generated by a pp formula (as opposed to an infinite set of pp formulas). It is tempting to define $\phi \otimes \phi'$ to be any generator of that pp-type. *But*, referring to the example above, namely $S \otimes S$ where $S = (\epsilon | x)$, a free realisation of $\epsilon|x$ is (R, ϵ) and the element $\epsilon \otimes \epsilon$ in $R \otimes R = R$ is 0, so this attempt at a definition would give the formula $x = 0$, not the correct formula $\epsilon|x$. (This recipe using free realisations does work for quantifier-free pp formulas - these correspond to the projective = representable functors - because in that case, the tuples chosen will (without loss of generality) generate the modules C_ϕ and C_ψ.)

14.5. Another monoidal structure $\mathrm{Ab}(K[\epsilon])$ **when** $\mathrm{char}(K) = 2$. Just to show that inducing tensor products on pp formulas is rather intriguing, we consider another monoidal structure on modules over $K[\epsilon : \epsilon^2 = 0]$ in the case that the characteristic of K is 2. In that case, $R = K[\epsilon; \epsilon^2 = 0]$ is also the group algebra over K of the cyclic group of order 2. Namely, if g is the non-identity element of that group then we set $\epsilon = 1 + g$. The tensor product of $A, B \in R$-mod as group representations is obtained by forming $A \otimes_K B$ and then having g act by $g.a \otimes b = ga \otimes gb$. This is quite different from the tensor product as R-modules: for example, this tensor product is exact, not just right exact, and one can compute, for example, that $R \otimes_K R \simeq R^2$, an R-basis for the latter being $1 \otimes \epsilon, (1 \otimes 1 + 1 \otimes \epsilon)$. We will write $\otimes_K$ in order to distinguish this from the tensor product over R.

We can repeat the same kinds of computations of the induced monoidal structure on the functor category as for the usual tensor product over R; we will see that the induced tensor product on the functor category is different from the previous one.

First we see that $K \otimes_K K = K$ and $R \otimes_K K \simeq R \simeq K \otimes_K R$. The trivial module K is a $\otimes_K$ unit, with $1_K \otimes_K j = j$ and $1_K \otimes_K p = p$, where j, p are as in the exact sequence

$$0 \to K \xrightarrow{j} R \xrightarrow{p} K \to 0.$$

Exactness of $R \otimes_K -$ applied to that sequence gives the exact sequence

$$0 \to R \xrightarrow{1_R \otimes_K j} R^2 \xrightarrow{1_R \otimes_K p} R \to 0$$

which must be split-exact, with $1_R \otimes_K j$ and $1_R \otimes_K p$ being respectively a split embedding and split epimorphism. We continue to use the notation from Section 14.

To compute $S \otimes_K S$:

$$\begin{array}{ccccccc}
(K \otimes_K K, -) & \xrightarrow{(p \otimes_K 1_K, -)} & (R \otimes_K K, -) & \xrightarrow{\pi_S \otimes_K (1_K, -)} & S \otimes_K (K, -) & \longrightarrow & 0 \\
\downarrow {\scriptstyle (1_K \otimes_K p,\ )} & & \downarrow {\scriptstyle (1_R \otimes_K p, -)} & & \downarrow {\scriptstyle 1_S \otimes_K (p, -)} & & \\
(K \otimes_K R, -) & \xrightarrow{(p \otimes_K 1_R, -)} & (R \otimes_K R, -) & \xrightarrow{\pi_S \otimes_K (1_R, -)} & S \otimes_K (R, -) & \longrightarrow & 0 \\
\downarrow {\scriptstyle (1_K, -) \otimes_K \pi_S} & & \downarrow {\scriptstyle (1_R, -) \otimes_K \pi_S} & & \downarrow {\scriptstyle 1_S \otimes_K \pi_S} & & \\
(K, -) \otimes_K S & \xrightarrow{(p, -) \otimes_K 1_S} & (R, -) \otimes_K S & \xrightarrow{\pi_S \otimes_K 1_S} & S \otimes_K S & \longrightarrow & 0 \\
\downarrow & & \downarrow & & \downarrow & & \\
0 & & 0 & & 0 & &
\end{array}$$

That is:

$$\begin{array}{ccccccc}
(K,-) & \xrightarrow{(p,-)} & (R,-) & \xrightarrow{\pi_S} & S & \longrightarrow & 0 \\
\downarrow{\scriptstyle (p,-)} & & \downarrow{\scriptstyle (1_R\otimes_K p,-)} & & \downarrow{\scriptstyle 1_S\otimes_K (p,-)} & & \\
(R,-) & \xrightarrow{(p\otimes_K 1_R,-)} & (R^2,-) & \xrightarrow{\pi_S\otimes_K (1_R,-)} & (R,-)\otimes_K S & \longrightarrow & 0 \\
\downarrow{\scriptstyle \pi_S} & & \downarrow{\scriptstyle (1_R,-)\otimes_K \pi_S} & & \downarrow{\scriptstyle 1_S\otimes_K \pi_S} & & \\
S & \xrightarrow{(p,-)\otimes_K 1_S} & S\otimes_K (R,-) & \xrightarrow{\pi_S\otimes_K 1_S} & S\otimes_K S & \longrightarrow & 0 \\
\downarrow & & \downarrow & & \downarrow & & \\
0 & & 0 & & 0 & &
\end{array}$$

Since $1_R \otimes_K p$ is a split epi, $(1_R \otimes_K p, -)$ is a split embedding, so we deduce $(R, -) \otimes_K S = (R, -)$ and $(1_R, -) \otimes_K \pi_S$ is a split epi. The question then is whether $1_S \otimes_K (p, -) : S \to (R, -) \otimes_K S$ is or is not 0, equivalently, since π_S is epi, whether, in the top left square, the composition down then across is 0, equivalently whether the image of $(1_R \otimes_K p, -)$ is the same as the image of $(p \otimes_K 1_R, -)$. We can compute the components of these natural transformations at the R-module R and see that the image of the first map is the vector subspace of $R \otimes_K R$ generated by $1_R \otimes_K \epsilon$ and $\epsilon \otimes_K \epsilon$ (where here ϵ means the endomorphism multiplication-by-ϵ of R). By symmetry, the image of the horizontal map is the subspace generated by $\epsilon \otimes_K 1_R$ and $\epsilon \otimes_K \epsilon$. They are not the same, hence $1_S \otimes_K (p, -)$ is monic. Therefore its cokernel, $S \otimes_K S = \overset{S}{T}$. Recall from before that $S \otimes S = S$ under the monoidal structure on $\mathrm{Ab}(R)$ induced from tensoring over R, so these structures are quite different and, with respect to this tensor, we have that $(\epsilon|x) \otimes_K (\epsilon|x)$ is the formula $x = x$ (as opposed to $\epsilon|x$ for the other structure).

To compute $T \otimes T$ we start with the same 3×3 diagram of right exact sequences as in Section 14 but with $\otimes_K$ replacing $\otimes$ and recalling that $(K, -)$ is the $\otimes_K$-identity on $\mathrm{Ab}(R)$. That gives:

$$\begin{array}{ccccccc}
(R^2,-) & \xrightarrow{(j\otimes_K 1_R,-)} & (R,-) & \xrightarrow{\pi_T\otimes_K (1_R,-)} & T\otimes_K (R,-) & \longrightarrow & 0 \\
\downarrow{\scriptstyle (1_R\otimes_K j,-)} & & \downarrow{\scriptstyle (j,-)} & & \downarrow{\scriptstyle 1_T\otimes_K (j,-)} & & \\
(R,-) & \xrightarrow{(j,-)} & (K,-) & \xrightarrow{\pi_T} & T & \longrightarrow & 0 \\
\downarrow{\scriptstyle (1_R,-)\otimes_K \pi_T} & & \downarrow{\scriptstyle \pi_T} & & \downarrow{\scriptstyle 1_T\otimes_K \pi_T} & & \\
(R,-)\otimes_K T & \xrightarrow{(j,-)\otimes_K 1_T} & T & \xrightarrow{\pi_T\otimes_K 1_T} & T\otimes_K T & \longrightarrow & 0 \\
\downarrow & & \downarrow & & \downarrow & & \\
0 & & 0 & & 0 & &
\end{array}$$

Since $(j \otimes_K 1_R, -)$ is a split projection, $T \otimes_K (R, -) = 0$ and $\pi_T \otimes_K (1_R, -) = 0$

$$
\begin{array}{ccccccc}
(R^2,-) & \xrightarrow{(j\otimes_K 1_R,-)} & (R,-) & \xrightarrow{0} & 0 & \longrightarrow & 0 \\
{\scriptstyle (1_R\otimes_K j,-)}\big\downarrow & & {\scriptstyle (j,-)}\big\downarrow & & \big\downarrow & & \\
(R,-) & \xrightarrow{(j,-)} & (K,-) & \xrightarrow{\pi_T} & T & \longrightarrow & 0 \\
{\scriptstyle \pi_T}\big\downarrow & & {\scriptstyle \pi_T}\big\downarrow & & \big\downarrow{\scriptstyle 1_T\otimes_K \pi_T} & & \\
0 & \longrightarrow & T & \xrightarrow{\pi_T\otimes_K 1_T} & T\otimes_K T & \longrightarrow & 0 \\
\big\downarrow & & \big\downarrow & & \big\downarrow & & \\
0 & & 0 & & 0 & &
\end{array}
$$

From which we see that $T \otimes_K T = T$ and $1_T \otimes_K \pi_T = 1_T$.

Finally, to compute $S \otimes_K T$ we obtain the diagram:

$$
\begin{array}{ccccccc}
(R,-) & \xrightarrow{(p\otimes_K 1_R,-)} & (R^2,-) & \xrightarrow{\pi_S\otimes_K(1_R,-)} & S\otimes_K (R,-) & \longrightarrow & 0 \\
{\scriptstyle (j,-)}\big\downarrow & & {\scriptstyle (1_R\otimes_K j,-)}\big\downarrow & & {\scriptstyle 1_S\otimes_K (j,-)}\big\downarrow & & \\
(K,-) & \xrightarrow{(p,-)} & (R,-) & \xrightarrow{\pi_S} & S & \longrightarrow & 0 \\
{\scriptstyle \pi_T}\big\downarrow & & {\scriptstyle (1_R,-)\otimes_K \pi_T}\big\downarrow & & {\scriptstyle 1_S\otimes_K \pi_T}\big\downarrow & & \\
T & \xrightarrow{(p,-)\otimes_K 1_T} & (R,-)\otimes_K T & \xrightarrow{\pi_S\otimes_K 1_T} & S\otimes_K T & \longrightarrow & 0 \\
\big\downarrow & & \big\downarrow & & \big\downarrow & & \\
0 & & 0 & & 0 & &
\end{array}
$$

We see (again) that $(R,-) \otimes_K T = 0$ and hence deduce $S \otimes_K T = 0$.

In summary: $S \otimes_K S = (R,-)$; $T \otimes_K T = T$; $S \otimes_K T = 0$.

Indeed, further computation (given after the table) gives us the following.

$\otimes_K$	S	T	$(K,-) = \begin{smallmatrix}T\\S\end{smallmatrix}$	$\begin{smallmatrix}S\\T\end{smallmatrix}$	$(R,-) = \begin{smallmatrix}S\\T\\S\end{smallmatrix}$
S	$\begin{smallmatrix}S\\T\end{smallmatrix}$	0	S	$\begin{smallmatrix}S\\T\end{smallmatrix}$	$(R,-)$
T	0	T	T	0	0
$(K,-)$	S	T	$(K,-)$	$\begin{smallmatrix}S\\T\end{smallmatrix}$	$(R,-)$
$\begin{smallmatrix}S\\T\end{smallmatrix}$	$\begin{smallmatrix}S\\T\end{smallmatrix}$	0	$\begin{smallmatrix}S\\T\end{smallmatrix}$	$\begin{smallmatrix}S\\T\end{smallmatrix}$	$(R,-)$
$(R,-)$	$(R,-)$	0	$(R,-)$	$(R,-)$	$(R,-)^2$

We can get the other entries from the exact sequence

$$0 \to T \to \begin{smallmatrix}S\\T\end{smallmatrix} \to S \to 0.$$

Tensoring with S gives the exact sequence

$$S \otimes_K T \to S \otimes_K \begin{smallmatrix}S\\T\end{smallmatrix} \to S \otimes_K S \to 0,$$

that is $0 \to S \otimes_K \begin{smallmatrix}S\\T\end{smallmatrix} \to \begin{smallmatrix}S\\T\end{smallmatrix} \to 0$ from which we deduce $S \otimes_K \begin{smallmatrix}S\\T\end{smallmatrix} = \begin{smallmatrix}S\\T\end{smallmatrix}$.

Next, since $(R,-)\to \overset{S}{T}$ is an epimorphism, so is $T\otimes_K (R,-)\to T\otimes_K \overset{S}{T}$ so, since $T\otimes_K (R,-)=0$ we have $T\otimes_K \overset{S}{T}=0$.

Then, tensoring the first exact sequence with $\overset{S}{T}$ gives

$$\overset{S}{T}\otimes_K T\to \overset{S}{T}\otimes_K \overset{S}{T}\to \overset{S}{T}\otimes_K S\to 0,$$

that is, $0\to \overset{S}{T}\otimes_K \overset{S}{T}\to \overset{S}{T}\to 0$, hence $\overset{S}{T}\otimes_K \overset{S}{T}=\overset{S}{T}$.

Finally, applying $(R,-)$ to the first exact sequence gives

$$(R,-)\otimes_K T\to (R,-)\otimes_K \overset{S}{T}\to (R,-)\otimes_K S\to 0,$$

that is, $0\to (R,-)\otimes \overset{S}{T}\to (R,-)\to 0$, so $(R,-)\otimes_K \overset{S}{T}=(R,-)$.

References

[1] J. Adámek and J. Rosický, *Locally presentable and accessible categories*, London Mathematical Society Lecture Note Series, vol. 189, Cambridge University Press, Cambridge, 1994. MR1294136

[2] M. Adelman, *Abelian categories over additive ones*, J. Pure Appl. Algebra **3** (1973), 103–117, DOI 10.1016/0022-4049(73)90026-1. MR0318265

[3] K. Arnesen, R. Laking, D. Pauksztello and M. Prest, *The Ziegler spectrum for derived-discrete algebras*, Adv. in Math. **319** (2017), 653–698. DOI:10.1016/j.aim.2017.07.016

[4] M. Artin, A. Grothendieck, and J. L. Verdier, *Théorie des topos et cohomologie étale des schémas. Tome 1: Théorie des topos* (French), Lecture Notes in Mathematics, Vol. 269, Springer-Verlag, Berlin-New York, 1972. MR0354652

[5] I. Assem, D. Simson and A. Skowroński, Elements of the Representation Theory of Associative Algebras. 1: Techniques of Representation Theory, London Math. Soc. Student Texts, Vol. 65, Cambridge University Press, 2006.

[6] M. Auslander, I. Reiten, and S. O. Smalø, *Representation theory of Artin algebras*, Cambridge Studies in Advanced Mathematics, vol. 36, Cambridge University Press, Cambridge, 1995. MR1314422

[7] L. Barbieri-Viale, *A pamphlet on motivic cohomology*, Milan J. Math. **73** (2005), 53–73, DOI 10.1007/s00032-005-0038-9. MR2175036

[8] L. Barbieri-Viale, $\mathbb{T}$*-motives*, J. Pure Appl. Algebra **221** (2017), no. 7, 1565–1588, DOI 10.1016/j.jpaa.2016.12.017. MR3614967

[9] L. Barbieri-Viale, O. Caramello, and L. Lafforgue, *Syntactic categories for Nori motives*, Selecta Math. (N.S.) **24** (2018), no. 4, 3619–3648, DOI 10.1007/s00029-018-0425-z. MR3848028

[10] L. Barbieri-Viale, $\mathbb{T}$*-motives*, J. Pure Appl. Algebra **221** (2017), no. 7, 1565–1588, DOI 10.1016/j.jpaa.2016.12.017. MR3614967

[11] L. Barbieri-Viale and M. Prest, *Definable categories and* $\mathbb{T}$*-motives*, Rend. Semin. Mat. Univ. Padova **139** (2018), 205–224, DOI 10.4171/RSMUP/139-8. MR3825187

[12] A. Beligiannis, *Relative homological algebra and purity in triangulated categories*, J. Algebra **227** (2000), no. 1, 268–361, DOI 10.1006/jabr.1999.8237. MR1754234

[13] A. Beligiannis, *Purity and almost split morphisms in abstract homotopy categories: a unified approach via Brown representability*, Algebr. Represent. Theory **5** (2002), no. 5, 483–525, DOI 10.1023/A:1020535022658. MR1935857

[14] D. J. Benson, *Representations and cohomology. I: Basic representation theory of finite groups and associative algebras*, Cambridge Studies in Advanced Mathematics, vol. 30, Cambridge University Press, Cambridge, 1991. MR1110581

[15] D. J. Benson and G. Ph. Gnacadja, *Phantom maps and purity in modular representation theory. I*, Fund. Math. **161** (1999), no. 1-2, 37–91. Algebraic topology (Kazimierz Dolny, 1997). MR1713200

[16] P. Bridge, Essentially Algebraic Theories and Localizations in Toposes and Abelian Categories, Doctoral Thesis, University of Manchester, 2012, *available at* `http://www.maths.manchester.ac.uk/~mprest/publications`.

[17] K. H. Burke, *Some model-theoretic properties of functor categories for modules*, ProQuest LLC, Ann Arbor, MI, 1994. Thesis (Ph.D.)–The University of Manchester (United Kingdom). MR3828642

[18] K. Burke and M. Prest, *Rings of definable scalars and biendomorphism rings*, Model theory of groups and automorphism groups (Blaubeuren, 1995), London Math. Soc. Lecture Note Ser., vol. 244, Cambridge Univ. Press, Cambridge, 1997, pp. 188–201, DOI 10.1017/CBO9780511629174.013. MR1689875

[19] S. Crivei, M. Prest, and G. Reynders, *Model theory of comodules*, J. Symbolic Logic **69** (2004), no. 1, 137–142, DOI 10.2178/jsl/1080938832. MR2039352

[20] P. Freyd, *Representations in abelian categories*, Proc. Conf. Categorical Algebra (La Jolla, Calif., 1965), Springer, New York, 1966, pp. 95–120. MR0209333

[21] G. Garkusha and M. Prest, *Injective objects in triangulated categories*, J. Algebra Appl. **3** (2004), no. 4, 367–389, DOI 10.1142/S0219498804000927. MR2114415

[22] G. Garkusha and M. Prest, *Triangulated categories and the Ziegler spectrum*, Algebr. Represent. Theory **8** (2005), no. 4, 499–523, DOI 10.1007/s10468-005-8147-2. MR2199207

[23] A. Grothendieck and J. A. Dieudonné, *Eléments de géométrie algébrique. I* (French), Grundlehren der Mathematischen Wissenschaften [Fundamental Principles of Mathematical Sciences], vol. 166, Springer-Verlag, Berlin, 1971. MR3075000

[24] I. Herzog, *Elementary duality of modules*, Trans. Amer. Math. Soc. **340** (1993), no. 1, 37–69, DOI 10.2307/2154545. MR1091706

[25] A. Huber and S. Müller-Stach, *Periods and Nori motives*, Ergebnisse der Mathematik und ihrer Grenzgebiete, Ser. 3, vol. 65, Springer, 2017. MR3618276

[26] S. L. Kleiman, *Motives*, Algebraic geometry, Oslo 1970 (Proc. Fifth Nordic Summer-School in Math., Oslo, 1970), Wolters-Noordhoff, Groningen, 1972, pp. 53–82. MR0382267

[27] H. Krause, *Exactly definable categories*, J. Algebra **201** (1998), no. 2, 456–492, DOI 10.1006/jabr.1997.7252. MR1612398

[28] H. Krause, *Smashing subcategories and the telescope conjecture—an algebraic approach*, Invent. Math. **139** (2000), no. 1, 99–133, DOI 10.1007/s002229900022. MR1728877

[29] T. G. Kucera and M. Prest, *Imaginary modules*, J. Symbolic Logic **57** (1992), no. 2, 698–723, DOI 10.2307/2275302. MR1169204

[30] D. Marker, *Model theory: An introduction*, Graduate Texts in Mathematics, vol. 217, Springer-Verlag, New York, 2002. MR1924282

[31] J. S. Milne, Motives - Grothendieck's dream, 2012 *available at* `http://www.jmilne.org/math/xnotes/MOT.pdf`.

[32] B. Mitchell, *Rings with several objects*, Advances in Math. **8** (1972), 1–161, DOI 10.1016/0001-8708(72)90002-3. MR0294454

[33] M. Nori, Lectures on mixed motives, TIFR, *notes by N. Fakhruddin*, 2000.

[34] S. Perera, Grothendieck Rings of Theories of Modules, Doctoral Thesis, University of Manchester, 2011, *available at* `http://www.maths.manchester.ac.uk/~mprest/publications`.

[35] M. Prest, *Model theory and modules*, London Mathematical Society Lecture Note Series, vol. 130, Cambridge University Press, Cambridge, 1988. MR933092

[36] M. Prest, *Purity, spectra and localisation*, Encyclopedia of Mathematics and its Applications, vol. 121, Cambridge University Press, Cambridge, 2009. MR2530988

[37] M. Prest, *Model theory in additive categories*, Models, logics, and higher-dimensional categories, CRM Proc. Lecture Notes, vol. 53, Amer. Math. Soc., Providence, RI, 2011, pp. 231–244. MR2867973

[38] M. Prest, *Definable additive categories: purity and model theory*, Mem. Amer. Math. Soc. **210** (2011), no. 987, vi+109, DOI 10.1090/S0065-9266-2010-00593-3. MR2791358

[39] M. Prest, *Modules as exact functors*, Surveys in representation theory of algebras, Contemp. Math., vol. 716, Amer. Math. Soc., Providence, RI, 2018, pp. 37–65, DOI 10.1090/conm/716/14426. MR3852399

[40] M. Prest, V. Puninskaya, and A. Ralph, *Some model theory of sheaves of modules*, J. Symbolic Logic **69** (2004), no. 4, 1187–1199, DOI 10.2178/jsl/1102022218. MR2135662

[41] G. Reynders, *Ziegler spectra over serial rings and coalgebras*, ProQuest LLC, Ann Arbor, MI, 1998. Thesis (Ph.D.)–The University of Manchester (United Kingdom). MR3768687

[42] M. Ziegler, *Model theory of modules*, Ann. Pure Appl. Logic **26** (1984), no. 2, 149–213, DOI 10.1016/0168-0072(84)90014-9. MR739577

Current address: School of Mathematics, Alan Turing Building, University of Manchester, Manchester M13 9PL, UK

Email address: mprest@manchester.ac.uk

Contemporary Mathematics
Volume **730**, 2019
https://doi.org/10.1090/conm/730/14715

Pure projective modules over non-singular serial rings

Pavel Příhoda

Abstract. We characterize non-singular serial rings possessing a pure projective module which is not a direct sum of finitely presented modules. As a consequence we obtain that every pure projective module over a non-singular serial ring with Krull dimension is a direct sum of finitely presented modules.

1. Introduction

A module is said to be uniserial if the lattice of its submodules is a chain. A module is called serial if it is a direct sum of uniserial modules. A ring is serial if its regular (left and right) modules are serial. A theorem of Drozd and Warfield (see Theorem 3.1) shows that finitely presented modules over serial rings have a reasonable structure. Indeed, every finitely presented module over a serial ring is a direct sum of cyclically presented uniserial modules. A natural question, posed by Warfield, was whether the Krull-Remak-Schmidt theorem holds in the category of finitely presented modules over a serial ring, i.e., whether every finitely presented module over a serial ring has up to isomorphism only one decomposition as a direct sum of indecomposable modules. Facchini [**3**] showed that the answer is 'no' in general. However, Facchini's work gives tools to understand when the Krull-Remak-Schmidt holds in the category of finitely presented modules over a serial ring. Puninski [**9**, Question 2.24] asked for a characterization of such serial rings. To answer this question it is enough to translate Facchini's conditions on homomorphisms between uniserial modules to conditions on elements of the ring.

Recall that a module is pure projective if it is a direct summand of a (possibly infinite) direct sum of finitely presented modules. Another natural question is to characterize serial rings whose category of pure projective modules consists only of direct sums of finitely presented modules. In [**13**] an answer was given for chain domains, i.e., domains having the (left and right) regular modules uniserial: If R is a chain domain, then every pure projective R-module is a direct sum of finitely presented modules if and only if there is no idempotent ideal of the form RrR, where $0 \neq r \in J(R)$. The main result of this paper extends this characterization to non-singular serial rings. Note that a similar question in terms of RD-projectivity has been considered in [**5**].

The paper is organized as follows. Section 2 is a collection of basic facts about uniserial modules. These facts are used frequently throughout the paper, often

2010 *Mathematics Subject Classification.* Primary 16D70,16P70 .
The author was partially supported by grant GA ČR P201/12/G028.

without giving detailed references to them. Section 3 characterizes basic serial rings with essentially unique indecomposable direct sum decompositions of finitely presented modules. Section 4 is a brief overview of the dimension theory established in [**8**]. This theory is then used in the last section where a characterization of nonsingular serial rings possessing only 'trivial' pure projective modules is given. To prove the result, we followed the approach of [**13**] closely, but we had to overcome several technical difficulties in the proof. Therefore we included almost complete proofs of Proposition 5.2 and Proposition 5.6 except for implications $(c) \Rightarrow (a)$ which have similar proofs as the corresponding implications in [**13**, Proposition 6.1, Proposition 7.3].

Throughout the paper a ring means an associative ring with unit, an R-module means a right module over a ring R, mod-R is the category of finitely presented R-modules, and $J(R)$ is the Jacobson radical of R.

2. Basics on uniserial modules

Let R be a ring. An R-module is *uniserial* if the lattice of its submodules is a chain. Below we list several important properties of uniserial modules. Usually we do not give a detailed reference when a result from this list is used.

(i) If U is a nonzero uniserial R-module, then the ring $S := \mathrm{End}_R(U)$ contains the following ideals $I := \{f \in S \mid f \text{ is not surjective}\}$ and $K := \{f \in S \mid f \text{ is not injective}\}$. If I and K are comparable in inclusion, then S is local and $J(S) = I \cup K$. If I and K are incomparable, then I, K are the only (left, right or two-sided) maximal ideals of S. In this case $J(S) = I \cap K = \{f \in S \mid f \text{ is neither injective nor surjective}\}$. If S is local, U is said to be *of type 1*, otherwise U is said to be *of type 2* (see [**4**, Theorem 9.1]).

(ii) If R is commutative or right noetherian, then every nonzero uniserial R-module is of type 1 (see [**4**, Proposition 9.24]).

(iii) Let U, V be uniserial R-modules such that there are $f, g \in \mathrm{Hom}_R(U, V)$, f injective and g surjective. Then $U \simeq V$, in fact, at least one of $f, g, f+g$ is an isomorphism (see [**4**, Lemma 9.2]).

(iv) Let U, V, W be nonzero uniserial modules, $f \in \mathrm{Hom}_R(U, V), g \in \mathrm{Hom}_R(V, W)$. Then gf is injective (surjective) if and only if f and g are both injective (surjective) (see [**4**, Lemma 6.26]).

Let U, V be uniserial modules. We say that U and V are of the same *monogeny class* ($[U]_m = [V]_m$) if there exist monomorphisms $f\colon U \to V$ and $g\colon V \to U$. We also use $[U]_m$ to denote the class of all modules having the same monogeny class as U. Similarly, U and V are of the same *epigeny class* ($[U]_e = [V]_e$) if there are epimorphisms $f\colon U \to V$ and $g\colon V \to U$. We also use $[U]_e$ to denote the class of modules having the same epigeny class as U. Observe that the property (iii) implies that $U \simeq V$ if and only if $[U]_m = [V]_m$ and $[U]_e = [V]_e$. The following theorem (known as the weak Krull-Schmidt theorem) clarifies the importance of monogeny and epigeny classes.

THEOREM 2.1. [**4**, Theorem 9.12] *Let $U_1, \dots, U_n, V_1, \dots, V_m$ be nonzero uniserial R-modules. Then $\oplus_{i=1}^n U_i \simeq \oplus_{j=1}^m V_j$ if and only if $n = m$ and there are permutations $\sigma, \tau \in S_n$ such that $[U_i]_m = [V_{\sigma(i)}]_m$ and $[U_i]_e = [V_{\tau(i)}]_e$ for every $i \in \{1, \dots, n\}$.*

Recall that an R-module N is said to be *quasi-small* if for every family $\{M_i, i \in I\}$ of R-modules such that N is isomorphic to a direct summand of $\oplus_{i\in I} M_i$ there exists a finite set $F \subseteq I$ such that N is isomorphic to a direct summand of $\oplus_{i\in F} M_i$. The existence of uniserial modules which are *not* quasi-small (proved by Puninski [**10**]) is the main obstacle for a straightforward generalization of the weak Krull-Schmidt theorem to infinite direct sums of uniserial modules. In this paper we study how the existence of a uniserial module which is not quasi-small is related to the existence of a pure projective module which is not a direct sum of finitely presented modules.

Let U be a nonzero uniserial R-module, S, I, K as in (i). Set $U_e := \sum_{f\in S\setminus I} \operatorname{Ker} f$, $U_m := \bigcap_{f\in S\setminus K} \operatorname{Im} f$. With this notation we add the following properties to the list:

(v) U_m and U_e are fully invariant submodules of U (see [**11**, Lemma 2.2,2.3]).
(vi) U_e is not finitely generated unless $U_e = 0$ (see [**11**, Lemma 2.3]).
(vii) If $U_e \neq 0$ and $V \subseteq U$, then $[U/V]_e = [U]_e$ if and only if $V \subsetneq U_e$ (see [**11**, Lemma 2.3]).
(viii) The monogeny class of U contains a uniserial module V which is not quasi-small if and only if $U_m \subsetneq U_e$. This V is unique up to isomorphism ([**12**, Theorem 1.1] and [**11**, Lemma 2.8]) and V is isomorphic to a direct summand of $uR^{(\omega)}$ for every $u \in U \setminus U_m$.

3. Finitely presented modules over serial rings

The aim of this section is to review basic facts about the category of finitely presented modules over a serial ring. As an easy consequence of known facts we derive a criterion to determine when the Krull-Remak-Schmidt theorem holds in the category mod-R (the category of finitely presented right R-modules) if R is a basic serial ring.

Recall that an R-module is called *serial* if it is a direct sum of uniserial modules. A ring R is *serial* if R_R and $_RR$ are (right and left) serial modules. Throughout the paper we consider a serial ring R *always* with a complete set of its primitive orthogonal idempotents $\{e_1, \dots, e_n\}$. An easy application of the Krull-Remak-Schmidt theorem gives that modules $e_1R, \dots, e_nR$ and $Re_1, \dots, Re_n$ are (right and left) uniserial modules.

The following result proved by Warfield [**14**] and Drozd [**2**] describes the objects of this category.

THEOREM 3.1. [**9**, Theorem 2.3] *Every finitely presented module over a serial ring R is isomorphic to a direct sum of modules of the form $e_iR/e_irR, r \in R$.*

In particular, every finitely presented module over a serial ring is serial. Note that $e_irR = e_ire_jR$ whenever $e_ire_jR = \max_{\subseteq}\{e_ire_kR \mid k = 1, \dots, n\}$. If e_iR/e_irR and e_jR/e_jsR are indecomposable finitely presented R-modules, the elements of $\operatorname{Hom}_R(e_iR/e_irR, e_jR/e_jsR)$ can be described by elements from $X = \{x \in e_jRe_i \mid xe_irR \subseteq e_jsR\}$. We say that $x \in X$ (or, more precisely, left multiplication by x) induces $f \in \operatorname{Hom}_R(e_iR/e_irR, e_jR/e_jsR)$ if $f(u + e_irR) = xu + e_jsR$ for every $u \in e_iR$.

The following lemma describes when a homomorphism induced by an element of the ring is injective or surjective.

LEMMA 3.2. *Let e_iR/e_irR and e_jR/e_jsR be indecomposable modules from* mod-R, *where R is a serial ring.*

(a) *If $e_js \neq 0$, then $x \in e_jRe_i$ induces an epimorphism if and only if $x \notin e_jJ(R)e_i$ and $xe_irR \subseteq e_jsR$.*
(b) *If $e_js \neq 0$, then $x \in e_jRe_i$ induces a monomorphism if and only if $xe_irR = e_jsR$*

PROOF. (a) is immediate since e_jR/e_jsR is a local module. To prove (b), we can follow the idea of [**13**, Corollary 2.2]: Suppose that $x \in e_jRe_i$ induces a homomorphism $f\colon e_iR/e_irR \to e_jR/e_jsR$, i.e., $xe_irR \subseteq e_jsR$. If $f \neq 0$, then $e_jsR \subsetneq xe_iR$. So if f is a monomorphism, then the inclusion $xe_irR \subseteq e_jsR$ cannot be strict. Conversely, assume that $xe_irR = e_jsR$. Then f is monic since if $e_it \in e_iR \setminus e_irR$ satisfies $xe_it \in e_jsR = xe_irR$, then $x(e_it - e_iru) = 0$ for some $u \in R$ and, consequently, the right annihilator of x contains e_irR. But then $xe_irR = e_jsR \neq 0$, a contradiction. □

Let R be a serial ring. If $e_ir, e_is \in e_iR$, then either $\mathrm{Hom}_R(e_iR/e_irR, e_iR/e_isR)$ or $\mathrm{Hom}_R(e_iR/e_isR, e_iR/e_irR)$ contains an epimorphism. This is immediate since e_iR is uniserial. Similarly, if e_ire_k and $e_jse_k \in Re_k$ then either $Re_ire_k \subseteq Re_jse_k$ or $Re_jse_k \subseteq Re_ire_k$. As a consequence of the previous lemma we get:

COROLLARY 3.3. *Let R be a serial ring, e_ire_k, e_jse_k nonzero elements of $J(R)$. Then either* $\mathrm{Hom}_R(e_iR/e_ire_kR, e_jR/e_jse_kR)$ *or* $\mathrm{Hom}_R(e_jR/e_jse_kR, e_iR/e_ire_kR)$ *contains a monomorphism.*

LEMMA 3.4. *Let R be a serial ring and let e_iR/e_irR, e_jR/e_jsR be indecomposable R-modules. If $e_js \neq 0$, then the image of every homomorphism $f\colon e_iR/e_irR \to e_jR/e_jsR$ is finitely presented.*

PROOF. Suppose that f is induced by $x \in e_jRe_i$ and $f \neq 0$. Then $xe_irR \subseteq e_jsR \subsetneq xe_iR$. Let $r' \in R$ be such that $xe_ir' = e_js$. Then it is easy to see that $\mathrm{Ker}\, f = e_ir'R/e_irR$, thus $\mathrm{Im}\, f \simeq e_iR/e_ir'R$. □

The serial ring R is called *basic* if $e_iR \simeq e_jR$ implies $i = j$ for every $i, j \in \{1, \dots, n\}$. The following lemma simplifies the description of basic serial rings with unique indecomposable decompositions of finitely presented modules.

LEMMA 3.5. *Assume that R is a serial ring and let $r, s \in R$.*

(a) *If there exists an epimorphism in* $\mathrm{Hom}_R(e_iR/e_irR, e_jR/e_jsR)$ *and $s \in J(R)$, then $e_iR \simeq e_jR$. If R is basic, then $i = j$.*
(b) *If there exists a monomorphism in* $\mathrm{Hom}_R(e_iR/e_ire_mR, e_jR/e_jse_nR)$ *and $r \in J(R)$, $0 \neq e_jse_n \in J(R)$, then $e_mR \simeq e_nR$. If R is basic, then $m = n$.*

PROOF. (a) Let $f \in \mathrm{Hom}_R(e_iR/e_irR, e_jR/e_jsR)$ be an epimorphism. If $s \in J(R)$, the modules $e_jR/e_jsR, e_iR/e_irR$ are nonzero. If $x \in e_jRe_i$ induces f, then $x \notin e_jJ(R)e_i$ and therefore $e_iR \simeq e_jR$. This implies $i = j$ if R is basic.

(b) Assume $f \in \mathrm{Hom}_R(e_iR/e_ire_mR, e_jR/e_jse_nR)$ is a monomorphism. Then $e_iR/e_ire_mR, e_jR/e_jse_nR$ are nonzero and f is induced by an element $x \in e_jRe_i$ such that $xe_ire_mR = e_jse_nR$. So there are $u, v \in R$ such that $xe_ire_mue_n = e_jse_n$ and $xe_ire_m = e_jse_nve_m$. Then $e_jse_n = e_jse_nve_mue_n$. Since $e_jse_n \neq 0$, $e_nve_mue_n \notin J(R)$. It follows that $e_mR \simeq e_nR$ and, if R is basic, $m = n$. □

Let R be a serial ring. Consider the following graph Γ_R. Its set of vertices is $\mathcal{V}_m \dot{\cup} \mathcal{V}_e$, where $\mathcal{V}_m$ is the set of all monogeny classes of indecomposable modules from mod-R and $\mathcal{V}_e$ is the set of all epigeny classes of indecomposable modules from mod-R. The graph Γ_R is bipartite, i.e., each edge connects a vertex from $\mathcal{V}_m$ and a vertex from $\mathcal{V}_e$: The vertices $v_m \in \mathcal{V}_m$ and $v_e \in \mathcal{V}_e$ are connected in Γ_R if and only if there exists an indecomposable module $U \in$ mod-R such that $[U]_m = v_m$ and $[U]_e = v_e$. Observe that there is a natural bijective correspondence between the isoclasses of indecomposable modules in mod-R and the edges of Γ_R, and we use this bijection to identify isoclasses of finitely presented indecomposables and edges of Γ_R. Repeating the argument of [**6**, Proposition 7.1], one can prove:

LEMMA 3.6. *Let R be a serial ring. Then every connected component of Γ_R is a complete bipartite graph.*

If U and V are indecomposable modules from mod-R, define $U \sim V$ if and only if the edges of Γ_R corresponding to $[U]$ and $[V]$ are in the same connected component of Γ_R. Since each connected component of Γ_R is a complete bipartite graph, $U \sim V$ if and only if there exists an indecomposable module $W \in$ mod-R such that $[W]_m = [U]_m$ and $[W]_e = [V]_e$, if and only if there exists an indecomposable module $W \in$ mod-R such that $[W]_e = [U]_e$ and $[W]_m = [V]_m$.

By [**7**, Lemma 5.2], if $U \sim V$ for $U, V \in$ mod-R indecomposable, then U is of type 1 if and only if V is of type 1. Therefore modules corresponding to edges of a connected component $\mathcal{C}$ are of the same type. The connected component $\mathcal{C}$ is said to be of *type 2* if all the modules corresponding to its edges are of type 2.

LEMMA 3.7. *Let $U \sim V$ be indecomposable finitely presented modules over a serial ring R.*

(a) *If* $\mathrm{Hom}_R(U, V)$ *contains a monomorphism, then* $[U]_m = [V]_m$.

(b) *If* $\mathrm{Hom}_R(U, V)$ *contains an epimorphism, then* $[U]_e = [V]_e$.

PROOF. (a) Consider a uniserial module W such that $[W]_m = [V]_m$ and $[W]_e = [U]_e$. The existence of a monomorphism in $\mathrm{Hom}_R(U, V)$ implies the existence of a monomorphism in $\mathrm{Hom}_R(U, W)$ and $[W]_e = [U]_e$ guarantees the existence of an epimorphism in $\mathrm{Hom}_R(U, W)$. Therefore $U \simeq W$ and $[U]_m = [V]_m$. (b) Consider a uniserial module W such that $[W]_e = [V]_e$ and $[W]_m = [U]_m$ and conclude making use of arguments dual to those in (a). □

We say that the Krull-Remak-Schmidt theorem holds in mod-R if every finitely presented R-module M can be written as $M = \oplus_{i=1}^n U_i$, where U_i are indecomposable and if $M = \oplus_{i=1}^m V_i$ is another such a decomposition, then $m = n$ and there exists a permutation $\pi \in S_n$ such that $U_i \simeq V_{\pi(i)}$ for every $i \in \{1, \dots, n\}$.

We can apply the weak Krull-Schmidt theorem to get a condition for unique decompositions in mod-R in terms of Γ_R.

COROLLARY 3.8. *Let R be a serial ring. The following statements are equivalent*

(a) *The Krull-Remak-Schmidt theorem holds in* mod-R.

(b) *The graph Γ_R does not contain a cycle.*

(c) *There are no indecomposable modules $U, V, W \in$ mod-R such that $[U]_m = [W]_m \neq [V]_m$ and $[V]_e = [W]_e \neq [U]_e$.*

PROOF. $(a) \Rightarrow (b)$. Assume (a) and consider a cycle in Γ_R. Since every connected component of Γ_R is a complete bipartite graph, there exists a cycle of length 4. Consider the edges $[U_1], [U_2], [U_3], [U_4]$ of such a cycle. Assume that $[U_1]_e = [U_2]_e, [U_2]_m = [U_3]_m, [U_3]_e = [U_4]_e$ and $[U_4]_m = [U_1]_m$. The isoclasses of U_1, U_2, U_3, U_4 are different edges of Γ_R, so these modules are pair-wise non-isomorphic. But the weak Krull-Schmidt theorem gives $U_1 \oplus U_3 \simeq U_2 \oplus U_4$ so the Krull-Remak-Schmidt theorem does not hold in mod-R.

$(b) \Rightarrow (a)$ Consider a connected component $\mathcal{C}$ of Γ_R. If there are no cycles in Γ_R, either $\mathcal{C} \cap \mathcal{V}_m$ or $\mathcal{C} \cap \mathcal{V}_e$ is a one-element set. If $|\mathcal{C} \cap \mathcal{V}_m| = 1$, then $[U]_e = [V]_e$ implies $U \simeq V$ whenever $[U], [V] \in \mathcal{C}$. In this case we say that $\mathcal{C}$ has one monogeny class. Similarly, if $|\mathcal{C} \cap \mathcal{V}_e| = 1$, we say that the component $\mathcal{C}$ has one epigeny class and, in this case $[U]_m = [V]_m \Rightarrow U \simeq V$ if $[U], [V] \in \mathcal{C}$.

Let $U_1, \dots, U_n, V_1, \dots, V_m$ be indecomposable modules in mod-R such that $\oplus_{i=1}^n U_i \simeq \oplus_{j=1}^m V_j$. By the weak Krull-Schmidt theorem, $n = m$ and there are $\sigma, \tau \in S_n$ such that $[U_i]_m = [V_{\sigma(i)}]_m$ and $[U_i]_e = [V_{\tau(i)}]_e$ for every $i \in \{1, \dots, n\}$. We have to show that there is $\pi \in S_n$ such that $U_i \simeq V_{\pi(i)}, i \in \{1, \dots, n\}$. For $i \in \{1, \dots, n\}$ define $\pi(i) := \tau(i)$ if $[U_i]$ is in a connected component having one monogeny class and $\pi(i) := \sigma(i)$ if $[U_i]$ is not in a connected component of Γ_R having one monogeny class. Thus if $[U_i]$ is in a connected component having one monogeny class, then $[U_i]_e = [V_{\pi(i)}]_e$, hence $V_{\pi(i)} \sim U_i$ and $U_i \simeq V_{\pi(i)}$. If $[U_i]$ is not in a connected component having one monogeny class, it is in a connected component having one epigeny class. Therefore $[U_i]_m = [V_{\pi(i)}]_m$, $V_{\pi(i)} \sim U_i$ and $U_i \simeq V_{\pi(i)}$. It remains to show that π is a bijection. But this is obvious since if $\pi(i) = \pi(i')$ for $i \neq i'$, then exactly one element from $\{[U_i], [U_{i'}]\}$ corresponds to an edge of a connected component having one monogeny class. Consequently, $[V_{\pi(i)}]$ and $[V_{\pi(i')}]$ correspond to edges in different connected components of Γ_R, thus $\pi(i) \neq \pi(i')$.

$(b) \Leftrightarrow (c)$ is an easy consequence of Lemma 3.6. □

In the remaining part of this section we rewrite (b),(c) in terms of the ring R and obtain a characterization of when the Krull-Remak-Schmidt theorem holds in mod-R provided R is a basic serial ring.

If R is a chain ring and $U = R/rR$ for some $r \in R$, then, by [**11**, Lemma 4.3], $U_e = RrR/rR$. The following lemma extends this result to the case of serial rings. Recall that, with any serial ring, we consider a complete set of indecomposable orthogonal idempotents $\{e_1, \dots, e_n\}$.

LEMMA 3.9. *Let R be a serial ring, $i \in \{1, \dots, n\}$ and $U = e_iR/e_irR$ for some $r \in J(R)$. Then $U_e = e_iRe_irR/e_irR$.*

PROOF. If $e_ir = 0$, then U is an indecomposable projective R-module, therefore $U_e = 0$. Assume $e_ir \neq 0$. Let $E := e_iRe_i$ be the endomorphism ring of e_iR. Then E is local and $J(E) = e_iJ(R)e_i$. It follows that every epimorphism from $\mathrm{End}_R(U)$ is induced by some $x \in E \setminus J(E)$ satisfying $xe_irR \subseteq e_irR$. Then it is easy to check that the kernel of every epimorphism in $\mathrm{End}_R(U)$ is contained in e_iRe_irR/e_irR, hence $U_e \subseteq e_iRe_irR/e_irR$.

Conversely, suppose that $u = e_ise_irt \in e_iRe_irR \setminus e_irR$. We want to check the existence of an epimorphism $f \in \mathrm{End}_R(U)$ satisfying $f(u + e_irR) = 0$. If $e_ise_i \notin J(E)$, then left multiplication by $(e_ise_i)^{-1} \in E$ induces such an epimorphism, so assume $e_ise_i \in J(E)$. Then $e_i + e_ise_i \in E \setminus J(E)$. Observe that $u + e_irR = (e_ise_i + e_i)e_irt + e_irR$. Therefore it suffices to find $e \in E \setminus J(E)$ such that $e(e_ise_i + e_i)e_irt \in$

e_irR. Obviously, the inverse of $e_ise_i+e_i$ in E satisfies this, therefore we can choose f to be the endomorphism of U induced by left multiplication by $(e_ise_i+e_i)^{-1}$. □

COROLLARY 3.10. *Let R be a serial ring, $i \in \{1,\dots,n\}$, $U = e_iR/e_irR$ and $V = e_iR/e_isR$ for some $r,s \in J(R)$. Then $[U]_e = [V]_e$ if and only if $e_iRe_irR = e_iRe_isR$, if and only if $Re_irR = Re_isR$.*

PROOF. We prove only the first equivalence, the proof of the second one is straightforward. Assume that $[U]_e = [V]_e$, i.e., there are epimorphisms in $\mathrm{Hom}_R(U,V)$ and $\mathrm{Hom}_R(V,U)$. If $e_ir = 0$, then U is projective and V contains a direct summand isomorphic to U. Since V is uniserial, $V \simeq U$, but then e_isR is a direct summand of e_iR, hence also $e_is = 0$. Similarly $e_is = 0$ implies $e_ir = 0$. Therefore we may assume $e_ir, e_is \neq 0$.

Without loss on generality suppose $0 \neq e_irR \subseteq e_isR \subsetneq e_iR$. We are left to check that $e_iRe_isR \subseteq e_iRe_irR$. Let $\pi\colon U \to V$ be the canonical projection and $g\colon V \to U$ be an epimorphism. Then $f = g\pi \in \mathrm{End}_R(U)$ is an epimorphism such that $f(e_is + e_irR) = 0$. By Lemma 3.9, $e_is \in e_iRe_irR$ and therefore $e_iRe_isR \subseteq e_iRe_irR$.

Now assume $e_iRe_irR = e_iRe_isR$. Obviously, $e_ir = 0$ if and only if $e_is = 0$. Therefore suppose $0 \neq e_irR \subseteq e_isR \subsetneq e_iR$. Since $e_irR \subseteq e_isR$, the canonical projection is an epimorphism $\pi\colon U \to V$. Since $e_is \in e_iRe_irR$ there exists an epimorphism $f \in \mathrm{End}_R(U)$ such that $f(e_is+e_irR) = 0$, i.e., $\mathrm{Ker}\,\pi \subseteq \mathrm{Ker}\,f$. Therefore there exists $g\colon V \to U$ such that $f = g\pi$. In particular, g is an epimorphism in $\mathrm{Hom}_R(V,U)$, thus $[U]_e = [V]_e$. □

We give a similar criterion for monogeny classes. Recall that if R is a serial ring, then every indecomposable finitely presented R-module is of the form e_iR/e_ire_jR, $i,j \in \{1,\dots,n\}$. As observed in Lemma 3.5, if R is basic and $e_ire_m, e_jse_n \in J(R)$ are nonzero, then $\mathrm{Hom}_R(e_iR/e_ire_mR, e_jR/e_jse_nR)$ can contain a monomorphism only if $m = n$.

LEMMA 3.11. *Suppose that R is a serial ring $0 \neq e_ire_k, e_jse_k \in J(R)$. Then $[e_iR/e_ire_kR]_m = [e_jR/e_jse_kR]_m$ if and only if $Re_ire_kRe_k = Re_jse_kRe_k$, if and only if $Re_ire_kR = Re_jse_kR$.*

PROOF. Suppose that $[e_iR/e_ire_kR]_m = [e_jR/e_jse_kR]_m$. Then there are $a \in e_jRe_i$ and $b \in e_iRe_j$ satisfying $ae_ire_kR = e_jse_kR$ and $be_jse_kR = e_ire_kR$. In particular, $e_ire_k \in Re_jse_kRe_k$ and $e_jse_k \in Re_ire_kRe_k$. This implies that $Re_ire_kRe_k = Re_jse_kRe_k$.

Conversely, assume $Re_ire_kRe_k = Re_jse_kRe_k$. Write e_jse_k as a sum of elements of the form $v'e_ire_ku'e_k$, for $u', v' \in R$. Since the left module Re_k is uniserial, there are $u \in R$ and $v \in e_jRe_i$ such that $e_jse_k = ve_ire_kue_k$. Left multiplication of v induces a monomorphism in $\mathrm{Hom}_R(e_iR/e_ire_kue_kR, e_jR/e_jse_kR)$. Observe that $e_iRe_ire_kue_kR$ contains e_ire_k. By Lemma 3.9, we get $[e_iR/e_ire_kR]_e = [e_iR/e_ire_kue_kR]_e$. Furthermore, $e_ire_kue_k$ and e_ire_k are nonzero elements of e_iRe_k. Therefore, by Corollary 3.3, at least one of $\mathrm{Hom}_R(e_iR/e_ire_kR, e_iR/e_ire_kue_kR)$, $\mathrm{Hom}_R(e_iR/e_ire_kue_kR, e_iR/e_ire_kR)$ contains a monomorphism. In any case the existence of such a monomorphism together with $[e_iR/e_ire_kR]_e = [e_iR/e_ire_kue_kR]_e$ implies $e_iR/e_ire_kR \simeq e_iR/e_ire_kue_kR$. So there exists an injective homomorphism from e_iR/e_ire_kR to e_jR/e_jse_kR. Similarly we prove that there exists a monomorphism from e_jR/e_jse_kR to e_iR/e_ire_kR. □

Combining the previous results, we obtain the following generalization of [**9**, Prop 2.21]. Observe that, for basic serial rings, it gives a classification of indecomposable finitely presented R-modules up to isomorphism.

COROLLARY 3.12. *Suppose that R is a serial ring and $0 \neq e_ire_j, e_ise_j \in e_iJ(R)e_j$. Then the following are equivalent*

(a) $[e_iR/e_ire_jR]_m = [e_iR/e_ise_jR]_m$
(b) $[e_iR/e_ire_jR]_e = [e_iR/e_ise_jR]_e$
(c) $e_iR/e_ire_jR \simeq e_iR/e_ise_jR$
(d) $Re_ire_jR = Re_ise_jR$.

We conclude the section giving an answer to [**9**, Question 2.24] for basic serial rings.

THEOREM 3.13. *Let $\{e_1, \dots, e_n\}$ be a complete set of orthogonal indecomposable idempotents of a basic serial ring R. The following statements are equivalent:*

(a) *The Krull-Remak-Schmidt theorem does not hold in* mod-R.
(b) *There are nonzero $e_ire_j, e_ise_k, e_lue_k, e_lte_j \in J(R)$ such that $j \neq k$, $i \neq l$ and $e_iRe_ire_jR = e_iRe_ise_kR$, $Re_ise_kRe_k = Re_lue_kRe_k$, $e_lRe_lue_kR = e_lRe_lte_jR$, $Re_lte_jRe_j = Re_ire_jRe_j$.*
(c) *There are nonzero $e_ire_j, e_ise_k, e_lue_k \in J(R)$ such that $j \neq k, i \neq l$ and $e_iRe_ire_jR = e_iRe_ise_kR$, $Re_ise_kRe_k = Re_lue_kRe_k$.*

PROOF. *(a)* $\Rightarrow$ *(b)* By Corollary 3.8, (a) is equivalent to the existence of a cycle in Γ_R. Since every connected component of Γ_R is a complete bipartite graph, (a) is in fact equivalent to the existence of a cycle of length 4. Let $[U_1], [U_2], [U_3], [U_4]$ be edges of such a cycle, say $[U_1]_e = [U_2]_e$, $[U_2]_m = [U_3]_m$, $[U_3]_e = [U_4]_e$ and $[U_4]_m = [U_1]_m$. Furthermore, $[U_1]_m \neq [U_3]_m$ and $[U_2]_e \neq [U_4]_e$. Note that if a connected component of Γ_R contains an edge corresponding to a projective module, then the component has only one monogeny class. Therefore U_1, U_2, U_3, U_4 are not projective. Let $U_1 = e_iR/e_ire_jR$, $U_2 = e_iR/e_ise_kR$, (to see that U_1 and U_2 have to be factors of the same e_iR use Lemma 3.5 (a)), $U_3 = e_lR/e_lue_kR$ (the occurence of the same primitive idempotent e_k in the presentations of U_2 and U_3 is a consequence of Lemma 3.5(b)), $U_4 = e_lR/e_lte_jR$. Since U_1, U_2, U_3, U_4 are not projective, $e_ire_j, e_ise_k, e_lue_k, e_lte_j$ are nonzero elements of $J(R)$. The relations between monogeny and epigeny classes of U_1, U_2, U_3, U_4 translate to relations in (b). Finally, if $j = k$ then, by Corollary 3.3, either $\mathrm{Hom}_R(U_1, U_3)$ or $\mathrm{Hom}_R(U_3, U_1)$ contains a monomorphism. But then $[U_1]_m = [U_3]_m$ follows from Lemma 3.7 since $U_1 \sim U_3$. This is not possible, so $j \neq k$. Similarly one can show that $[U_2]_e \neq [U_4]_e$ implies $i \neq l$.

(b) $\Rightarrow$ *(a)* Using the same arguments as in the proof of the previous implication one can see that $[e_iR/e_ire_jR], [e_iR/e_ise_kR], [e_lR/e_lue_kR], [e_lR/e_lte_jR]$ give a cycle in Γ_R.

The proof of *(a)* $\Leftrightarrow$ *(c)* is similar. □

Note that [**9**, Question 2.24] also aks whether the condition (a) of Theorem 3.13 is left-right symmetric. It is possible to use condition (b) to check that this is indeed the case. However, it is much easier to apply Corollary 3.8 and the Auslander-Bridger transpose to see that (cf. [**1**, Proposition 7.2]).

4. Dimension theory for pure projective modules over serial rings

In this section we briefly review the dimension theory developed in [**8**] to study direct summands of serial modules. We are interested only in a particular case of pure projective modules over serial rings.

Let R be a serial ring, $\mathcal{P} := \text{Add}(\text{mod-}R)$ the full subcategory of Mod-R whose objects are pure projective R-modules. Let U be an indecomposable finitely presented R-module of type 1. Let $\mathcal{J}_U$ be the largest ideal of $\mathcal{P}$ such that $\mathcal{J}_U(U,U) = J(\text{End}_R(U))$. If $X, Y \in \mathcal{P}$, then

$$\mathcal{J}_U(X,Y) := \{f \in \text{Hom}_R(X,Y) \mid \forall \alpha \in \text{Hom}_R(U,X),\\ \forall \beta \in \text{Hom}_R(Y,U)\ \beta f \alpha \in J(\text{End}_R(U))\}.$$

The factor category $\mathcal{P}/\mathcal{I}_U$ is equivalent to the category of right modules over the division ring $\text{End}_R(U)/J(\text{End}_R(U))$ and therefore there exists a natural way to assign a cardinal $\dim_U(M)$ to every pure projective module M (see [**8**, page 30] for details). Note that if $V \in \text{mod-}R$ is indecomposable and $U \not\simeq V$, then $1_V \in \mathcal{J}_U$, so V is a zero object in $\mathcal{P}/\mathcal{I}_U$. Thus $\dim_U(V) = 0$. On the other hand, if $V \simeq U$, then $\dim_U(V) = 1$.

Similarly, if U is an indecomposable finitely presented R-module of type 2 and $I \subseteq \text{End}_R(U)$ ($K \subseteq \text{End}_R(U)$) is the ideal of $\text{End}_R(U)$ consisting of nonmonomorphisms (nonepimorphisms), let $\mathcal{I}_U$ ($\mathcal{K}_U$) be the ideals of $\mathcal{P}$ such that $\mathcal{I}_U$ ($\mathcal{K}_U$) is the largest ideal of $\mathcal{P}$ satisfying $\mathcal{I}_U(U,U) = I$ ($\mathcal{K}_U(U,U) = K$). The category $\mathcal{P}/\mathcal{I}_U$ (resp. $\mathcal{P}/\mathcal{K}_U$) is equivalent to the category of modules over the division ring $\text{End}_R(U)/I$ ($\text{End}_R(U)/K$). Therefore we may assign a cardinal $\text{mdim}_U(M)$ ($\text{edim}_U(M)$) to every pure projective R-module M. Let $V \in \text{mod-}R$ be indecomposable. Similarly as above, it is possible to check that $\text{mdim}_U(V) = 0$ ($\text{edim}_U(V) = 0$) if $[U]_m \neq [V]_m$ ($[U]_e \neq [V]_e$) and $\text{mdim}_U(V) = 1$ ($\text{edim}_U(V) = 1$) if $[U]_m = [V]_m$ ($[U]_e = [V]_e$).

The cardinal numbers defined by indecomposable objects of mod-R determine objects of $\mathcal{P}$ up to isomorphism, because from [**8**, Theorem 7.4] it is easy to derive:

THEOREM 4.1. *Let R be a serial ring, M, N pure projective R-modules. Then $M \simeq N$ if and only if* $\dim_U(M) = \dim_U(N)$ *for every indecomposable $U \in$* mod-*R of type* 1*, and* $\text{mdim}_V(M) = \text{mdim}_V(N)$, $\text{edim}_V(M) = \text{edim}_V(N)$ *for every indecomposable $V \in$* mod-*R of type 2.*

We need this theorem only for countably generated pure projective modules M, N. In this case the theorem is a particular case of [**8**, Lemma 7.2].

If $U, V \in \text{mod-}R$ are of type 2 and $[U]_m = [V]_m$, then $\text{mdim}_U(M) = \text{mdim}_V(M)$ for every $M \in \mathcal{P}$. If v is a vertex of a connected component of type 2 in Γ_R contained in $\mathcal{V}_m$, we define a dimension function mdim_v on objects of $\mathcal{P}$ by $\text{mdim}_v(M) := \text{mdim}_U(M)$, where $U \in \text{mod-}R$ is indecomposable and v consists of modules having the same monogeny class as U. Similarly we define a dimension function edim_v for every vertex $v \in \Gamma_R$ of a connected component of type 2 which is contained in $\mathcal{V}_e$.

The cardinal invariants are compatible with direct sums, i.e., if $*\dim_U$ is a function described as above and $M_i, i \in I$, is a family of modules in $\mathcal{P}$, then $*\dim_U(\oplus_{i\in I} M_i) = \sum_{i\in I} *\dim_U(M_i)$ (see [**8**, Corollary 2.7]).

If $U_i, i \in I$, is a family of indecomposable modules from mod-R and N is a direct summand of $\oplus_{i\in I} U_i$, then N is an image of an idempotent endomorphism

$\pi \in \operatorname{End}_R(\oplus_{i\in I} U_i)$. We display π as a column-finite matrix $\pi = (p_{i,j})_{i,j\in I}$, where $p_{i,j} \in \operatorname{Hom}_R(U_j, U_i)$.

If $V \in \text{mod-}R$ is indecomposable and of type 2, it is easy to see that $\operatorname{mdim}_V(N) > 0$ if and only if there are $i, j \in I$, such that $[U_i]_m = [U_j]_m = [V]_m$ and $p_{i,j}$ is a monomorphism. Similarly, $\operatorname{edim}_V(N) > 0$ if and only if there are $i, j \in I$ such that $[U_i]_e = [U_j]_e = [V]_e$ and $p_{i,j}$ is an epimorphism.

5. The main result

Using the dimension theory it is easy to prove the following criterion.

LEMMA 5.1. *Let M be a pure projective module over a serial ring R. The following are equivalent:*

(a) *M is a direct sum of finitely presented modules.*
(b) *The equality $\sum_{v_m\in\mathcal{V}_m\cap\mathcal{C}} \operatorname{mdim}_{v_m}(M) = \sum_{v_e\in\mathcal{V}_e\cap\mathcal{C}} \operatorname{edim}_{v_e}(M)$ holds for every connected component $\mathcal{C} \subseteq \Gamma_R$ of type 2.*

PROOF. $(a) \Rightarrow (b)$ Suppose $M = \oplus_{i\in I} U_i$, where every $U_i \in \text{mod-}R$ is indecomposable. Then (b) follows from $\sum_{v_m\in\mathcal{V}_m\cap\mathcal{C}} \operatorname{mdim}_{v_m}(M) = |\{i \in I \mid [U_i] \in \mathcal{C}\}| = \sum_{v_e\in\mathcal{V}_e\cap\mathcal{C}} \operatorname{edim}_{v_e}(M)$.

$(b) \Rightarrow (a)$ For every vertex $v \in \mathcal{V}_m$ in a connected component of type 2 set $\kappa_v := \operatorname{mdim}_v(M)$ and for every vertex $v \in \mathcal{V}_e$ in a connected component of type 2 let $\lambda_v := \operatorname{edim}_v(M)$. If $\mathcal{C}$ is a connected component of type 2, set $\kappa_{\mathcal{C}} := \dot\bigcup_{v\in\mathcal{C}\cap\mathcal{V}_m} \kappa_v$ and $\lambda_{\mathcal{C}} := \dot\bigcup_{v\in\mathcal{C}\cap\mathcal{V}_e} \lambda_v$. (b) implies the existence of a bijection $b_{\mathcal{C}} \colon \kappa_{\mathcal{C}} \to \lambda_{\mathcal{C}}$ for every connected component $\mathcal{C}$ of type 2. Using $b_{\mathcal{C}}$ we construct a module $M_{\mathcal{C}}$ which is a direct sum of indecomposable modules from mod-R, $\operatorname{mdim}_v(M_{\mathcal{C}}) = \kappa_v$ for every $v \in \mathcal{C} \cap \mathcal{V}_m$ and $\operatorname{edim}_v(M_{\mathcal{C}}) = \lambda_v$ for every $v \in \mathcal{C} \cap \mathcal{V}_e$. Let $i \in \kappa_{\mathcal{C}} = \dot\bigcup_{v\in\mathcal{C}\cap\mathcal{V}_m} \kappa_v$. Then $i \in \kappa_v$ for unique $v \in \mathcal{C} \cap \mathcal{V}_m$ and $b_{\mathcal{C}}(i) \in \lambda_w$ for unique $w \in \mathcal{C} \cap \mathcal{V}_e$. Let U_i be a module in mod-R corresponding to the edge of Γ_R connecting v and w. Define $M_{\mathcal{C}} := \oplus_{i\in\kappa_{\mathcal{C}}} U_i$ and $M_2 := \oplus_{\mathcal{C}} M_{\mathcal{C}}$, where the direct sum is over all connected components of type 2.

Finally, let $\mathcal{I}$ be the set of isoclases of indecomposable finitely presented modules of type 1, and let $M_1 := \oplus_{[U]\in\mathcal{I}} U^{(\operatorname{dim}_U(M))}$. Consider $M' := M_1 \oplus M_2$. It is easy to check that for every indecomposable $W \in \text{mod-}R$ the dimensions related to W coincides on M and M'. Therefore $M \simeq M'$ and M is isomorphic to a direct sum of finitely presented modules. □

PROPOSITION 5.2. *Let R be a serial ring. The following are equivalent*

(a) *There exists a countably generated pure projective module M and a connected component $\mathcal{C}$ of Γ_R of type 2 such that $\sum_{v_m\in\mathcal{V}_m\cap\mathcal{C}} \operatorname{mdim}_{v_m}(M) < \sum_{v_e\in\mathcal{V}_e\cap\mathcal{C}} \operatorname{edim}_{v_e}(M)$.*
(b) *There exists a countably generated pure projective module M and a connected component $\mathcal{C}$ of Γ_R of type 2 such that $0 = \sum_{v_m\in\mathcal{V}_m\cap\mathcal{C}} \operatorname{mdim}_{v_m}(M)$ and $0 < \sum_{v_e\in\mathcal{V}_e\cap\mathcal{C}} \operatorname{edim}_{v_e}(M)$.*
(c) *There exist $U, W \in \text{mod-}R$ indecomposable, U of type 2 and $W \not\sim U$, and homomorphisms $\alpha\colon U \to W, \beta\colon W \to U$ such that $U \not\sim W$ and $\operatorname{Ker} \beta\alpha \subsetneq U_e$.*

PROOF. Suppose (a) is true and $0 < \sum_{v_m\in\mathcal{V}_m\cap\mathcal{C}} \operatorname{mdim}_{v_m}(M)$. Since M is countably generated, it is a direct summand of a countable direct sum of finitely presented modules. Therefore, $\sum_{v_e\in\mathcal{V}_e\cap\mathcal{C}} \operatorname{edim}_{v_e}(M)$ is at most countable and

$\sum_{v_m\in\mathcal{V}_m\cap\mathcal{C}} \mathrm{mdim}_{v_m}(M)$ is finite. Since components of Γ_R are complete bipartite, there exists $U \in$ mod-R indecomposable such that $[U] \in \mathcal{C}$, $\mathrm{mdim}_U(M) > 0$ and $\mathrm{edim}_U(M) > 0$. Then, using the same arguments as in [**13**, Lemma 4.3], $M \simeq U \oplus M'$ and $0 \leq \sum_{v_m\in\mathcal{V}_m\cap\mathcal{C}} \mathrm{mdim}_{v_m}(M') = \sum_{v_m\in\mathcal{V}_m\cap\mathcal{C}} \mathrm{mdim}_{v_m}(M) - 1$ and $1+\sum_{v_e\in\mathcal{V}_e\cap\mathcal{C}} \mathrm{edim}_{v_e}(M') = \sum_{v_e\in\mathcal{V}_e\cap\mathcal{C}} \mathrm{edim}_{v_e}(M)$. Repeating this procedure, after finitely many steps we get a pure projective module satisfying (b).

Suppose that (b) is true. Let M be a direct summand of $N = \oplus_{i\in\mathbb{N}} U_i$, where the modules $U_i \in$ mod-R are indecomposable. Let $\pi = (p_{i,j})_{i,j\in\mathbb{N}} \in \mathrm{End}_R(N)$ be a projection of N onto M, so $\pi^2 = \pi$ and $\pi(N) = M$. Since $0 < \sum_{v_e\in\mathcal{V}_e\cap\mathcal{C}} \mathrm{edim}_{v_e}(M)$, there are $i, j \in \mathbb{N}$ such that $[U_i]_e = [U_j]_e$, $[U_i], [U_j] \in \mathcal{C}$ and $p_{i,j}\colon U_j \to U_i$ is surjective. Fix such i, j and consider the set $I := \{t \in \mathbb{N} \mid [U_t] \in \mathcal{C}\}$. Observe that for every $t \in I$ the homomorphism $p_{t,j}\colon U_j \to U_t$ cannot be a monomorphism, since otherwise, according to Lemma 3.7(a), $[U_j]_m = [U_t]_m$ would hold and $\mathrm{mdim}_{U_j}(M) > 0$ would contradict the assumption (b). Let $i_0 \in I$ be such that $\mathrm{Ker}\ p_{i_0,j} \subseteq \mathrm{Ker}\ p_{t,j}$ for every $t \in I$. Observe that $i \in I$ and $p_{i,j}$ onto imply $\mathrm{Ker}\ p_{i_0,j} \subsetneq (U_j)_e$.

Now use the identity $\pi^2 = \pi$, in particular $p_{i_0,j} = \sum_{k\in\mathbb{N}} p_{i_0,k}p_{k,j}$. Of course, there are only finitely many nonzero summands on the right hand side. Then there exists k such that $\mathrm{Ker}\ p_{i_0,k}p_{k,j} \subseteq \mathrm{Ker}\ p_{i_0,j}$. If $U_k \in \mathcal{C}$, i.e., $k \in I$, then $\mathrm{Ker}\ p_{k,j} = \mathrm{Ker}\ p_{i_0,j}$ and $p_{i_0,k}$ is a monomorphism. Since $U_{i_0} \sim U_k$, $[U_{i_0}]_m = [U_k]_m$ and, again, we get a contradiction since $\mathrm{mdim}_{U_{i_0}}(M) > 0$. Therefore $W := U_k$ is not in $\mathcal{C}$.

Consider the homomorphism $\gamma = p_{i_0,k}p_{k,j}$. Since $\mathrm{Ker}\ \gamma \subsetneq (U_j)_e$, $[\mathrm{Im}\ \gamma]_e = [U_j]_e$. In particular, $U := \mathrm{Im}\ \gamma \sim U_{i_0}$ and since $U \subseteq U_{i_0}$, an application of Lemma 3.7(a) gives $[U_{i_0}]_m = [U]_m$. Let $\mu\colon U_{i_0} \to U$ be a monomorphism and $\delta\colon U \to U_j$ be an epimorphism. Then it is enough to put $\alpha := p_{k,j}\delta$ and $\beta := \mu p_{i_0,k}$. This proves (c) since, by [**11**, Lemma 2.3(4)], $\delta^{-1}((U_j)_e) = U_e$ and all modules corresponding to edges of the connected component $\mathcal{C}$ are of type 2.

Finally suppose (c) is true. Consider an epimorphism $f \in \mathrm{End}_R(U)$ such that $\mathrm{Ker}\ \beta\alpha \subsetneq \mathrm{Ker} f$. Then $\mathrm{Ker}\ (f^2 + \beta\alpha) \subsetneq \mathrm{Ker}\ f$, so there exists an epimorphism $\varphi \in \mathrm{End}_R(U)$ such that $f = \varphi f^2 + (\varphi\beta)\alpha$. Now since f is onto and not mono, $W \not\sim U$ one can construct as in [**13**, Lemma 5.2] a countably generated pure projective module M such that $\mathrm{mdim}_U(M) = 0$, $\mathrm{edim}_U(M) = 1$ and dimensions related to W are infinite (in the construction [**13**, page 201] we set $M := U$, $N := W$, $f_i := f$ for every $i \in \mathbb{N}$ and the arguments of [**13**, Lemma 5.2] apply also in this more general case). In fact, if $\mathcal{C}$ is the connected component of Γ_R containing the edge $[U]$, then $\sum_{v_m\in\mathcal{V}_m\cap\mathcal{C}} \mathrm{mdim}_{v_m}(M) = 0$ and $\sum_{v_e\in\mathcal{V}_e\cap\mathcal{C}} \mathrm{edim}_{v_e}(M) = 1$ since M is a direct summand of $U^{(\omega)} \oplus W^{(\omega)}$. Therefore (a) and (b) hold. □

LEMMA 5.3. *Let R be a serial ring and let U, V be indecomposable modules of* mod-R *such that $U \sim V$. If $[U]_m$ contains a module which is not quasi-small, then the same holds for $[V]_m$.*

PROOF. In other words, we have to prove that $U_m \subsetneq U_e$ implies $V_m \subsetneq V_e$. The claim is obvious if $[U]_m = [V]_m$, so it is enough to prove it under the assumption $[U]_e = [V]_e$. Let $f \in \mathrm{End}_R(U)$ and $g \in \mathrm{End}_R(U)$ be a monomorphism and an epimorphism such that $gf = 0$. Furthermore, let $\alpha\colon V \to U$, $\beta\colon U \to V$ be epimorphisms. Consider $V' = \alpha^{-1}(\mathrm{Im}\ f)$. It is a submodule of V, $\alpha(V') \simeq U$

implies that there exists an epimorphism in $\mathrm{Hom}_R(V', V)$. Hence $V' \simeq V$. If $\gamma \in \mathrm{Hom}_R(V, V')$ is a monomorphism, then $(\beta g\alpha)\gamma = 0$, where $\beta g\alpha \in \mathrm{End}_R(V)$ is an epimorphism. Therefore $V_m \subsetneq V_e$. $\square$

PROPOSITION 5.4. *Let R be a serial ring. Then there exists a pure projective R-module which is uniserial and not quasi-small if and only if there exists $0 \neq e_ire_j \in J(R)$ such that the ideal Re_ire_jR is idempotent.*

PROOF. Suppose that $I = Re_ire_jR$ is a nonzero idempotent ideal contained in $J(R)$. Observe that, by Nakayama's lemma, e_iI is not a finitely generated right ideal. Consider $U := e_iR/e_ire_jR$. It is a uniserial module with $U_e = e_iI/e_ire_jR \neq 0$. Since $I = I^2$, there are $a \in e_iI$, $b \in Ie_j$ such that $e_ire_j = ab$. Write $a = se_ire_jv$, $b = ue_ire_jw$, where $s \in e_iR, v, u \in R, w \in Re_j$. Then $e_ire_j = (se_ire_jvue_i)(e_ire_jw)$. Set $V := e_iR/e_ire_jwR$ and note $V_e = e_iI/e_ire_jwR$, hence $[V]_e = [U]_e$. Consider the homomorphism $f \colon V \to U$ induced by $se_ire_jvue_i$. Since $e_ire_jR = (se_ire_jvue_i)(e_ire_jw)R$, f is injective and $U \simeq V$ follows. On the other hand $se_ire_jvue_i \in e_iI$, so there exists an epimorphism $g \in \mathrm{End}_R(U)$ such that $gf = 0$. As $U \simeq V$, we conclude $U_m \subsetneq U_e$ and therefore $\mathrm{Add}(U)$ contains a uniserial module which is not quasi-small.

Conversely, assume that there exists a pure projective R-module X which is uniserial and not quasi-small. Then X contains a finitely generated submodule W' such that $W'_m \subsetneq W'_e$. Consider W' of the form e_iR/e_iI, where I is a right ideal of R. Since $W'_e \neq 0$ there exists $W = e_iR/e_ire_jR$ such that $[W]_e = [W']_e$, for example choose e_ire_j such that $e_ire_j + e_iI$ is a nonzero element of W'_e. The argument from the proof of Lemma 5.3 shows that $W_m \subsetneq W_e$, so there are a monomorphism $f \in \mathrm{End}_R(W)$, an epimorphism $g \in \mathrm{End}_R(W)$ such that $gf = 0$. Suppose that g is induced by $u \in e_iRe_i$ and f is induced by $v \in e_iRe_i$. Then $ve_ire_jR = e_ire_jR$ and $uv \in e_ire_jR$. But u is invertible in e_iRe_i, so $v \in Re_ire_jR$ and consequently the ideal Re_ire_jR is idempotent. Finally note that e_ire_j is has to be an element of $\mathrm{rad}(e_iR) = e_iJ(R)$, so $Re_ire_jR \subseteq J(R)$. $\square$

LEMMA 5.5. *Let W be a non-projective indecomposable finitely presented module over a serial ring R, and $f, g \in \mathrm{End}_R(W)$ monomorphisms. Then there exists a monomorphism $h \in \mathrm{End}_R(W)$ such that $f = hg$ or $g = hf$.*

PROOF. Let $W = e_kR/e_kwR$ for some primitive idempotent $e_k \in R$, and let $u, v \in e_kRe_k$ be elements inducing f, g. Since $0 \neq e_kw \in J(R)$, we have $ue_kwR = e_kwR$ and $ve_kwR = e_kwR$. Since e_kRe_k is a chain ring, there exists $t \in e_kRe_k$ such that either $tu = v$ or $u = tv$. In the former case we have $te_kwR = tue_kwR = ve_kwR = e_kwR$. Therefore t induces $h \in \mathrm{End}_R(W)$ such that $hf = g$. Similarly, in the latter case t induces $h \in \mathrm{End}_R(W)$ such that $f = hg$. $\square$

PROPOSITION 5.6. *Let R be a serial ring such that there is no pure projective R-module which is uniserial and not quasi-small. The following are equivalent*

(a) *There exists a countably generated pure projective module M and a connected component $\mathcal{C}$ of Γ_R of type 2 such that $\sum_{v_m \in \mathcal{V}_m \cap \mathcal{C}} \mathrm{mdim}_{v_m}(M) > \sum_{v_e \in \mathcal{V}_e \cap \mathcal{C}} \mathrm{edim}_{v_e}(M)$.*

(b) *There exists a countably generated pure projective module M and a connected component $\mathcal{C}$ of Γ_R of type 2 such that $\sum_{v_m \in \mathcal{V}_m \cap \mathcal{C}} \mathrm{mdim}_{v_m}(M) > 0$ and $\sum_{v_e \in \mathcal{V}_e \cap \mathcal{C}} \mathrm{edim}_{v_e}(M) = 0$.*

(c) *There exist* $U, W \in \text{mod-}R$ *indecomposable,* U *of type* 2 *and* $W \not\sim U$, *and homomorphisms* $\alpha\colon U \to W, \beta\colon W \to U$, $f, g\colon U \to U$ *such that* f *is mono but not epi and* $gf^2 - f = \beta\alpha$.

PROOF. $(a) \Rightarrow (b)$: It can be proved as $(a) \Rightarrow (b)$ in Proposition 5.2.

$(b) \Rightarrow (c)$: Suppose there exists M satisfying (b). Then M is a direct summand of $N = \oplus_{i\in\mathbb{N}} U_i$, where the modules $U_i := e_{l_i}R/e_{l_i}s_ie_{r_i}R$ are indecomposable and finitely presented. Let $\pi = (p_{i,j}) \in \text{End}_R(N)$ be an idempotent endomorphism with image M. Then there are $i, j \in \mathbb{N}$ such that $[U_i]_m = [U_j]_m$, $p_{i,j}$ is a monomorphism and $[U_i], [U_j] \in \mathcal{C}$. Moreover, if $[U_k]_e = [U_l]_e$ and $[U_k], [U_l] \in \mathcal{C}$, then $p_{k,l}$ is not onto.

Fix i, j such that $[U_i], [U_j] \in \mathcal{C}$ and $p_{i,j}$ is a monomorphism. By Lemma 3.7, $[U_i]_m = [U_j]_m$. Let $x_{i,j} \in e_{l_i}Re_{l_j}$ be an element inducing $p_{i,j}$. We claim that the left ideal $Rx_{i,j}$ is independent of the choice of $x_{i,j}$, i.e., if $y_{i,j} \in e_{l_i}Re_{l_j}$ is another element inducing $p_{i,j}$, then $Rx_{i,j} = Ry_{i,j}$. Indeed, in this case $x_{i,j} - y_{i,j} \in e_{l_i}s_ie_{r_i}Re_{l_j}$. If $x_{i,j} \in Re_{l_i}s_ie_{r_i}R$, the image of $p_{i,j}$ would be contained in $(U_i)_e$ and it is not possible since in this case $\text{Add}(U_i)$ would contain a uniserial module which is not quasi-small. Therefore the submodule $Rx_{i,j} \subseteq Re_{l_j}$ is strictly bigger than $R(x_{i,j} - y_{i,j})$ and, since Re_{l_j} is uniserial, $Rx_{i,j} = Ry_{i,j}$ follows.

Consider the set $I = \{k \in \mathbb{N} \mid [U_k] \in \mathcal{C}\}$ and its subset $I' = \{k \in I \mid p_{k,j}$ can be induced by $x_{k,j} \in e_{l_k}Re_{l_j}$ such that $Rx_{i,j} \subseteq Rx_{k,j}\}$. Observe that $i \in I'$, so I' is nonempty. We claim that if $k \in I'$, then $Rx_{k,j}$ does not depend on the choice of the element inducing $p_{k,j}$. Assume that $x_{k,j}$ induces $p_{k,j}$ and $Rx_{i,j} \subseteq Rx_{k,j}$. Let $y_{k,j} \in e_{l_k}Re_{l_j}$ be another element inducing $p_{k,j}$, again $y_{k,j} - x_{k,j} \in e_{l_k}s_ke_{r_k}R$ and therefore it suffices to prove $x_{k,j} \notin Re_{l_k}s_ke_{r_k}R$. Since $U_k \sim U_j$ there exists a finitely presented module $W := e_{l_k}R/e_{l_k}wR$ such that $[W]_e = [U_k]_e$ and $[W]_m = [U_j]_m = [U_i]_m$. Let $\nu\colon W \to U_j$, $\mu\colon U_i \to W$ be monomorphisms and let $\varrho\colon U_k \to W$ be an epimorphism. Suppose μ is induced by $u \in e_{l_k}Re_{l_i}$, ν is induced by $v \in e_{l_j}Re_{l_k}$ and ϱ is induced by $s \in e_{l_k}Re_{l_k}$. Let s' be the inverse of s in $e_{l_k}Re_{l_k}$. If $x_{k,j} \in Re_{l_k}s_ke_{r_k}R = Re_{l_k}s'se_{l_k}s_ke_{r_k}R \subseteq Re_{l_k}wR$, then $ux_{i,j}v \in e_{l_k}Re_{l_k}wR$. Therefore the image of the monomorphism $\mu p_{i,j}\nu \in \text{End}_R(W)$ is contained in W_e, a contradiction to the non existence of a non quasi-small uniserial pure projective R-module.

For any $a, b \in \mathbb{N}$ let $x_{a,b} \in e_{l_a}Re_{l_b}$ be an element inducing $p_{a,b}$, and let $k_0 \in I'$ be such that $Rx_{k_0,j}$ is the largest element of the set $\{Rx_{k,j} \mid k \in I'\}$. Consider the equality $p_{k_0,j} = \sum_n p_{k_0,n}p_{n,j}$. Observe that $\sum_n x_{k_0,n}x_{n,j}$ is an element inducing $p_{k_0,j}$, therefore there exists $n \in \mathbb{N}$ such that $Rx_{k_0,j} \subseteq Rx_{k_0,n}x_{n,j}$. If $[U_n] \in \mathcal{C}$, then $n \in I$ and, in fact, $n \in I'$. By the choice of k_0, we have $Rx_{k_0,n}x_{n,j} = Rx_{n,j}$ and $p_{k_0,n}$ is an epimorphism. By Lemma 3.7, $[U_{k_0}]_e = [U_n]_e$ and hence $\text{edim}_{U_n}(M) > 0$. It is not possible since we assume $\text{edim}_{U_n}(M) = 0$. Therefore $n \notin I$.

Let $U \in \text{mod-}R$ be indecomposable such that $[U]_e = [U_{k_0}]_e$ and $[U]_m = [U_j]_m$. Let $\nu\colon U \to U_j$, $\mu\colon U_i \to U$ be non-epic monomorphisms and $\varrho\colon U_{k_0} \to U$ an epimorphism. Suppose μ is induced by u, ν is induced by v and ϱ is induced by s. Observe that we may assume that s is an invertible element of $e_{l_{k_0}}Re_{l_{k_0}}$. Then $ux_{i,j}v$ induces $\mu p_{i,j}\nu$ and $sx_{k_0,n}x_{n,j}v$ induces $\varrho p_{k_0,n}p_{n,j}\nu$. Since $Rux_{i,j}v \subseteq Rsx_{k_0,n}x_{n,j}v$, Lemma 5.5 gives the existence of $g \in \text{End}_R(U)$ such that $\mu p_{i,j}\nu = g((\mu p_{i,j}\nu)^2 + \varrho p_{k_0,n}p_{n,j}\nu)$. So we may set $W := U_n$, $\alpha := p_{n,j}\nu$, $\beta := -g\varrho p_{k_0,n}$, $f := \mu p_{i,j}\nu$.

$(c) \Rightarrow (a)$ Having such $U, W, f, g, \alpha, \beta$ it is possible to apply the construction from [**13**, page 201] to get a countably generated module $M \in \mathrm{Add}(U \oplus W)$. The same arguments as in [**13**, Lemma 5.2] give $\mathrm{mdim}_U(M) = 1, \mathrm{edim}_U(M) = 0$, $*\mathrm{dim}_W(M) = \omega$, where $*\mathrm{dim}_W$ is any dimension function related to W, and $*\mathrm{dim}_V(M) = 0$ if $V \in \text{mod-}R$ is indecomposable such that the edge $[V]$ does not intersect neither with $[W]$ nor with $[U]$. □

LEMMA 5.7. *Let R be a serial ring, $U, V \in \text{mod-}R$ be indecomposable modules such that U is of type 2, $U \not\sim V$ and there are homomorphisms $\alpha \colon U \to V$, $\beta \colon V \to U$ with $\beta\alpha(U_e) \neq 0$. If $U = e_iR/e_ire_jR$, $V = e_kR/e_kse_lR$, $a \in e_kRe_i$ induces α and $b \in e_iRe_k$ induces β, then $e_ire_j \notin e_iRe_ibae_ire_jR$.*

PROOF. Observe that U is of type 2, hence U is not projective, so $e_ire_j \neq 0$. If $e_kse_l = 0$, then $ae_ire_j = 0$ and the statement is obvious. Therefore we may assume $e_kse_l \neq 0$.

Since $\alpha(U_e) \neq 0$, $[\mathrm{Im}\ \alpha]_e = [U]_e$. Since $e_kse_l \neq 0$, Im α is finitely presented by Lemma 3.4. Therefore there exists $z \in e_iRe_ire_jR$ such that Ker $\alpha = zR/e_ire_jR$ and Im α is isomorphic to e_iR/zR. There exists an embedding of e_iR/zR to V, therefore, by Lemma 3.5, it is possible to choose $z \in e_iRe_l$, let $z = e_ir'e_l$.

If $e_kte_i \in e_kRe_i$ is an element inducing a monomorphism of $e_iR/e_ir'e_lR$ to $V = e_kR/e_kse_lR$, then $e_kte_ir'e_lR = e_kse_lR$. Observe that we may choose $e_kte_i = a$.

Consider Im β. Again, by Lemma 3.4, it is a finitely presented module, so it is isomorphic to a module of the form e_kR/wR, where $wR \subseteq e_kR$ contains e_kse_lR. There exists a monomorphism from e_kR/wR to U, so we may choose $w \in e_kRe_j$. Let $w = e_ks'e_j$ for some $s' \in R$. Moreover, observe that $[\mathrm{Im}\ \beta]_m = [U]_m$: Of course, Im β is a submodule of U containing Im $\beta\alpha$ and, since $\beta\alpha(U_e) \neq 0$, $[\mathrm{Im}\ \beta\alpha]_e = [U]_e$. In particular, $\mathrm{Hom}_R(\mathrm{Im}\ \beta\alpha, U)$ contains a monomorphism and an epimorphism, therefore Im $\beta\alpha \simeq U$ and also $[U]_m = [\mathrm{Im}\ \beta]_m$.

We assume $V \not\sim U$ and therefore $[V]_e \neq [\mathrm{Im}\ \beta]_e$. Since Im $\beta \simeq e_kR/e_ks'e_jR$, we have $e_ks'e_j \notin e_kRe_kse_lR$.

Let $e_iue_k \in e_iRe_k$ be an element inducing a monomorphism from $e_kR/e_ks'e_jR$ to U. Again we may choose $e_iue_k = b$. Now assume that the statement of the Lemma is not true, that is, $e_ire_j \in e_iRe_ibae_ire_jR$. Since $ae_ire_j \in e_kse_lR$, we have $e_ire_j \in e_iRe_ibe_kse_lR$. Furthermore, by Lemma 3.11, $[U]_m = [e_kR/e_ks'e_jR]_m$ implies $Re_ire_jRe_j = Re_ks'e_jRe_j$. In particular, $e_ks'e_j = ce_ire_jd$ for some $c \in e_kRe_i, d \in e_jRe_j$. Therefore $e_ks'e_j = ce_ire_jd \in (ce_iRe_ibe_k)(se_lRd) \subseteq e_kRe_kse_lR$. This contradiction concludes the proof. □

LEMMA 5.8. *Let R be a serial ring such that there exist $U, V \in \text{mod-}R$ indecomposable modules, $U \not\sim V$, U is of type 2, monomorphisms $g, f \in \mathrm{End}_R(U)$, with f not onto, and homomorphisms $\alpha \colon U \to V, \beta \colon V \to U$ such that $gf^2 - f = \beta\alpha$. Then there exists a pure projective uniserial R-module which is not quasi-small.*

PROOF. Let e_i, e_k be primitive idempotents of R such that $U = e_iR/e_irR$, $V = e_kR/e_ksR$. Consider elements $x, y \in e_iRe_i$ inducing f, g and elements $a \in e_kRe_i, b \in e_iRe_k$ inducing α, β.

First assume that $\beta\alpha(U_e) \neq 0$. Let us choose $r \in e_iRe_j, s \in e_kRe_l$, then, by Lemma 5.7, $e_iRe_ibae_irR$ does not contain e_ir. The relation $gf^2 - f = \beta\alpha$ can be written as $yx^2 - x - ba \in e_irR$. Therefore $(yx - e_i)xe_ir - bae_ir \in e_irRe_ir$. Observe that $xe_irR = e_irR$, since f is a monomorphism. As f is not onto, $yx - e_i$ is invertible in e_iRe_i. If u is its inverse, then $xe_ir - ubae_ir \in e_iRe_irRe_irR$. Note that $e_irR = xe_irR$ and $e_ir \notin e_iRe_ibae_irR$ implies $e_irR = (xe_ir - ubae_ir)R \subseteq e_iRe_irRe_irR$. Therefore $(Re_irR)^2 = Re_irR$ and, by Proposition 5.4, there exists a pure projective uniserial R-module which is not quasi-small.

Now suppose that $\beta\alpha(U_e) = 0$. Recall U_e is an invariant submodule of U, so we may consider restrictions $f_1 = f|_{U_e}$ and $g_1 = g|_{U_e}$ as elements of $\mathrm{End}_R(U_e)$. Then f_1, g_1 are monomorphisms such that $(g_1f_1 - 1)f_1 = 0$. If we prove that f_1 is not onto, then the monogeny class of U_e contains a uniserial module which is not quasi-small (cf Section 2, property (viii)). Then the same arguments as in the proof of Proposition 5.4 show that there exists a pure projective R-module which is uniserial and not quasi-small.

Assume that f_1 is onto, that is, $xe_iRe_irR = e_iRe_irR$. From $\beta\alpha(U_e) = 0$ we have $(yx - e_i)xe_iRe_irR \subseteq e_irR$. We have already observed that $yx - e_i$ is invertible in e_iRe_i and if u is its inverse then $xe_iRe_irR \subseteq ue_irR \subseteq e_iRe_irR$. But if $xe_iRe_irR = e_iRe_irR$ we get $ue_irR = e_iRe_irR$ and this is a contradiction with U_e not finitely generated. □

By [**9**, Lemma 1.33], a serial ring R is right non-singular if and only if $0 \neq r \in e_iRe_j$ and $0 \neq s \in e_jRe_k$ imply $rs \neq 0$. Observe that, in particular, if R is a right non-singular serial ring, then e_iRe_i is a chain domain for every $i \in \{1, \dots, n\}$. It follows that a serial ring is right non-singular if and only if it is left non-singular. Such a ring is called a *non-singular serial* ring.

LEMMA 5.9. *Let R be a non-singular serial ring such that there is no pure projective R-module which is uniserial and not quasi-small. Then there are no indecomposable $U, V \in$ mod-R, U of type 2, $U \not\sim V$ and homomorphisms $\alpha\colon U \to V$, $\beta\colon V \to U$ such that $\beta\alpha(U_e) \neq 0$.*

PROOF. Suppose that the statement is not true. Let $U = e_iR/e_ire_jR, V = e_kR/e_kse_lR$, $\alpha \in \mathrm{Hom}_R(U, V)$, and $\beta \in \mathrm{Hom}_R(V, U)$ be such that U is of type 2, $U \not\sim V$, and $\beta\alpha(U_e) \neq 0$. Let $a \in e_kRe_i$, $b \in e_iRe_k$ be elements inducing α, β. By Lemma 5.7, $e_ire_j \notin e_iRe_ibae_ire_jR$.

Since R is non-singular, $R_j := e_jRe_j$ is a chain domain. Observe that R_j does not contain nontrivial idempotent ideals of the form R_jxR_j, otherwise there would exist a pure projective R-module which is uniserial and not quasi-small.

Let $X := \{x \in J(R_j) \mid \exists y \in e_iRe_i\ ye_ire_jx = e_ire_j\}$. For every $x \in X$ set $Q_x := \bigcap_{t\in\mathbb{N}}(R_jxR_j)^t$. Since R_jxR_j is not idempotent, $x \notin Q_x$. By [**4**, Proposition 5.2(d)], Q_x is a completely prime ideal of R_j. In particular, $Q_x = zQ_x = Q_xz$ for every $z \in R_j \setminus Q_x$. Observe that $bae_ire_j \in e_ire_jx^tR_j$ for every $x \in X$ and every $t \in \mathbb{N}$. Otherwise $e_ire_jx^t \in bae_ire_jR_j$ for some $x \in X$ and $t \in \mathbb{N}$, and if $y \in e_iRe_i$ is such that $ye_ire_jx = e_ire_j$, then $e_ire_j = y^te_ire_jx^t \in e_iRe_ibae_ire_jR$, which is not possible by Lemma 5.7. It follows that $bae_ire_j \in e_ire_jQ_x$ for every $x \in X$. Indeed, since $e_ire_j \notin bae_ire_jR$, we have $bae_ire_j \in e_ire_jR_j$. As R is right non-singular, left multiplication by $e_ire_j\colon R_j \to e_ire_jR_j$ is an isomorphism of right R_j-modules by [**9**, Lemma 1.33]. Let $c \in R_j$ be such that $bae_ire_j = e_ire_jc$. We observed $c \in (R_jxR_j)^t$ for every $t \in \mathbb{N}$ and therefore $c \in Q_x$.

The assumption $\beta\alpha(U_e) \neq 0$ can be rewritten as $e_ire_jR \subsetneq bae_iRe_ire_jR$. Let $y \in e_iRe_i$ be such that $baye_ire_j \notin e_ire_jR$. Obviously $ye_ire_j \notin e_ire_jR$ therefore there exists $x \in X$ such that $ye_ire_jx = e_ire_j$. Then $baye_ire_jx = bae_ire_j \in e_ire_jQ_x$. Since $x \notin Q_x$, there is $q \in Q_x$ such that $bae_ire_j = e_ire_jqx$. Then $baye_ire_jx =$

$e_i r e_j q x$ and we may cancel x because R is non-singular (apply [**9**, Lemma 1.33] again). Then we get $baye_i re_j \in e_i re_j R$, a contradiction. □

Now we can formulate the main result of the paper.

THEOREM 5.10. *Let R be a non-singular serial ring. Then the following are equivalent*

(a) *There exists a pure projective R-module which is not a direct sum of finitely presented modules.*
(b) *There exists a pure projective R-module which is uniserial and not quasi-small.*
(c) *There exists $0 \neq e_i r e_j \in J(R)$ such that $(Re_i re_j R)^2 = Re_i re_j R$.*

PROOF. Suppose (a) holds and there is no pure projective R-module which is uniserial and not quasi-small. By Kaplansky's theorem (see [**9**, Corollary 14.3]), there exists a countably generated pure projective R-module which is not a direct sum of finitely presented modules. Then Lemma 5.1 implies that the equivalent conditions of Proposition 5.2 or Proposition 5.6 are satisfied. Lemma 5.8 imply that the condition (c) of Proposition 5.6 cannot be satisfied. If the equivalent conditions of Proposition 5.2 are true, we get a contradiction with Lemma 5.9. Therefore (a) implies (b).

(b) implies (a) is obvious.

(b) and (c) are equivalent by Proposition 5.4. □

By [**4**, Lemma 7.17], the Jacobson radical of a ring with right Krull dimension does not contain nonzero idempotent ideals. Therefore

COROLLARY 5.11. *Let R be a non-singular serial ring with right Krull dimension. Then every pure projective R-module is a direct sum of finitely presented modules.*

EXAMPLE 5.12. Recall Facchini's example [**4**, Example 9.20]: Let p, q be different primes, and let $\mathbb{Z}_{(p)}, \mathbb{Z}_{(q)}$ be the localizations of $\mathbb{Z}$ at these primes. Let R be a subring of $M_4(\mathbb{Q})$ consisting of matrices of the form

$$\begin{pmatrix} \mathbb{Z}_{(p)} & \mathbb{Z}_{(p)} & 0 & 0 \\ p\mathbb{Z}_{(p)} & \mathbb{Z}_{(p)} & 0 & 0 \\ \mathbb{Q} & \mathbb{Q} & \mathbb{Z}_{(q)} & \mathbb{Z}_{(q)} \\ \mathbb{Q} & \mathbb{Q} & q\mathbb{Z}_{(q)} & \mathbb{Z}_{(q)} \end{pmatrix}$$

Then R is a serial ring such that the Krull-Remak-Schmidt theorem does not hold in mod-R. Observe that R is non-singular and, by [**4**, Proposition 7.29], R has right Krull dimension 1. Therefore every pure projective R-module is a direct sum of finitely presented modules.

References

[1] B. Amini, A. Amini, and A. Facchini, *Equivalence of diagonal matrices over local rings*, J. Algebra **320** (2008), no. 3, 1288–1310, DOI 10.1016/j.jalgebra.2008.04.008. MR2427644

[2] Ju. A. Drozd, *Generally uniserial rings* (Russian), Mat. Zametki **18** (1975), no. 5, 705–710. MR0404325

[3] A. Facchini, *Krull-Schmidt fails for serial modules*, Trans. Amer. Math. Soc. **348** (1996), no. 11, 4561–4575, DOI 10.1090/S0002-9947-96-01740-0. MR1376546

[4] A. Facchini, *Module theory: Endomorphism rings and direct sum decompositions in some classes of modules*, Progress in Mathematics, vol. 167, Birkhäuser Verlag, Basel, 1998. MR1634015

[5] A. Facchini and A. Moradzadeh-Dehkordi, *Rings over which every RD-projective module is a direct sums of cyclically presented modules*, J. Algebra **401** (2014), 179–200, DOI 10.1016/j.jalgebra.2013.11.018. MR3151254

[6] A. Facchini and P. Příhoda, *Monogeny dimension relative to a fixed uniform module*, J. Pure Appl. Algebra **212** (2008), no. 9, 2092–2104, DOI 10.1016/j.jpaa.2007.11.014. MR2422193

[7] A. Facchini and P. Příhoda, *Representations of the category of serial modules of finite Goldie dimension*, Models, modules and abelian groups, Walter de Gruyter, Berlin, 2008, pp. 463–486. MR2513260

[8] A. Facchini and P. Příhoda, *Factor categories and infinite direct sums*, Int. Electron. J. Algebra **5** (2009), 135–168. MR2471385

[9] G. Puninski, *Serial rings*, Kluwer Academic Publishers, Dordrecht, 2001. MR1855271

[10] G. Puninski, *Some model theory over a nearly simple uniserial domain and decompositions of serial modules*, J. Pure Appl. Algebra **163** (2001), no. 3, 319–337, DOI 10.1016/S0022-4049(00)00140-7. MR1852123

[11] P. Příhoda, *On uniserial modules that are not quasi-small*, J. Algebra **299** (2006), no. 1, 329–343, DOI 10.1016/j.jalgebra.2005.11.010. MR2225779

[12] P. Příhoda, Add(U) *of a uniserial module*, Comment. Math. Univ. Carolin. **47** (2006), no. 3, 391–398. MR2281002

[13] P. Příhoda and G. Puninski, *Pure projective modules over chain domains with Krull dimension*, J. Algebra **459** (2016), 189–212, DOI 10.1016/j.jalgebra.2016.04.010. MR3503971

[14] R. B. Warfield Jr., *Serial rings and finitely presented modules*, J. Algebra **37** (1975), no. 2, 187–222, DOI 10.1016/0021-8693(75)90074-5. MR0401836

Charles University, Faculty of Mathematics and Physics, Department of Algebra, Sokolovská 83, 18675 Prague 8, Czech Republic

Email address: prihoda@karlin.mff.cuni.cz

Contemporary Mathematics
Volume **730**, 2019
https://doi.org/10.1090/conm/730/14716

Mittag-Leffler modules and definable subcategories

Philipp Rothmaler

To Mike

ABSTRACT. We study (relative) $\mathcal{K}$-Mittag-Leffler modules as was done in the author's habilitation thesis, rephrase old, unpublished results in terms of definable subcategories, and present newer ones, culminating in a characterization of countably generated $\mathcal{K}$-Mittag-Leffler modules.

1. Introduction

We call a module M *countably pure-separable by pure-projectives* if every countable subset of M is contained in a pure submodule of M that is itself pure-projective. (A module is said to be *pure-projective* if it is projective with respect to pure-exact sequences, which is equivalent to being a direct summand of a direct sum of finitely presented modules.) More than four decades ago, Raynaud and Gruson showed:

(1) *A module is Mittag-Leffler iff it is countably pure-separable by (in fact, countably generated) pure-projectives.*

One may take this as a definition. It implies that a countably generated module is Mittag-Leffler if and only if it is pure-projective. So 'Mittag-Leffler' is a generalization of 'pure-projective' (and 'flat Mittag-Leffler' of 'projective'). The author, more than two decades ago, proved:

(2) *A module is Mittag-Leffler iff it is positively atomic,*

in the sense that the positive primitive (henceforth *pp*) type of every finite tuple in M be finitely generated. This model-theoretic condition—to be explained below—makes many of the algebraic properties of Mittag-Leffler modules very transparent and easy to prove.

For instance, since the pp type of a tuple from M does not change when passing to a pure submodule containing it or when passing to a pure supermodule of M, one sees at once:

(3) *The union of a pure chain of Mittag-Leffler modules is Mittag-Leffler.*

Results (1) and (3) imply:

(4) *The countably generated (automatically pure-projective) pure submodules of a Mittag-Leffler module form an $\aleph_1$-dense system,*

2010 *Mathematics Subject Classification.* 16D40, 16B70, 16D80, 16S90.
Supported in part by PSC CUNY Award 64595-00 42.

in the sense of [**HT**, Def. 2.4]:

(5) *An $\aleph_1$-dense system of a module M is an upward directed system $\mathcal{C}$ of submodules of M satisfying*
   (a) *every countable subset of M is contained in a member of $\mathcal{C}$;*
   (b) *$\mathcal{C}$ is closed under unions of countable chains.*

Note, this definition is a slight weakening of [**EM**, Def.IV.1.1], where, as an additional condition, every member of $\mathcal{C}$ is required to be countably generated (in which case directedness of $\mathcal{C}$ is entailed by (a)).

A Mittag-Leffler module M trivially has $\mathcal{C} = \{M\}$ as an $\aleph_1$-dense system of Mittag-Leffler submodules, but the one given in (4) is much more meaningful and, by (2), the converse of (4) is true as well:

(6) *A module M is a Mittag-Leffler module iff it has an $\aleph_1$-dense system of Mittag-Leffler modules each of which is* pure *in M iff it has an $\aleph_1$-dense system of (countably generated) pure-projective modules each of which is* pure *in M.*

Next we see how to drop 'pure' in (6). In view of (2), a straightforward argument about finite generation of types in unions of directed systems of submodules yields:

(7) *Suppose M is the union of a directed system of submodules, $\mathcal{C}$, all of whose members are Mittag-Leffler. Then M is Mittag-Leffler iff every finite tuple $\overline{m}$ of M is contained in a member $C_{\overline{m}}$ of $\mathcal{C}$ such that, for every $C \in \mathcal{C}$ containing $C_{\overline{m}}$, the pp type of $\overline{m}$ is the same whether considered in $C_{\overline{m}}$ or in C.*

This condition requires precisely what is needed to make up for the dropped purity condition on the members of $\mathcal{C}$. Namely, it says that if a finite system of linear equations, given in matrix form as $\mathfrak{B}\overline{y} = \mathfrak{A}\overline{m}$, has a solution in M (that is, in some $C \supseteq C_{\overline{m}}$), it already has one in $C_{\overline{m}}$. (A more general statement for direct limits, but essentially the same simple argument, has been made explicit in [**PR**, Lemma 3.6], see Lemma 3.8 below.)

If such a union $M = \bigcup \mathcal{C}$ is not Mittag-Leffler, there must be, by (7), a tuple $\overline{m}$ in M and an ω-chain $C_0 \subset C_1 \subset C_2 \subset \ldots$ in $\mathcal{C}$ such that the pp type of $\overline{m}$ keeps increasing (i.e., more and more equations $\mathfrak{B}\overline{y} = \mathfrak{A}\overline{m}$ have a solution) when going up the chain. Then, by (7) applied to the directed system of the C_i and their union M', the module M' is not Mittag-Leffler, that is:

(8) *If $\mathcal{C}$ is a directed system of submodules of a module, then its union is Mittag-Leffler provided the union of every ω-chain in $\mathcal{C}$ is.*

This yields the missing direction of the strengthening of (6) where purity is no longer required, see [**HT**, Cor.2.6] (whose proof is purely algebraic):

(9) *A module is Mittag-Leffler iff it has an $\aleph_1$-dense system of Mittag-Leffler modules iff it has an $\aleph_1$-dense system of (countably generated) pure-projective modules.*

The flat case. One can specialize the above to flat Mittag-Leffler modules by replacing 'pure-projective' by 'projective' and 'finitely presented' by 'finitely generated projective' (see [**R**] for this), because *flat + pure-projective = projective* and hence *flat + finitely presented = finitely generated projective*. In a 'pure situation' this is immediate. For dense systems of not necessarily pure submodules, one may either repeat the above argument under the extra assumption of flatness, or, more elegantly, simply apply the following to the corresponding statements above.

(10) *A module is flat if and only if it has an $\aleph_0$-dense system of flat submodules.*

Here, naturally, $\aleph_0$-dense systems are defined like $\aleph_1$-dense systems in (5) above with 'countable' replaced by 'finite.' Then condition (*b*) becomes vacuously true and so an $\aleph_0$-dense system in a module M is simply an upward directed collection $\mathcal{C}$ of submodules of M such that every finite subset of M is contained in a member of $\mathcal{C}$. (Again, one direction is trivial: take $\mathcal{C} = \{M\}$.)

Either way, one deduces at once the following flat versions of (9), which was given an algebraic proof in [**HT**, Thm.2.9]:

(11) *A module is flat and Mittag-Leffler iff it has an $\aleph_1$-dense system of flat Mittag-Leffler modules iff it has an $\aleph_1$-dense system of (countably generated) projective modules.*

Contents. In this paper, all of the above, except the previously mentioned flat case, will be done for relativized Mittag-Leffler modules as introduced and for the first time systematically studied in [**R**], and continued in [**AH**], [**HT**] and others. The concordance for the relativized versions is: (1) is Cor. 3.5(1) and Cor. 6.5, (2) is Thm. 3.1, (3) is (part of) Cor. 3.5(1), (7) is Lemma 3.8, (8) is Cor. 3.11, and (9) is Cor. 3.12 and Prop. 6.4.

In Section 3, $\mathcal{K}$-Mittag-Leffler modules are introduced and their main properties developed. The classical case is that of $\mathcal{K}$ = Mod-R. This and two others, $\mathcal{K}$, the class $\flat_R$ of flat right R-modules, and $\mathcal{K}$, the class $\sharp_R$ of absolutely pure right R-modules, are considered in Section 4. The fundamental countable separation result (1) (and thus the left-to-right direction of (9)) above is proved for arbitrary $\mathcal{K}$-Mittag-Leffler modules in Section 6, see Prop. 6.4 and Cor. 6.5. For this, one needs to relativize purity as well, which is done in the first part of Section 5, the rest of which is preparation for the final Section 7, where countably generated $\mathcal{K}$-Mittag-Leffler modules are characterized.

A general remark. The point of the model-theoretic perspective taken here and in the previous work [**R**] is not to make possible all sorts of relativizations—it has been subsequently shown, for instance in [**AH**], that this can be done completely algebraically. The point rather is, and this applies to Mittag-Leffler modules in general—relativized or not—that this perspective reveals some features of the concept that have gone unnoticed previously. Take, for instance, the ease with which direct limits of Mittag-Leffler modules can be treated, as in (8) above (and Section 3.4 below for relativizations), or the role of definable subcategories as revealed by the relativized version of (2), Theorem 3.1 below (and Sections 3.3 and 3.6), which showed—if stated in different terms—that the very concept of $\mathcal{K}$-Mittag-Leffler module depended only on the definable subcategory $\mathcal{K}$ generates.

2. Preliminaries

Most of the preliminaries can be found in Mike Prest's two monographs, [**P1**] and [**P2**], where the latter is closer in style to current usage. I will largely use its notation—with a few exceptions that are rather self-explanatory. [**R**] may also serve as an introduction with all necessary detail.

I freely use the fundamental tool of *pp formulas* (called *pp condition* in [**P2**])[1] and the lattice order $\leq$ between them, often adorned by a class $\mathcal{L} \subseteq R$-Mod.

2.1. Modules and tuples. Module means *left module* (unless otherwise indicated) over an associative ring R with 1.

A *tuple* is a finite sequence. Given an i-tuple $\overline{r} = (r_0, \dots, r_{i-1})$ of ring elements and an i-tuple $\overline{m} = (m_0, \dots, m_{i-1})$ of elements of a module M, I use $\overline{r}\,\overline{m}$ to denote the corresponding linear combination, i.e., $\overline{r}\,\overline{m} = \sum_{j<i} r_i m_i$. In doing so I assume that those tuples match and that the tuple $\overline{m}$ is tacitly transposed to a column vector. (This is for left modules. For right modules it would, naturally, be the other way around.)

It is convenient to expand (this is the syntactic term) a module M by a tuple $\overline{m}$ from M. The resulting object is denoted $(M, \overline{m})$, an *i-pointed module.* Morphisms in the pointed category are supposed to send each entry of a (source) tuple to the corresponding entry of the (target) tuple.

I use $\langle\,\rangle$ for generation of various objects. If $C \subseteq M$, then , or more precisely, $\langle C \rangle_M$, denotes the submodule of M generated by C. In the lattices of pp formulas (see below), specifically, for finite generation of types, these symbols are used thus: $\langle \varphi \rangle_{\mathcal{L}} = \{\psi \mid \varphi \leq_{\mathcal{L}} \psi\}$. For a class $\mathcal{K}$ of modules, $\langle \mathcal{K} \rangle$ denotes the definable subcategory generated by $\mathcal{K}$, see Section 2.4 below.

Add($\mathcal{K}$) (resp., add($\mathcal{K}$)) denotes the smallest class of modules containing $\mathcal{K}$ that is closed under direct summands and (resp., finite) direct sum. Zg_R stands for the Ziegler spectrum of R-Mod, that is the space of indecomposable pure-injective left R-modules as introduced in [**Zie**], cf. [**P1**] or [**P2**].

2.2. Formulas and types. The atomic formulas (in the common first-order language of R-modules where the scalars figure as unary function symbols) are precisely the linear equations over R. But for us the more important syntactic object are the finite conjunctions of atomic formulas, which are also known as *quantifier-free pp formulas*, the term I use. These are the finite (homogeneous) systems of linear equations (over the ring). We write them in matrix form with two kinds of variables, $\mathfrak{A}\overline{x} \doteq \mathfrak{B}\overline{y}$, where $\mathfrak{A}$ and $\mathfrak{B}$ are matching matrices over R. Given such a quantifier-free pp formula $\alpha = \alpha(\overline{x}, \overline{y})$, we obtain a typical *pp formula* by existentially quantifying out $\overline{y}$, that is, $\exists\overline{y}\alpha$, or, more explicitly, $\exists\overline{y}\alpha(\overline{x}, \overline{y})$, which we also write as $\mathfrak{B}|\mathfrak{A}\overline{x}$.

Denote this formula by φ or $\varphi(\overline{x})$. In any given module M it defines the set, $\varphi(M)$, of all $l(\overline{x})$-tuples $\overline{a}$ for which the inhomogeneous system $\alpha(\overline{a}, \overline{y})$ has a solution in M. Obviously, $\varphi(M)$ forms an additive group, in fact, a subgroup of $M^{l(\overline{x})}$. It is called a(n *$l(\overline{x})$-place*) *pp subgroup* of M. Clearly, $\varphi(M)$ is the projection of $\alpha(M)$ onto the first $l(\overline{x})$ coordinates.

It is easily verified that the n-place pp subgroups of M form a sublattice, denoted by $\Lambda^n(M)$, of the subgroup lattice of M^n.

In fact, the n-place pp formulas themselves form a lattice, whose meet is conjunction $\wedge$ and whose join is $+$. We denote this lattice by ${}_R\Lambda^n$ (and for right R-modules by Λ^n_R). It turns into $\Lambda^n(M)$ when specialized at M. When we specialize it at a class $\mathcal{L}$, we obtain a lattice whose order plays a major role in this work and is discussed in the next subsection.

[1]One may work with Zimmermann's (equivalent) concept of *finite matrix group* and the correponding *p-functors* or the subgroups of finite definition of Gruson and Jensen [**GJ**] instead.

A *pp n-type* is a subset of ${}_R\Lambda^n$. Of particular importance are those that form a filter in ${}_R\Lambda^n$, as the *pp type* of a given n-tuple $\overline{a}$ in M, denoted $\mathrm{pp}_M(\overline{a})$, which is, by definition, the collection of all n-place pp formulas $\overline{a}$ satisfies in M. The *quantifier-free pp type* of $\overline{a}$ in M, denoted $\mathrm{qf}_M(\overline{a})$, is the subset consisting of all quantifier-free pp formulas in $\mathrm{pp}_M(\overline{a})$.

It is easily seen that every pp formula (and thus every pp type) constitutes a subfunctor of the forgetful functor $R\text{-Mod} \longrightarrow Ab$, from left R-modules to abelian groups.

While often times the actual lengths of tuples are irrelevant, care has to be taken, however, that they match. It is assumed that they (as well as sizes of corresponding matrices) *always* do.

2.3. The lattice order. Given pp formulas φ and ψ, I use $\varphi \leq_{\mathcal{L}} \psi$ to mean that the sentence $\forall\overline{x}(\varphi(\overline{x}) \rightarrow \psi(\overline{x}))$ is true in all members of $\mathcal{L}$. We say ψ is *$\mathcal{L}$-implied* by φ. When the class $\mathcal{L}$ is a singleton L, I simple write $\varphi \leq_L \psi$ for $\varphi \leq_{\{L\}} \psi$. When $\mathcal{L} = R\text{-Mod}$, the subscript will be dropped.

Equivalence is denoted by $\sim$ (which stands for '$\leq$ *and* $\geq$'), adorned or unadorned. So to be *$\mathcal{L}$-equivalent*, i.e., $\varphi \sim_{\mathcal{L}} \psi$, means that φ and ψ are equivalent everywhere in $\mathcal{L}$ (hence everywhere in $\langle\mathcal{L}\rangle$, see Section 2.4).

All this extends to types in an obvious fashion. Given two pp-n-types p and q, I say q is *$\mathcal{L}$-implied* by p and write $p \leq_{\mathcal{L}} q$ if, in every member L of $\mathcal{L}$, every realization (= solution) of p realizes q as well. (That is, $p \leq_{\mathcal{L}} q$ really means $\bigwedge p \leq_{\mathcal{L}} \bigwedge q$.) Note, if $p = \mathrm{pp}_L(\overline{a})$ with $L \in \mathcal{L}$, then $p \leq_{\mathcal{L}} q$ if and only if $q \subseteq p$ (iff $p \leq q$), but if $p = \mathrm{pp}_M(\overline{a})$ with M *not* in $\mathcal{L}$, no such relationship is entailed in general. The proper context for this relation between types is therefore that of the definable subcategory generated by $\mathcal{L}$.

2.4. Definable subcategories. A *definable subcategory* of left R-modules is a full subcategory of R-Mod whose object class is the model class of a set of pp implications, i.e., of statements of the form $\forall\overline{x}(\varphi(\overline{x}) \rightarrow \psi(\overline{x}))$ where φ and ψ are pp formulas whose free variables are among those of $\overline{x}$. This is equivalent to being closed under direct product, direct limit and pure submodule, cf. [**Rot2**] or [**P2**, Thm.3.4.7].

It is clear from the definition that every class $\mathcal{L}$ of left R-modules is contained in a smallest definable subcategory (namely the intersection of all of those that contain $\mathcal{L}$), which is denoted by $\langle\mathcal{L}\rangle$ and called the *definable subcategory generated by $\mathcal{L}$*. Here is a concrete way to obtain it: if $\mathrm{T}_{\mathcal{L}}$ denotes the set of axioms $\{\forall\overline{x}(\varphi(\overline{x}) \rightarrow \psi(\overline{x})) \mid \varphi \leq_{\mathcal{L}} \psi\}$ (together with the axioms for left R-modules), $\langle\mathcal{L}\rangle$ is the class of models of $\mathrm{T}_{\mathcal{L}}$, i.e., the class of modules for which all of those implications hold. That is, $C \in \langle\mathcal{L}\rangle$ iff $\varphi \leq_{\mathcal{L}} \psi$ implies $\varphi \leq_C \psi$.

Note, two classes $\mathcal{L}$ and $\mathcal{L}'$ generate the same definable subcategory if and only if the corresponding lattice orderings on pp formulas, $\leq_{\mathcal{L}}$ and $\leq_{\mathcal{L}'}$, are the same. In particular, $\leq_{\mathcal{L}}$ and $\leq_{\langle\mathcal{L}\rangle}$ are the same—at least on formulas.

As a special case we have that $\leq_{R\text{-Mod}}$, $\leq_{R\text{-mod}}$ and $\leq_{{}_R\mathrm{Zg}}$ (with ${}_R\mathrm{Zg}$ the left Ziegler spectrum of R) are the same. More cases are listed in Corollary 3.6 below.

The lattice orderings of $\mathcal{L}$-implication between types as introduced above behave better when the context indicated by the subscript is a definable subcategory (as then compactness of first order logic is available).

Remark 2.1. Let p and q be pp types of the same arity with p closed under finite conjunction. Then $p \leq_{\langle \mathcal{L} \rangle} q$ if and only if for all $\psi \in q$ there is $\varphi \in p$ such that $\varphi \leq_{\mathcal{L}} \psi$.

First of all, $p \leq_{\mathcal{L}} q$ if and only if $p \leq_{\mathcal{L}} \psi$ for all $\psi \in q$ (and this is true for *any* $\mathcal{L}$). But as $\langle \mathcal{L} \rangle$ is first order axiomatizable, the statement $p \leq_{\langle \mathcal{L} \rangle} \psi$ is equivalent to the inconsistency of $p \cup \{\neg\psi\}$ (with $\mathrm{T}_{\mathcal{L}}$). By compactness, we can replace p by a finite subset. Finite conjunctions of pp formulas being pp, a single φ from p will do. But inconsistency of $\{\varphi \wedge \neg\psi\}$ with $\mathrm{T}_{\mathcal{L}}$ means $\varphi \leq_{\mathcal{L}} \psi$.

Remark 2.2. This may be a good opportunity to emphasize why sometimes $\langle \mathcal{L} \rangle$ and sometimes only $\mathcal{L}$. In order to apply compactness to a statement $p \leq_{\mathcal{X}} \varphi$ when p is an infinite type, we need $\mathcal{X}$ to be axiomatizable (so we take $\mathcal{X} = \langle \mathcal{L} \rangle$). In contrast, for $\varphi \leq_{\mathcal{X}} q$ it doesn't matter, because all that says is that $\varphi \leq_{\mathcal{X}} \psi$ for every $\psi \in q$, so we can go interchangeably with $\mathcal{X} = \langle \mathcal{L} \rangle$ or $\mathcal{X} = \mathcal{L}$.

Remark 2.3.

(1) [**Rot2**, Fact 7.7.] about pure embeddability of any structure from $\langle \mathcal{K} \rangle$ in a reduced product of structures from $\mathcal{K}$ shows that in order to generate the definable subcategory, one may first take all direct products, then all direct limits, and finally all pure substructures; see also the note before [**P2**, Prop.3.4.9]. This is true for any first-order structures, not just modules, cf. [**Rot2**].

(2) It is easy to check that for $\mathcal{K} \subseteq \mathcal{K}' \subseteq \mathrm{Add}(\mathcal{K})$, the preorders $\leq_{\mathcal{K}}$ and $\leq_{\mathcal{K}'}$ are the same, hence $\mathcal{K}$, $\mathcal{K}'$, and $\mathrm{Add}(\mathcal{K})$ generate the same definable subcategories.

2.5. Elementary duality. *Elementary duality* as introduced by Prest and further developed by Herzog is a major tool to switch sides. Namely, for all n, this duality, denoted by D, constitutes an anti-isomorphism between the lattices ${}_R\Lambda^n$ and Λ^n_R, see [**P1**], [**P2**], [**Her**], [**Z-HZ**], or [**R**] for an introduction.

Elementary duality turns the relation $\leq_{\mathcal{K}}$ upside down in $\mathrm{D}\mathcal{K}$, the class of all character duals of members of $\mathcal{K}$ (here we use character modules obtained by hom-ing into $\mathbb{Q}/\mathbb{Z}$). More precisely, $\varphi \leq_{\mathcal{K}} \psi$ if and only if $\mathrm{D}\psi \leq_{\mathrm{D}\mathcal{K}} \mathrm{D}\varphi$.

The use of D is illustrated by the following elegant criterion for a tensor to be zero. Note that we take the same liberty as in linear combinations and write $\overline{n} \otimes \overline{m}$ instead of $\sum_{j<i} n_i \otimes m_i$ (assuming, as always, that the tuples match).

Fact 2.4 (Herzog's Criterion). Given a k-tuple $\overline{n}$ in a right R-module N and a k-tuple $\overline{m}$ (of the same length) in a left R-module M, the following are equivalent.

(i) $\overline{n} \otimes \overline{m} = 0$
(ii) There is a (left) k-place pp formula φ such that $\overline{n} \in \mathrm{D}\varphi(\mathrm{N})$ and $\overline{m} \in \varphi(M)$.
(iii) There is a (right) k-place pp formula φ such that $\overline{n} \in \varphi(N)$ and $\overline{m} \in \mathrm{D}\varphi(\mathrm{M})$.

Definition 2.5. By *definably dual classes* we mean classes $\mathcal{K} \subseteq \text{Mod-}R$ and $\mathcal{L} \subseteq R\text{-Mod}$ such that $\mathrm{D}\langle \mathcal{K} \rangle = \langle \mathcal{L} \rangle$.

Since elementary duality goes both ways and is an involution, this definition is completely symmetric. That is, $\mathcal{K}$ and $\mathcal{L}$ are definably dual iff $\langle \mathcal{K} \rangle = \mathrm{D}\langle \mathcal{L} \rangle$.

COROLLARY 2.6. *Two families, $\mathcal{K}$ and $\mathcal{L}$, of right, resp. left R-modules are definably dual, provided any of the following holds.*

(a) $\varphi \leq_{\mathcal{L}} \psi$ *iff* $\mathrm{D}\psi \leq_{\mathcal{K}} \mathrm{D}\varphi$, *for all left pp formulas* φ *and* ψ.
(b) $\mathcal{L}$ *consists of all character duals of members of* $\mathcal{K}$.
(c) *Any of the corresponding symmetric statements.*

CONVENTION 2.7. Unless indicated otherwise, $\mathcal{K}$ and $\mathcal{L}$ are **definably dual** classes of R-modules in the sense of the above definition.

Prominent examples of this situation are the flat right R-modules $\mathcal{K} = \flat_R$ and the absolutely pure left R-modules $\mathcal{L} = {}_R\sharp$ [**P2**, Prop.3.4.26], and the symmetric situation $\mathcal{K} = \sharp_R$ and $\mathcal{L} = {}_R\flat$, considered in Sections 4.2 and 4.3 respectively. See [**Rot3**] for other examples.

2.6. Finite generation of types. Following [**P1**], a pp type p is called *finitely generated* if there is a pp formula φ with $\langle\varphi\rangle = p$. I emphasize that this means that $\varphi \leq p$ *and* $\varphi \in p$. In [**R**, Sect.2.1] this was relativized to any context $\mathcal{L} \subseteq R$-Mod. Since there is a subtlety with partial types here (even when the context $\mathcal{L}$ is all of R-Mod), let me be very precise. Consider p, the pp type of a tuple $\overline{a}$ in a module M, i.e., $p = \mathrm{pp}_M(\overline{a})$. A subtype $q \subseteq p$ is said to be *$\mathcal{L}$-finitely generated* if q is $\mathcal{L}$-implied by some $\varphi \in p$, i.e., when there is $\varphi \in p$ such that $q \subseteq \langle\varphi\rangle_{\mathcal{L}}$, where $\langle\varphi\rangle_{\mathcal{L}} = \{\psi \mid \varphi \leq_{\mathcal{L}} \psi\}$.

Note, p is $\mathcal{L}$-finitely generated iff it is $\mathcal{L}$-implied by some formula *it contains.* No such assumption is made if q is partial. The 'generator' is only required to be in p, not in q. This will be of importance below where we will encounter finitely generated quantifier-free types, i.e., the case where $q = \mathrm{qf}_M(\overline{a})$. According to the definition just made, such q is said to be $\mathcal{L}$-finitely generated when there is a pp formula true of $\overline{a}$—quantifier-free or not!—that $\mathcal{L}$-implies q. Certainly q is so generated when p is. But we will see cases where q is finitely generated, while p is not.

Consider types $p = \mathrm{pp}_M(\overline{a})$ and $q = \mathrm{pp}_N(\overline{b})$ with $p \leq_{\mathcal{L}} q$. If p is $\mathcal{L}$-implied by a single formula φ, so is q. But φ may not be in q and thus q may not be $\mathcal{L}$-finitely generated in the sense of the definition above.

However, if the types in question are $\mathcal{L}$-equivalent, $p \sim_{\mathcal{L}} q$, *and* $\mathcal{L} = \langle\mathcal{L}\rangle$, then, by compactness, every formula in p is $\mathcal{L}$-implied by some formula in q (see Rem. 2.1 above). Hence, if p is $\mathcal{L}$–implied by $\varphi \in p$, this formula must be $\mathcal{L}$-implied by a formula $\psi \in q$, which then $\mathcal{L}$-implies all of q, making q in turn $\mathcal{L}$-finitely generated as desired. Let us summarize.

REMARK 2.8. Let $p = \mathrm{pp}_M(\overline{a})$ and $q = \mathrm{pp}_N(\overline{b})$ be $\langle\mathcal{L}\rangle$-equivalent. If p is $\mathcal{L}$-finitely generated, then so is q.

I close this section with the simple fact (known at least since [**R**, Rem.1.6]) that to check $\mathcal{L}$-finite generation of types in a finitely generated module it suffices to consider a generating tuple (cf. Lemma 5.5 for a similar phenomenon).

LEMMA 2.9. *If the pp type of some generating tuple of a finitely generated module is $\mathcal{L}$-finitely generated, then the pp type of every other tuple in that module is $\mathcal{L}$-finitely generated as well.*

PROOF. Suppose A is generated by $\overline{a}$ and $\mathrm{pp}_A(\overline{a})$ is $\mathcal{L}$-generated by $\varphi \in \mathrm{pp}_A(\overline{a})$. Any other tuple $\overline{b}$ in A can be written as $\mathfrak{B}\overline{a}$ for some matrix $\mathfrak{B}$ over the ring. Let

$\varphi_{\mathfrak{B}}$ be the pp formula $\exists\overline{x}(\overline{z} = \mathfrak{B}\overline{x} \wedge \varphi(\overline{x}))$ (in the free variables $\overline{z}$). Clearly, this formula is in $\mathrm{pp}_A(\overline{b})$, and I claim it $\mathcal{L}$-generates this type. Indeed, for any other $\psi = \psi(\overline{z}) \in \mathrm{pp}_A(\overline{b})$, the formula $\psi(\mathfrak{B}\overline{x})$ is in $\mathrm{pp}_A(\overline{a})$, hence $\mathcal{L}$-implied by φ. It is, finally, straightforward to verify that $\varphi \leq_{\mathcal{L}} \psi(\mathfrak{B}\overline{x})$ entails $\varphi_{\mathfrak{B}} \leq_{\mathcal{L}} \psi$. □

3. Mittag-Leffler modules

Here we state the main theorem of [**R**]—and draw some more or less immediate consequences—about relative Mittag-Leffler modules. It is the basis for everything in this and other work of mine on Mittag-Leffler modules, yet it has, though widely circulated, never been published other than in the thesis [**R**, Thm.2.2]. (It appeared in the classical case (with no relativization) as the main theorem of [**Rot1**], cf. [**P2**, Thm.1.3.22].)

DEFINITION 3.1. [**R**, Ch. 2]

(1) Given a family $\mathcal{K}$ of right R-modules, a *$\mathcal{K}$-Mittag-Leffler module* is a left R-module M such that the canonical map
$$\tau^I : (\prod_I N_i) \otimes M \longrightarrow \prod_I (N_i \otimes M)$$
is monic for all subfamilies $\{N_i : i \in I\}$ of $\mathcal{K}$.

(2) A right R-module N is called a *test module* for $\mathcal{K}$-Mittag-Leffler if, in the previous definition, it suffices to let all the N_i be equal to N. We then write τ_N^I for the transformation $N^I \otimes - \longrightarrow (N \otimes -)^I$.

(3) Given a test module N as before, a *test transformation* for $\mathcal{K}$-Mittag-Leffler is a transformation τ_N^I as before for a particular set I that suffices to test the $\mathcal{K}$-Mittag-Leffler condition above.

(4) We omit '$\mathcal{K}$-' if $\mathcal{K} = \text{Mod-}R$ and write 'N-' if $\mathcal{K} = \{N\}$.

The two cases of $\mathcal{K}$ that led to this definition in [**R**] were R-Mod, the case of (classical) Mittag-Leffler modules as introduced in [**RG**], and $\{R_R\}$, which was investigated in [**Goo**].

REMARK 3.2. Clearly, a direct sum of modules is $\mathcal{K}$-Mittag Leffler if and only if every summand is. Since the canonical map $(\prod_I \prod_J N_{ij}) \otimes M \longrightarrow \prod_I \prod_J (N_{ij} \otimes M)$ factors through the canonical map $(\prod_I \prod_J N_{ij}) \otimes M \longrightarrow \prod_I (\prod_J N_{ij} \otimes M)$, we see that $\mathcal{K}$-Mittag-Leffler implies $\mathcal{K}'$-Mittag-Leffler whenever $\mathcal{K}'$ consists of products of members of $\mathcal{K}$.

3.1. Atomic modules. The main theorem below describes $\mathcal{K}$-Mittag-Leffler modules in terms of the following concept, which is a variant of a classical notion from first-order model theory.

DEFINITION 3.3. Suppose $\mathcal{L}$ is a family of left R-modules. An *$\mathcal{L}$-atomic*[2] module is a left R-module M such that the pp type of every tuple in M is $\mathcal{L}$-finitely generated.

In case $\mathcal{L}$ is the class of all R-modules (or all finitely presented ones), this is what was called *positively atomic* in [**R**] and [**Rot1**].

[2] Be aware of different usage of this term in [**R**, §3.3].

The definition can be reformulated in terms of linear algebra as follows. For every tuple $\overline{a}$ in M there is a tuple $\overline{b}$ in M and a finite system of linear equations, $\alpha(\overline{x}, \overline{y})$ or $\mathfrak{A}\overline{x} \doteq \mathfrak{B}\overline{y}$, with solution $(\overline{a}, \overline{b})$ and such that

if $\mathfrak{C}\overline{x} \doteq \mathfrak{D}\overline{z}$ is a finite system of linear equations whose inhomogeneous instance $\mathfrak{C}\overline{a} \doteq \mathfrak{D}\overline{z}$ has a solution in M, then for every $L \in \mathcal{L}$ and every matching tuple $\overline{c}$ in L, if the inhomogeneous system $\mathfrak{A}\overline{c} \doteq \mathfrak{B}\overline{y}$ has a solution in L, so does the inhomogeneous system $\mathfrak{C}\overline{c} \doteq \mathfrak{D}\overline{z}$.

REMARK 3.4. Lemma 2.9 above says that a finitely generated module is $\mathcal{L}$-atomic if and only if the pp type of some *generating* tuple is $\mathcal{L}$-finitely generated.

Clearly, $\mathcal{L}$-atomicity depends only on $\leq_{\mathcal{L}}$, so if this relation is the same as $\leq_{\mathcal{L}'}$, then $\mathcal{L}$-atomic is the same as $\mathcal{L}'$-atomic. (This shows that the concept can be extended to theories, as was done in [**R**]).

3.2. The main theorem.

THEOREM 3.1. [**R**, Thm.2.2] *Let $\mathcal{K}$ and $\mathcal{L}$ be definably dual classes of right and left R-modules (i.e., such that $\varphi \leq_{\mathcal{L}} \psi$ iff $\mathrm{D}\psi \leq_{\mathcal{K}} \mathrm{D}\varphi$, for all left pp formulas φ and ψ).*

Then a module is $\mathcal{K}$-Mittag-Leffler if and only if it is $\mathcal{L}$-atomic.

PROOF. The case $\mathcal{K} = R$-Mod is contained in [**Rot1**, Thm.2.2] and in [**P2**, Thm.1.3.22]. The proof of the general case is almost verbatim the same (and carried out in detail in [**R**]). The main adjustment to be made is to replace every unrelativized occurrence of an implication $\leq$ between left pp formulas by the relativized implication $\leq_{\mathcal{L}}$ and every implication between right pp formulas by the relativized implication $\leq_{\mathcal{K}}$. □

Because of space concerns I state the following more or less immediate consequences without proof.

COROLLARY 3.5. [**R**, Cor.2.4]

(1) *A module M is $\mathcal{K}$-Mittag-Leffler iff every finite subset of M is contained in a pure submodule of M which is $\mathcal{K}$-Mittag-Leffler. In particular, the class of $\mathcal{K}$-Mittag-Leffler modules is closed under taking unions of pure chains.*
(2) *The class of $\mathcal{K}$-Mittag-Leffler modules is closed under pure submodules and pure extensions.*
(3) *A direct sum is $\mathcal{K}$-Mittag-Leffler iff all summands are $\mathcal{K}$-Mittag-Leffler.*
(4) *A finitely generated module is finitely presented iff it is Mittag-Leffler.*
(5) *If N is a finitely generated pure submodule of a $\mathcal{K}$-Mittag-Leffler module M (hence N is finitely presented), then M/N is $\mathcal{K}$-Mittag-Leffler too.*

3.3. Definable subcategories. As was stressed in its original form—albeit in different terminology—one of the most powerful direct consequences of this theorem is that the concept of $\mathcal{K}$-Mittag-Leffler module is invariant under passing to the generated definable subcategory. Many a preservation result for classes $\mathcal{K}$ found in the literature follow directly from this. We make it explicit in (1) below. The other statements are special cases contained in one way or another in [**R**, Ch. 2], for (3), see [**R**, Rem.2.3(b)].

COROLLARY 3.6. *Let $\mathcal{K}$ and $\mathcal{K}'$ be families of right R-modules. Then a module is $\mathcal{K}$-Mittag-Leffler if and only if it is $\mathcal{K}'$-Mittag-Leffler, provided any of the following holds.*

(a) *$\mathcal{K}$ and $\mathcal{K}'$ generate the same definable subcategory.*
(b) $\mathrm{Add}(\mathcal{K}) = \mathrm{Add}(\mathcal{K}')$ *(or* $\mathrm{add}(\mathcal{K}) = \mathrm{add}(\mathcal{K}')$*).*
(c) *$\mathcal{K}$ and $\mathcal{K}'$ have the same associated Ziegler-closed set, i.e.,*

$$\mathrm{Zg}_R \cap \langle \mathcal{K} \rangle = \mathrm{Zg}_R \cap \langle \mathcal{K}' \rangle.$$

(d) *$\mathcal{K}'$ consists of all direct products of members of $\mathcal{K}$.*
(e) *Every member of $\mathcal{K}$ is elementarily equivalent to a member of $\mathcal{K}'$, and vice versa.* □

The converses do not hold. For example, over a left noetherian ring, every left module is $\flat$-Mittag-Leffler, cf. Cor. 4.3(2) below. Hence the latter is the same as $\{0\}$-Mittag-Leffler, while $\langle \flat \rangle$ and $\{0\}$ are two different definable subcategories.

Since the definable subcategory generated by R-mod is R-Mod, as a special case of (1) we get that R-mod-Mittag-Leffler modules are Mittag-Leffler [**Fac**, Lemma 3.1].

3.4. Direct limits. Consider an upward directed poset $I = (I, \leq)$ and an I-directed system $\mathcal{M} = \{M_i\}_I$ of modules with connecting maps $f_i^j : M_i \longrightarrow M_j$ for all $i \leq j$ in I (and $f_j^k f_i^j = f_i^k$ for $i \leq j \leq k$). Let M be its direct limit with canonical maps $f_i : M_i \longrightarrow M$ (such that $f_j f_i^j = f_i$ for $i \leq j$).

A *tail* in $\mathcal{M}$ is a set of the form $\{f_i^j(\overline{m}_i) | i \leq j \in I\}$, where $i \in I$ and $\overline{m}_i$ is a tuple in M_i. Clearly, the canonical maps f_j send each entry $\overline{m}_j = f_i^j(\overline{m}_i)$ in this tail to the same tuple, $\overline{m}$ say, in M. All entries in this tail being preimages of $\overline{m}$ under the corresponding canonical maps, we also call this tail a *tail of* $\overline{m}$. Clearly, any two, and hence any finitely many tails of a given tuple merge at some jth entry. So, as far as eventual behavior in tails is concerned, it makes no difference with which tail we start. Let's make this explicit.

REMARK 3.7. A tuple $\overline{m}$ in the limit M satisfies a pp formula φ if and only if a(ny) tail of $\overline{m}$ eventually does.

Suppose now the M_i are $\mathcal{K}$-Mittag-Leffler, hence $\mathcal{L}$-atomic. We want to know what makes the limit $\mathcal{K}$-Mittag-Leffler, i.e., $\mathcal{L}$-atomic, too. For this, as we will see, it suffices to consider countable chains in the direct system at hand.

By atomicity, the pp type of the jth entry $\overline{m}_j$ in a tail $\{\overline{m}_j | j \geq i\}$ is $\mathcal{L}$-generated by some pp formula φ_j. For every $k \geq j$, the kth entry is an image of the jth and thus satisfies φ_j. Therefore these formulas form a descending chain, $\varphi_i \geq_{\mathcal{L}} \varphi_j \geq_{\mathcal{L}} \varphi_k$ for all $i \leq j \leq k$ in I. And they are all contained in the pp type of the limit $\overline{m} = f_i(\overline{m}_i)$.

We say a tail as above *$\mathcal{L}$-stabilizes at* φ_j if $\varphi_k \sim_{\mathcal{L}} \varphi_j$ for all $k \geq j$.

LEMMA 3.8. (Essentially [**PR**, Lemma 3.6]) *Given a limit tuple $\overline{m}$ and a pp formula φ, the following are equivalent.*

(i) $\mathrm{pp}_M(\overline{m})$ *is $\mathcal{L}$-generated by φ.*
(ii) *A(ny) tail $\{f_i^j(\overline{m}_i) | i \leq j \in I\}$ of $\overline{m}$ $\mathcal{L}$-stabilizes at some φ_j (in which case $\varphi_j \sim_{\mathcal{L}} \varphi$ and hence $\varphi_k \sim_{\mathcal{L}} \varphi$ for all $k \geq j$).*

PROOF. Let $\{\overline{m}_j | j \geq i\}$ be a tail of the tuple $\overline{m}$.

(i) $\Rightarrow$ (ii): For $\overline{m}$ to satisfy φ in M, the tail must eventually satisfy it in the respective M_j's, Rem. 3.7. As φ_k $\mathcal{L}$-generates $\text{pp}_{M_k}(\overline{m}_k)$, this implies $\varphi_k \leq_{\mathcal{L}} \varphi$ (from some j on). But all these formulas are contained in $\text{pp}_M(\overline{m})$ and are therefore, conversely, $\mathcal{L}$-implied by φ, so the tail stabilizes at φ_j and $\varphi_k \sim_{\mathcal{L}} \varphi$ from j on.

(ii) $\Rightarrow$ (i). Suppose the tail stabilizes at φ_j. To verify that $\text{pp}_M(\overline{m})$ is $\mathcal{L}$-generated by φ_j, let $\psi \in \text{pp}_M(\overline{m})$. Then the tail eventually satisfies ψ. In particular, there is M_k with $k \geq j$ where $\overline{m}_k$ satisfies ψ. Then $\varphi_k \leq_{\mathcal{L}} \psi$ and so $\varphi_j \sim_{\mathcal{L}} \varphi_k$ $\mathcal{L}$-implies ψ as well. □

DEFINITION 3.9.

(1) By an *ω-limit* we mean a direct limit of a chain of order type ω.
(2) By an *ω-sublimit* of the system $\mathcal{M} = \{M_i, f_i^j\}_{i \leq j \in I}$ we mean an ω-limit of the form $M_{i_0} \xrightarrow{f_{i_0}^{i_1}} M_{i_1} \xrightarrow{f_{i_1}^{i_2}} M_{i_2} \xrightarrow{f_{i_2}^{i_3}} \dots \xrightarrow{f_{i_{k-1}}^{i_k}} M_{i_k} \xrightarrow{f_{i_k}^{i_{k+1}}} \dots$ $(k < \omega)$, where $i_0 < i_1 < i_2 < \dots < i_k < \dots$ $(k < \omega)$ in I.
(3) Call the directed system $\mathcal{M}$ (or its limit) *ω-closed* if every ω-sublimit (as a structure) is itself a member of $\mathcal{M}$.

PROPOSITION 3.10. [**HT**, Prop.2.2] *A direct limit of $\mathcal{K}$-Mittag-Leffler modules is $\mathcal{K}$-Mittag-Leffler provided all ω-sublimits are. Thus, an ω-closed direct limit of $\mathcal{K}$-Mittag-Leffler modules is $\mathcal{K}$-Mittag-Leffler.*

PROOF. In the notation introduced in the beginning of this subsection, assume the limit M is *not* $\mathcal{K}$-Mittag-Leffler. By Thm. 3.1, there's a tuple $\overline{m}$ in M whose pp type is not $\mathcal{L}$-finitely generated. Then, by the previous lemma, some (hence every) tail $\{f_i^j(\overline{m}_i) | i \leq j \in I\}$ of $\overline{m}$ does not stabilize. Hence there's a sequence $i \leq i_0 < i_1 < i_2 < \dots < i_k < \dots$ $(k < \omega)$ in I for which the pp formulas corresponding to this subtail form a strictly descending chain: $\varphi_{i_0} >_{\mathcal{L}} \varphi_{i_1} >_{\mathcal{L}} \varphi_{i_2} >_{\mathcal{L}} \dots >_{\mathcal{L}} \varphi_{i_k} >_{\mathcal{L}} \dots$ $(k < \omega)$ in I.

Consider the sublimit N of the M_{i_k} (an ω-limit) with canonical maps $g_k : M_{i_k} \longrightarrow N$ $(k < \omega)$ and the tuple $\overline{n} = g_k(f_i^{i_k}(\overline{m}_i))$ therein. Its tail $\{f_i^{i_k}(\overline{m}_i) | k < \omega\}$ does not stabilize and so its pp type is, by the previous lemma, not $\mathcal{L}$-finitely generated and N, by Thm. 3.1, not $\mathcal{K}$-Mittag-Leffler, as desired. □

We show in Section 7.2 below what it takes to make an ω-limit $\mathcal{K}$-Mittag-Leffler.

COROLLARY 3.11. *The union of an ω-closed direct system of of $\mathcal{K}$-Mittag-Leffler submodules is $\mathcal{K}$-Mittag-Leffler.* □

3.5. Density. In the spirit of [**EM**], Herbera and Trlifaj [**HT**, Def.2.5] call a collection, $\mathcal{C}$, of submodules of a module M a *κ-dense system* (in M) provided that

(a) every subset of M of power less than κ is contained in some $C \in \mathcal{C}$;
(b) $\mathcal{C}$ is closed under chains of length less than κ;
(c) $\mathcal{C}$ is directed, i.e., every two members of $\mathcal{C}$ are contained in a member of $\mathcal{C}$.

The case of interest here is $\kappa = \aleph_1$ (but the general definition reveals the role of "$<$" more visibly).

COROLLARY 3.12. [**HT**, Thm.2.6] *A module is $\mathcal{K}$-Mittag-Leffler if and only if it possesses an $\aleph_1$-dense system of $\mathcal{K}$-Mittag-Leffler submodules. Furthermore, in that case these submodules can always be chosen to be countably generated.*

PROOF. A module with an $\aleph_1$-dense system of $\mathcal{K}$-Mittag-Leffler submodules is the union of that system, by (a). This union is ω-closed by (b), and so the module is $\mathcal{K}$-Mittag-Leffler by the previous corollary.

That, conversely, every $\mathcal{K}$-Mittag-Leffler module has such an $\aleph_1$-dense system will be immediate from Prop. 6.4 below, which shows that one can take for $\mathcal{C}$ the set of all countably generated $\mathcal{L}$-pure submodules. □

3.6. Test modules and test maps. It follows from the Main Theorem that a test module for $\mathcal{K}$-Mittag-Leffler in the sense of Definition 3.1 above is a module that generates the same definable subcategory as does $\mathcal{K}$. Another way of saying this is (1) of the following corollary. The rest are immediate consequences of that—for (4) invoke the Löwenheim-Skolem Theorem.

COROLLARY 3.13. [**R**, Cor.2.6] *Test modules for $\mathcal{K}$-Mittag-Leffler are:*

(1) *Any right module N such that $\leq_{\mathcal{K}}$ and $\leq_N$ are the same.*
(2) *N the direct sum (or product) of all pure-injective indecomposables in the Ziegler-closed set corresponding to $\mathcal{K}$.*
(3) *N the direct sum (or product) of one model of every completion of the first-order (or the implicational) theory of $\mathcal{K}$.*
(4) *Any module elementarily equivalent to a test module; so there are test modules of power $|R| + \aleph_0$.* □

In the remainder of this subsection I discuss, without detailed proofs, a bound on the *exponent* I occurring in a test transformation τ_N^I, see Definition 3.1. To minimize this exponent, we need some more notation. Given a lattice Λ, let $\triangle(\Lambda)$ be the smallest cardinal κ such that every ideal is generated by κ elements. Note that $\triangle(\Lambda)$ is finite if and only if it is 1 if and only if all ideals in Λ are principal iff Λ has no infinite (strictly) ascending chains, in which case Λ is called a *noetherian lattice*, cf. [**S**, Ch.III, ex.10]. Since finitely generated left ideals are pp definable in R_R, we see that $\triangle(R_R) = 1$ iff R is left noetherian. For a module N, let $\triangle(N)$ denote the supremum of the cardinals $\triangle(\Lambda^n(N))$, where, recall, $\Lambda^n(N)$ is the lattice of n-place pp subgroups of N and n runs over the natural numbers. Since there are no more than $|R|+\aleph_0$ pp formulas altogether, $\triangle(N)$ is always bounded by this number.

FACT 3.14. [**R**, Cor.2.6].

(1) Given a test module N for $\mathcal{K}$-Mittag-Leffler, $\tau_N^{\triangle(N)+|N|}$ is a test transformation.
(2) If this test module N has power $|R|+\aleph_0$, then τ_N^N is a test transformation.

It was noted in [**R**, Rem.2.3(c),(d)] that in case $\triangle(N) = 1$, the entire exponent in $\tau_N^{\triangle(N)+|N|}$ can be taken to be 1, and *every* left module M is N-Mittag-Leffler. In particular, over a left noetherian ring R all left modules are R-Mittag-Leffler. This goes back to [**Goo**, p.1]; see Cor. 4.3(2) below for the converse.

4. Special cases

Here we specialize some of the previous results to the following three cases: $\mathcal{K}$ = Mod-R and $\mathcal{L}$ = R-Mod, $\mathcal{K} = \flat_R$ and $\mathcal{L} = {}_R\sharp$, and $\mathcal{K} = \sharp_R$ and $\mathcal{L} = {}_R\flat$ (notation as introduced after CONVENTION 2.7).

4.1. (Classical) Mittag-Leffler modules, i.e., $\mathcal{K} = \text{Mod-}R$. Corollary 3.6 yields at once:

COROLLARY 4.1. [**R**, Rem.2.3] *Given a left R-module M, the following are equivalent.*

(i) *M is Mittag-Leffler.*
(ii) *M is $\mathcal{K}$-Mittag-Leffler and $\langle\mathcal{K}\rangle = \text{Mod-}R$.*
(iii) *M is Zg_R-Mittag-Leffler.*
(iv) *M is $(\text{mod-}R)$-Mittag-Leffler.*

COROLLARY 4.2. *Besides the test modules one obtains by specializing Cor. 3.13 to the case $\mathcal{K} = \text{Mod-}R$, another test module is P_*, the direct sum of one representative of each isomorphism type from mod-R. (This module has power at most $|R| + \aleph_0$.)*

4.2. $\flat$-Mittag-Leffler modules. The first concrete instance of relativized Mittag-Leffler module in the literature seems to have been considered in [**Goo**] (not under this name) with $\mathcal{K} = \flat_R$. By a result of Zimmermann [**Zim**], $\leq_{\flat_R}$ and $\leq_{R_R}$ are the same (i.e., $\flat_R$ and R_R generate the same definable subcategory), see also [**P2**, Thm.2.3.9]. Thus $\flat$-Mittag-Leffler is the same as R_R-Mittag-Leffler, or R-Mittag-Leffler for short. The following basic properties of these modules are immediate from the main theorem. All but condition (v) (from [**R**, Cor.2.7]) are contained in [**Goo**].

COROLLARY 4.3.

(1) *For any left R-module M the following are equivalent.*
 (i) *M is $\flat$-Mittag-Leffler.*
 (ii) *M is R_R-Mittag-Leffler, i.e., R_R is a test module.*
 (iii) *$R^R \otimes M \longrightarrow M^R$ is monic (i. e. $R^R \otimes - \longrightarrow (R \otimes -)^R$ is a test transformation for $\flat$-Mittag-Leffler).*
 (iv) *For every finitely generated submodule A of M, the inclusion $A \subseteq M$ factors through a finitely presented module.*
 (v) *All quantifier-free types realized in M are finitely generated.*
(2) [**Goo**, p.1] *Every left R-module is R-Mittag-Leffler iff R is left noetherian.*

PROOF. The second part follows from (iv) of the first.

Since $\leq_{\flat_R}$ and $\leq_{R_R}$ are the same, we see from the theorem that R_R is a test module, hence we have (i) $\Leftrightarrow$ (ii). To prove (ii) $\Leftrightarrow$ (iii) we are thus left with checking that the exponent can be taken to be R. If R is infinite, $|R| + \aleph_0 = |R| = \triangle(R_R) \cdot |R_R|$, hence we get the desired exponent from this and the second part of the above corollary. If R is finite, $\triangle(R_R) = 1$ and we get exponent 1 from the discussion after Fact 3.14.

(iv) $\Rightarrow$ (v): Let $A = \langle\overline{a}\rangle_M$ and $(A, \overline{a}) \longrightarrow (P, \overline{c}) \longrightarrow (M, \overline{a})$ the factorization given, with P finitely presented. Then $\text{qf}_M(\overline{a}) = \text{qf}_A(\overline{a}) \subseteq \text{pp}_P(\overline{c}) \subseteq \text{pp}_M(\overline{a})$, so the pp formula that generates $\text{pp}_P(\overline{c})$ also generates $\text{qf}_M(\overline{a})$.

(Of course this suffices for the proof, but it's interesting to note a direct proof of the converse: if $\varphi \in \text{pp}_M(\overline{a})$ generates $\text{qf}_M(\overline{a})$, then any finitely presented free realization $(P, \overline{c})$ of φ maps to $(M, \overline{a})$; it remains to find $(A, \overline{a}) \longrightarrow (P, \overline{c})$, which is easily done by extending the map $\overline{a} \longrightarrow \overline{c}$ by linearity; this is well defined, for if $\overline{ra} = 0$, then $\overline{rx} \doteq 0$ is in $\text{qf}_A(\overline{a})$, hence implied by φ, and so $\overline{rc} = 0$.)

(i) $\Leftrightarrow$ (v). By the theorem, (i) is equivalent to every pp type in M being $\mathrm{D}(\flat_{\mathrm{R}})$-finitely generated. We will show that this is the same as every quantifier-free type in M being (R-Mod-) finitely generated. First of all, by the duality of flat and absolutely pure [**Her**] (for another reference, if somewhat in disguise, see [**P2**, Prop.3.4.26]), $\leq_{\mathrm{D}(\flat_{\mathrm{R}})}$ is the same as $\leq_{R\sharp}$. So we have to show that $\mathrm{pp}_M(\overline{a})$ is $\sharp$-finitely generated iff the type $\mathrm{qf}_M(\overline{a})$ is finitely generated. Because of its significance, we make this claim a separate lemma. □

LEMMA 4.4. *Given a tuple $\overline{a}$ in a module M, the following are equivalent.*

(i) $\mathrm{qf}_M(\overline{a})$ *is finitely generated.*
(ii) $\mathrm{qf}_M(\overline{a})$ *is $\sharp$-finitely generated.*
(iii) $\mathrm{pp}_M(\overline{a})$ *is $\sharp$-finitely generated.*

Moreover, the same pp (not necessarily quantifier-free!) formula works in all three cases.

PROOF. For the nontrivial implications, first suppose $\psi \in \mathrm{pp}(\overline{a})$ generates $\mathrm{qf}_M(\overline{a})$. To show this formula $\sharp$-generates $\mathrm{pp}_M(\overline{a})$, let $\varphi \in \mathrm{pp}(\overline{a})$ and N be absolutely pure. We have to show that $\psi(N) \subseteq \varphi(N)$. Since $\overline{a} \in \varphi(M) \subseteq \mathrm{ann}_M \mathrm{D}\varphi(\mathrm{R_R})$, for all $\overline{r} \in \mathrm{D}\varphi(\mathrm{R_R})$ the formula $\overline{r}\,\overline{x} \doteq 0$ is in $\mathrm{qf}_M(\overline{a})$, hence $\psi \leq \overline{r}\,\overline{x} \doteq 0$. In particular, $\psi(N) \subseteq \mathrm{ann}_N \mathrm{D}\varphi(\mathrm{R_R})$. But as N is absolutely pure, by [**P2**, Prop.2.3.3], this latter set *is* $\varphi(N)$.

Now assume $\mathrm{qf}_M(\overline{a})$ to be $\sharp$-generated by $\psi \in \mathrm{pp}_M(\overline{a})$. To show ψ generates $\mathrm{qf}_M(\overline{a})$, let $\alpha \in \mathrm{qf}_M(\overline{a})$ and N be any module and $\overline{b} \in \psi(N)$. We have to show $\overline{b} \in \alpha(N)$. Let $N \subseteq E$ be an injective envelope. Being existential, the formula ψ is satisfied by $\overline{b}$ also in E. Since $\psi \leq_\sharp \alpha$, therefore $\overline{b} \in \alpha(E)$. But α is quantifier-free and $\overline{b}$ is in $N \subseteq E$, hence $\overline{b} \in \alpha(N)$, as desired. □

4.3. $\sharp$-Mittag-Leffler modules. Another instance is $\mathcal{K} = \sharp_R$. As $\mathrm{D}({}_{\mathrm{R}}\flat)$ and $\sharp_R$ generate the same definable subcategory [**P2**, Prop.3.4.26], we may then take $\mathcal{L} = {}_R\flat$, and the theorem shows that a module is $\sharp$-Mittag-Leffler iff it is $\flat$-atomic. Symmetrically, $\mathrm{D}(\flat_{\mathrm{R}})$ and ${}_R\sharp$ generate the same definable subcategory, hence a module is $\flat$-Mittag-Leffler iff it is $\sharp$-atomic, a fact we haven't exploited in the previous subsection.

The next example, though rather trivial, may be instructive as to what the role of $\mathcal{K}$ and $\mathcal{L}$.

EXAMPLE 4.5. Let R be a (not necessarily commutative) domain and let **Div** denote the class of divisible right R-modules. Then any torsion left R-module is **Div**–Mittag-Leffler (hence also $\sharp$-Mittag-Leffler, for every absolutely pure module over a domain is divisible): as a divisible module tensored with a torsion module is zero and direct products of divisible modules are divisible, the canonical map figuring in the definition of Mittag-Leffler module is vacuously monic, cf. [**Rot3**].

EXAMPLE 4.6. Let R be a (not necessarily commutative) domain with a two-sided ideal I that is not finitely generated as a left ideal. Then the cyclic left R-module R/I is not finitely presented. Yet it is $\sharp$-Mittag-Leffler (even **Div**–Mittag-Leffler), for it is torsion, see Example 4.5.

In order to elucidate the concept, let us give another proof via atomicity. We have to verify that R/I is $\flat$-atomic. We know from Lemma 2.9 that it is enough to verify that the generator $1/I$ has $\flat$-finitely generated pp type. Now, this type is equivalent to saying $Ix = 0$ or, more precisely, to the set of all formulas $rx = 0$

with $r \in I$. Let s be any non-zero element of I. I claim that the pp type of $1/I$ in R/I is $\flat$-generated by the formula $sx = 0$. First of all, of course this formula is in the type, for $s(1/I) = 0$ in R/I. To see, it $\flat$-implies the the rest of it, we have to verify that $sx = 0 \leq_\flat rx = 0$ for all $r \in I$. It satisfies to verify this implication in the left regular module R. But there, the only element satisfying $sx = 0$ is 0. And 0 realizes *any* pp type. (This shows that, given any pp type p in any left R-module M, if p contains a formula $sx = 0$ for some non-zero $s \in R$, then this formula $\flat$-generates p.)

For other relativized Mittag-Leffler modules see [**Rot3**].

5. Purity

Here we investigate three kinds of relativized purity. The first of these is needed in the next section, the other two in the last.

5.1. Left purity: pure monomorphisms. To prove the relativized version, Corollary 6.5 below, of Raynaud and Gruson's separability result (1) at the top of this paper, one needs relativized versions of purity. An algebraic one was given in [**HT**, Thm.2.5(ii)] and a rather model-theoretic one in [**Rot3**, Prop.3.3]. Here we show that they are the same. As it turns out, this purity goes back to Herzog's thesis [**Her**, Sect.8], where it was called *local.*

DEFINITION 5.1 (Herzog). An *$\mathcal{L}$-pure monomorphism* is a monomorphism $f : N \longrightarrow M$ (of left R-modules) such that $K \otimes f$ is a monomorphism for every $K \in \mathrm{D}\langle\mathcal{L}\rangle$ (the dual of the definable category generated by $\mathcal{L}$). An *$\mathcal{L}$-pure submodule* of a module M is a submodule N whose inclusion in M is $\mathcal{L}$-pure, i.e., such that the corresponding map $K \otimes N \longrightarrow K \otimes M$ is monic for every $K \in \mathrm{D}\langle\mathcal{L}\rangle$. As usual, we drop curly brackets around singletons.

REMARK 5.2. It is an easy exercise using Herzog's criterion (and elementary duality) to show that it suffices to check the above monomorphism for all $K \in \mathcal{K}$, where $\mathcal{K}$ is any class such that $\langle\mathcal{K}\rangle = \mathrm{D}\langle\mathcal{L}\rangle$. In particular, $\mathcal{L}$-purity is the same as $\langle\mathcal{L}\rangle$-purity.

The next result is essentially [**Her**, Prop.8.3]. Herzog's original proof was in terms of closed sets of the Ziegler spectrum. The above remark allows us to pass to a class that is closed under product and thus to apply McKinsey's Lemma, see [**H**, Cor.9.1.7, p.417] (or [**Rot2**, Cor.6.3]), instead of Ziegler's indecomposability calculus. I include such a proof.

LEMMA 5.3. *Given a submodule N of a module M, the following are equivalent.*

(i) *N is an $\mathcal{L}$-pure submodule of M.*
(ii) *For every tuple $\overline{n}$ in N, its pp type in N is $\langle\mathcal{L}\rangle$-equivalent to its pp type in M.*

PROOF. Consider a tuple $\overline{n}$ in N and denote its pp type in M by p and its pp type in N by q. Clearly, $q \subseteq p$, hence, by compactness, these types are $\langle\mathcal{L}\rangle$-equivalent precisely when every $\varphi \in p$ is $\langle\mathcal{L}\rangle$-implied by some $\psi \in q$. As $\leq_{\mathcal{L}}$ is the same as $\leq_{\langle\mathcal{L}\rangle}$, the last two conditions are indeed equivalent.

It is an(other) easy exercise using Herzog's criterion to show that any of these two conditions implies the first. So we are left with the less straightforward implication (i) $\Rightarrow$ (ii). We have to show that for every tuple $\overline{n}$ in N and every pp formula φ it satisfies in M, there is a pp formula $\psi \leq_{\mathcal{L}} \varphi$ that $\overline{n}$ satisfies in N. To this end, consider $\varphi \in p$ and let let $\mathcal{K}$ be any class closed under product such that $\langle\mathcal{K}\rangle = \mathrm{D}\langle\mathcal{L}\rangle$. Notice, this latter equality means that $\psi \leq_{\mathcal{L}} \varphi$ iff $\mathrm{D}\varphi \leq_{\mathcal{K}} \mathrm{D}\psi$.

First note that for every choice of $\overline{k} \in \mathrm{D}\varphi(\mathrm{K})$ in any $K \in \mathcal{K}$, there is $\psi \in q$ such that $\overline{k} \in \mathrm{D}\psi(\mathrm{K})$. Indeed, as $\overline{n}$ satisfies φ, we have $\overline{k} \otimes \overline{n} = 0$ in $K \otimes M$, hence, by (i), also in $K \otimes N$, which yields $\psi \in q$ such that $\overline{k}$ satisfies $\mathrm{D}\psi$ in K.

This shows that $\forall\overline{x}(\mathrm{D}\varphi \to \bigvee_{\psi\in q} \mathrm{D}\psi)$ holds in every member of $\mathcal{K}$. McKinsey's Lemma yields a single $\psi \in q$ such that $\forall\overline{x}(\mathrm{D}\varphi \to \mathrm{D}\psi)$ holds in every member of $\mathcal{K}$, i.e., $\mathrm{D}\varphi \leq_{\mathcal{K}} \mathrm{D}\psi$. By duality, $\psi \leq_{\mathcal{L}} \varphi$, as desired. □

We see immediately how this looks when all the types p are $\mathcal{L}$-finitely generated.

COROLLARY 5.4. *Suppose N is a submodule of an $\mathcal{L}$-atomic module M. Given a tuple $\overline{n}$ in N, let $\varphi_{\overline{n}}$ be an $\mathcal{L}$-generator of* $\mathrm{pp}_M(\overline{n})$.

Then the following are equivalent.

(i) *N is an $\mathcal{L}$-pure submodule of M.*

(ii) *For every tuple $\overline{n}$ in N, the type $\mathrm{pp}_N(\overline{n})$ is $\langle\mathcal{L}\rangle$-equivalent to $\varphi_{\overline{n}}$.*

(iii) *For every tuple $\overline{n}$ in N, the type $\mathrm{pp}_N(\overline{n})$ contains a pp formula $\psi_{\overline{n}} \leq_{\mathcal{L}} \varphi_{\overline{n}}$ (and hence $\psi_{\overline{n}} \sim_{\mathcal{L}} \varphi_{\overline{n}}$).*

In this case N is $\mathcal{L}$-atomic too.

PROOF. Just note that $\varphi_{\overline{n}}$ is $\mathcal{L}$- (hence $\langle\mathcal{L}\rangle$-) equivalent to $\mathrm{pp}_M(\overline{n})$. □

To check $\mathcal{L}$-purity of a finitely generated submodule it suffices to consider a generating tuple in Lemma 5.3. The argument is standard and similar to the one in Lemma 2.9:

LEMMA 5.5. *Let $\overline{a}$ be a tuple in M and N the submodule it generates. Then the following are equivalent.*

(i) *N is an $\mathcal{L}$-pure submodule of M.*

(ii) *$\mathrm{pp}_N(\overline{a})$ is $\langle\mathcal{L}\rangle$-equivalent to $\mathrm{pp}_M(\overline{a})$.*

(iii) *For every pp formula φ in $\mathrm{pp}_M(\overline{a})$, there is a pp formula $\psi \leq_{\mathcal{L}} \varphi$ (and hence $\psi \sim_{\mathcal{L}} \varphi$) in $\mathrm{pp}_N(\overline{a})$.*

PROOF. As $\overline{a}$ generates N, for every tuple $\overline{n}$ in N there is a matrix $\mathfrak{B}$ such that $\overline{n} = \mathfrak{B}\overline{a}$. For every pp formula φ in the type of $\overline{n}$ (in N or M), the pp formula $\varphi(\mathfrak{B}\overline{x})$ is in the corresponding type of $\overline{a}$. Using this it is straightforward to verify that (ii) entails $\langle\mathcal{L}\rangle$-equivalence of $\mathrm{pp}_N(\overline{n})$ and $\mathrm{pp}_M(\overline{n})$. □

The next, trivial example may serve to clarify some of the purity issues involved in relative implication (localized at a definable subcategory).

EXAMPLE 5.6. Consider $\mathcal{K}$ the class of divisible (=injective) and $\mathcal{L}$ the class of torsionfree (=flat) abelian groups. Recall, they form definably dual definable subcategories of $\mathrm{A}b = \mathbb{Z}$-Mod.

Let M be the cyclic abelian group of order 4 and N its subgroup of order 2. Of course, there is nothing pure about the inclusion of N in M. Yet, it is trivially $\mathcal{L}$-pure, for tensoring with a divisible group (with $\mathcal{K}$) annihilates every torsion group.

What's happening on the level of pp formulas? Let n be the generator of N and consider its pp type p in M and q in N. Clearly $q \subseteq p$. Consider the formula $2|x \wedge 2x = 0$. Call it φ. Clearly, it is a generator of the type p. Let ψ be the formula $2x = 0$. This is a generator of the type q. So, in particular, they are both

$\mathcal{L}$-generators of their corresponding types. But they are also $\mathcal{L}$-equivalent, in fact, to $x = 0$ (as $\mathbb{Z}$ is a domain), whence ψ $\mathcal{L}$-implies *any* pp formula. So ψ is also an $\mathcal{L}$-generator of p, but φ is not an $\mathcal{L}$-generator of the type q for the mere reason that it is not contained in it.

5.2. Partial/Restricted left purity. We need a restricted form of purity in Section 7.2 below.

DEFINITION 5.7. Let C be a submodule (or even subset) of the module N and $f : N \longrightarrow M$ a homomorphism.

(1) The map f is said to be *$\mathcal{L}$-pure on C* if for every tuple $\bar{c}$ in C, its pp type in N is $\langle\mathcal{L}\rangle$-equivalent to the pp type of $f(\bar{c})$ in M.
(2) f is said to be *monic on C* if $\ker f \cap C = \emptyset$.

REMARK 5.8.

(1) Clearly, $f : N \longrightarrow M$ is $\mathcal{L}$-pure iff it is $\mathcal{L}$-pure on N.
(2) It is easy to see that the versions of Lemma 5.3 and Cor. 5.4 hold for $\mathcal{L}$-purity on C when the tuple $\bar{n}$ runs only over C (instead of over all of N).

We are not trying to study this restricted concept in generality here, but rather make use of it in the special context of $\mathcal{L}$-ω-limits below.

5.3. A special assumption. The relevance of R_R being a member of $\mathcal{K}$ was pointed out in [**HT**]. The underlying reason is the following. While, in general, under an $\mathcal{L}$-pure homomorphism, a given tuple in the domain and its image have $\mathcal{L}$–equivalent pp types, under the assumption $R_R \in \langle\mathcal{K}\rangle$, the quantifier-free parts of those pp types are actually *equal* (or R-Mod–equivalent). Note, as $\langle R_R\rangle = \langle\flat_R\rangle$, the clause $R_R \in \langle\mathcal{K}\rangle$ is, by elementary duality, equivalent to ${}_R\sharp \subseteq \langle\mathcal{L}\rangle$.

LEMMA 5.9. *$\sharp$-equivalent types of tuples (in any modules) have identical quantifier-free part.*

PROOF. Let $p = \mathrm{pp}_N(\bar{c})$ and $q = \mathrm{pp}_M(\bar{b})$. We are going to show that $p \leq_\sharp q$ implies that the quantifier-free part of q is in p, i.e., $\mathrm{qf}_M(\bar{b}) \subseteq \mathrm{qf}_N(\bar{c})$. To this end, embed N in an injective module E. Then $\bar{c}$ satisfies p also in E. As $E \in \sharp$, $\bar{c}$ satisfies also q in E. The quantifier-free part $\mathrm{qf}_M(\bar{b})$ passes down to the submodule N, so $\bar{c}$ satisfies $\mathrm{qf}_M(\bar{b})$ in N, whence $\mathrm{qf}_M(\bar{b}) \subseteq \mathrm{qf}_N(\bar{c})$, as desired. □

REMARK 5.10. The same proof shows that $\alpha \leq_\sharp \varphi$ implies $\alpha \leq \varphi$ provided α is quantifier-free.

LEMMA 5.11. *Suppose $R_R \in \langle\mathcal{K}\rangle$ or, equivalently, ${}_R\sharp \subseteq \langle\mathcal{L}\rangle$.*

(1) *Every map $\mathcal{L}$-pure on a submodule preserves quantifier-free parts of pp types of tuples from that submodule.*
(2) *Every map $\mathcal{L}$-pure on a submodule is monic on it.*
(3) *${}_R\sharp$-pure maps are monomorphisms.*

PROOF. The last statement is a special case of the second, where the subset in question is the entire module, while the second is a special case of the first, for the (quantifier-free) formula $\bar{x} \doteq \bar{y}$.

For the first statement, let $f : N \longrightarrow M$ be $\mathcal{L}$-pure on a set $C \subset N$ and let $\bar{c}$ be a tuple in C and $\bar{b}$ the image of $\bar{c}$ in M. The assertion now is $\mathrm{qf}_N(\bar{c}) = \mathrm{qf}_M(\bar{b})$, which readily follows from the previous lemma. □

5.4. Right purity: pure epimorphisms. Dually to the lemma characterizing $\mathcal{L}$-pure monomorphisms, we make the following definition.

DEFINITION 5.12. A *uniform $\mathcal{L}$-pure epimorphism* is a homomorphism $g : N \longrightarrow M$ (of left R-modules) such for every tuple $\overline{m}$ in M there is a g-preimage $\overline{n}$ in N such that their corresponding pp types are $\mathcal{L}$-equivalent. We drop the prefix $\mathcal{L}$ when it is all of R-Mod.

We will encounter uniform pure epimorphisms in Thm. 7.1. For reference there, we state the following easy fact.

LEMMA 5.13. *Every uniformly $\mathcal{L}$-pure epimorphic image of a $\mathcal{K}$-Mittag-Leffler module is $\mathcal{K}$-Mittag-Leffler.*

PROOF. Let $g : N \longrightarrow M$ be a uniform $\mathcal{L}$-pure epimorphism and $\overline{m}$ a tuple in M. Let p be its pp type. To see that p is finitely $\mathcal{L}$-generated, choose a preimage $\overline{n}$ with $\mathcal{L}$-equivalent pp type, q say. By hypothesis, the pp type q contains a pp formula φ that $\mathcal{L}$-generates it. This formula is contained in its image p too, and as p and q are $\mathcal{L}$-equivalent, the former is $\mathcal{L}$-generated by φ as well. □

6. Pure separation

In the first part I prove the relativized version of Raynaud and Gruson's separability result (1) from the introduction. Previous, and equivalent versions are [**HT**, Thm.2.5(ii)] and [**Rot3**, Prop.3.3]. But with a new and more systematic view on relativized purity, it seems justified to include a proof of this result, especially since that very proof goes back again to the unpublished [**R**].[3] Another variant of this phenomenon is considered in the second part. I make a uniform definition.

DEFINITION 6.1. Let κ be a cardinal and $\mathcal{C} \subseteq R$-Mod. A module M is said to be *κ-$\mathcal{L}$-pure separable by (modules from) $\mathcal{C}$* if every subset A of M of cardinality $< \kappa$ is contained in an $\mathcal{L}$-pure submodule of M that is a member of $\mathcal{C}$.

I say *countably $\mathcal{L}$-pure separable* for $\aleph_1$-$\mathcal{L}$-pure separable and *finitely $\mathcal{L}$-pure separable* for $\aleph_0$-$\mathcal{L}$-pure separable.

The focus here lies in classes $\mathcal{C}$ of $\mathcal{K}$-Mittag-Leffler (=$\mathcal{L}$-atomic) modules.

REMARK 6.2. (1) By the Main Theorem, a module $\mathcal{L}$-pure separable by $\mathcal{K}$-Mittag-Leffler modules is certainly $\mathcal{K}$-Mittag-Leffler.
(2) Since unions of chains of $\mathcal{L}$-pure submodules are $\mathcal{L}$-pure, it is easy to see that a module finitely $\mathcal{L}$-pure separable by finitely generated $\mathcal{K}$-Mittag-Leffler modules is also countably $\mathcal{L}$-pure separable by countably generated $\mathcal{K}$-Mittag-Leffler modules.

6.1. Countable separation. First of all, Corollary 5.4 implies

COROLLARY 6.3. *$\mathcal{L}$-pure submodules of $\mathcal{K}$-Mittag-Leffler modules are $\mathcal{K}$-Mittag-Leffler.* □

That there are enough such $\mathcal{L}$-pure submodules can be viewed as a version of the downward Löwenheim-Skolem Theorem, where 'elementary submodule' is weakened to 'pure submodule' (which means that the realm of formulas is restricted to pp formulas) and 'cardinality of the language' is strengthened to 'countably generated'

[3] Where it was the definition that was missing, see the introduction to Section 5.

(no matter what that cardinality). Accordingly, a proof of (1) using pp formulas was given in [**R**, Lemma 3.11][4] and [**Rot1**]. A version of this proof closer to the original Löwenheim-Skolem Theorem, [**P2**, Thm.1.3.27], carries over one-to-one to the more general case of $\mathcal{K}$-Mittag-Leffler modules.

PROPOSITION 6.4. *Every countable subset of a $\mathcal{K}$-Mittag-Leffler module is contained in a countably generated $\mathcal{L}$-pure submodule that is $\mathcal{K}$-Mittag-Leffler.*

PROOF. Suppose M is a $\mathcal{K}$-Mittag-Leffler module and C a countable subset. We are going to construct an $\mathcal{L}$-pure submodule N of M that contains C (which then is $\mathcal{K}$-Mittag-Leffler).

First, list C as $(c_i)_{i<\omega}$. Then successively extend this to a list of tuples as follows. Starting with $\overline{a}_0$ the singleton c_0, (or any other tuple containing it), in the ith step ($i \geq 0$), choose a pp formula $\varphi_i(\overline{x}_0, \dots, \overline{x}_i)$ in $p_i := \mathrm{pp}_M(\overline{a}_0, \dots, \overline{a}_i)$ that $\mathcal{L}$-generates p_i. Write φ_i as $\exists \overline{x}'_{i+1} \alpha_i(\overline{x}_0, \dots, \overline{x}_i, \overline{x}'_{i+1})$ with α_i quantifier-free. Choose a witness $\overline{a}'_{i+1}$ for $\overline{x}'_{i+1}$ in M, concatenate it with c_{i+1} and call the resulting tuple $\overline{a}_{i+1}$. Accordingly, let $\overline{x}_{i+1}$ be the tuple of variables obtained from $\overline{x}'_{i+1}$ by concatenating one more variable for c_{i+1}. We add a dummy variable for that also in α_i so that we may think of its variables as being $\overline{x}_0, \dots, \overline{x}_i, \overline{x}_{i+1}$ (rather than ending in $\overline{x}'_{i+1}$). (In case, φ_i is quantifier-free, $\overline{a}_{i+1}$ would be the singleton c_{i+1}.)

Now, the module, N, generated by the $\overline{a}_i$ is $\mathcal{L}$-pure by construction. Indeed, since φ_i $\mathcal{L}$-generates p_i while its witnesses are inside N, $\mathcal{L}$-purity is guaranteed for the generators of N. But it's a simple exercise to verify the same for any other tuple in N (this was done in [**R**, Lemma 1.5], cf. [**PR**, Fact 2.4]). □

COROLLARY 6.5. *A module is $\mathcal{K}$-Mittag-Leffler iff it is countably $\mathcal{L}$-pure separable by countably generated $\mathcal{K}$-Mittag-Leffler modules.* □

6.2. Finite separation. Finitely pure separated modules were characterized in [**R**, Rem.3.13].[5] We desribe the $\mathcal{L}$-pure variant of this particular kind of $\mathcal{K}$-Mittag-Leffler module in the same way now. Key is the following.

LEMMA 6.6. [**R**, Rem.3.13] *If the pp type of a tuple $\overline{a}$ in a module M is $\mathcal{L}$-generated by a quantifier-free pp formula, then the submodule generated by $\overline{a}$ is $\mathcal{L}$-pure in M.*

If M is $\mathcal{K}$-Mittag-Leffler, the converse is true.

PROOF. Suppose, $\mathrm{pp}_M(\overline{a})$ is $\mathcal{L}$-generated by the quantifier-free formula α. Then $\alpha \leq_{\mathcal{L}} \varphi$ for every $\varphi \in \mathrm{pp}_M(\overline{a})$. But being quantifier-free, α is already in $\mathrm{pp}_A(\overline{a})$, so α makes (iii) of Lemma 5.5 true uniformly, and A is $\mathcal{L}$-pure in M.

If, conversely, A is $\mathcal{L}$-pure in the $\mathcal{K}$-Mittag-Leffler M, then $\mathrm{pp}_M(\overline{a})$ is $\mathcal{L}$-generated by a single pp formula φ and it suffices that this be $\mathcal{L}$-implied by a quantifier-free formula $\alpha \in \mathrm{pp}_M(\overline{a})$. By purity, φ is $\mathcal{L}$-implied by some ψ in $\mathrm{pp}_A(\overline{a})$. Write ψ as $\exists \overline{y} \theta(\overline{x}, \overline{y})$, where θ is quantifier-free, and pick a witness $\overline{b} \in A$ for $\theta(\overline{a}, \overline{y})$. As $\overline{a}$ generates A, we can write $\overline{b}$ as a term in $\overline{a}$ or, put in more linear algebraic terminology, there's a matrix $\mathfrak{B}$ over the ring such that $\overline{b} = \mathfrak{B}\overline{a}$. Let $\alpha(\overline{x})$ be the formula $\theta(\overline{x}, \mathfrak{B}\overline{x})$. It is quantifier-free and in $\mathrm{pp}_A(\overline{a}) \subseteq \mathrm{pp}_M(\overline{a})$. The sequence of implications $\alpha \sim \theta(\overline{x}, \mathfrak{B}\overline{x}) \leq \exists \overline{y} \theta(\overline{x}, \overline{y}) \sim \psi \leq_{\mathcal{L}} \varphi$ concludes the proof. □

---

[4]For another generalization of Mittag-Leffler module, where purities given by classes of finitely presented modules other than R-mod are considered.

[5]and called *strong Mittag-Leffler modules*—for purities as mentioned in the previous footnote.

The moreover clause follows from Lemma 5.11 and Remark 6.7. □

REMARK 7.6. (1) To avoid any misconception, while the $\mathcal{L}$-ω-limit M_ω is always the direct limit of the chain of monomorphisms $g_0(M_0) \hookrightarrow g_1(M_1) \hookrightarrow g_2(M_2)\ldots$, these images may no longer be finitely presented.
(2) If $\mathcal{S}$ is an ascending chain of coherent modules that is $\mathcal{L}$-pure on images, then its direct limit is a direct limit of an $\mathcal{L}$-pure chain of finitely presented modules. Namely, the images of all the maps along the chain are themselves finitely presented and $\mathcal{L}$-pure in the next member of the chain. In particular, countably generated left $\mathcal{K}$-Mittag-Leffler modules over left coherent rings are 'unions' of $\mathcal{L}$-pure chains of finitely presented modules (and conversely).

LEMMA 7.7. *Every $\mathcal{L}$-ω-limit is $\mathcal{L}$-atomic, hence $\mathcal{K}$-Mittag-Leffler.*

PROOF. Consider a chain as in 7.4(1) where all M_i are finitely presented. Every tuple in its limit M_ω is the image under a canonical map of a g_i-image in M_{i+1} for some i. By the previous lemma, its pp type, q, in M_ω is $\mathcal{L}$-equivalent to that of the image tuple in M_{i+1}. The latter module being $\mathcal{L}$-atomic, we see that also q is $\mathcal{L}$-finitely generated. □

REMARK 7.8. The proof shows the same for a limit of a chain $\mathcal{L}$-pure on images of *arbitrary* $\mathcal{K}$-Mittag-Leffler modules.

Next we establish a connection between $\mathcal{L}$-ω-limits and $\mathcal{L}$-chains of pp formulas.

NOTATION 7.9. Suppose $\Phi = (\varphi_i : i < \omega)$ is an $\mathcal{L}$-chain with corresponding chain of quantifier-free formulas $(\alpha_i : i < \omega)$ as in Remark 7.1. Every quantifier-free pp formula α_i has a free realization $(M_i, \overline{p}_0^i, \ldots, \overline{p}_{i+1}^i)$ with M_i finitely presented and generated by the realizations $\overline{p}_0^i, \ldots, \overline{p}_{i+1}^i$ of α_i. Then $(M_{i+1}, \overline{p}_0^{i+1}, \ldots, \overline{p}_{i+1}^{i+1})$ realizes α_i, hence there is a map

$$g_i : (M_i, \overline{p}_0^i, \ldots, \overline{p}_{i+1}^i) \longrightarrow (M_{i+1}, \overline{p}_0^{i+1}, \ldots, \overline{p}_{i+1}^{i+1}).$$

Note, $(M_i, \overline{p}_0^i, \ldots, \overline{p}_i^i)$ is a free realization of φ_i for all i.

Define the connecting maps f_i^j and the canonical maps $f_i : M_i \longrightarrow M_\omega$ as in 7.4(3). To indicate the origin of this limit, we may denote M_ω by M_Φ. By the *ith tail* in M_Φ we mean the tail $\{p_i^j | j \geq i\}$ and denote its limit in M_Φ by $\overline{q}_i$, i.e., $\overline{q}_i = f_i(\overline{p}_i^i)$ for $i < \omega$ (hence $\overline{q}_i = f_j(\overline{p}_i^j)$ for any $j \geq i$).

For every i, let $\Phi_i := \{\exists \overline{x}_{i+1} \ldots \exists \overline{x}_j \varphi_j : i \leq j < \omega\}$ (this is also a chain of pp formulas, but in finitely many free variables; for $j = i$, the formula is taken to be just φ_i).

LEMMA 7.10.

(1) *The tuples $\overline{q}_0, \overline{q}_1, \overline{q}_2, \ldots$ generate the module M_Φ.*
(2) *For every i, the type $\mathrm{pp}_{M_\Phi}(\overline{q}_0, \ldots, \overline{q}_i)$ is equivalent to Φ_i.*
(3) $\mathrm{pp}_{M_\Phi}(\overline{q}_0, \ldots, \overline{q}_i)$ *is $\mathcal{L}$-generated by φ_i.*

PROOF. (1) is clear from the construction. (3) follows from (2), since all the formulas in Φ_i are $\mathcal{L}$-equivalent to φ_i in an $\mathcal{L}$-chain.

To prove (2), first note that, by the defining properties of direct limit, $\Phi_i \subseteq \mathrm{pp}_{M_\Phi}(\overline{q}_0, \ldots, \overline{q}_i)$, for $f_j(\overline{p}_0^j, \ldots, \overline{p}_i^j) = (\overline{q}_0, \ldots, \overline{q}_i)$ for every $j \geq i$ (remember, pp formulas are preserved under homomorphisms).

Now let $\varphi = \varphi(\overline{x}_0, \ldots, \overline{x}_i) \in \mathrm{pp}_{M_\Phi}(\overline{q}_0, \ldots, \overline{q}_i)$. Then there must be $j \geq i$ such that $(P_j, \overline{p}_0^j, \ldots, \overline{p}_i^j) \models \varphi$. But then, as $(P_j, \overline{p}_0^j, \ldots, \overline{p}_j^j)$ freely realizes φ_j, the formula $\exists \overline{x}_{i+1} \ldots \exists \overline{x}_j \varphi_j$ is freely realized by $(P_j, \overline{p}_0^j, \ldots, \overline{p}_i^j)$ and thus implies φ, as desired. □

REMARK 7.11. If $\sharp \subseteq \langle \mathcal{L} \rangle$, then, by Lemma 4.4, φ_i generates the corresponding quantifier-free type in all of R-Mod.

LEMMA 7.12. *For every $\mathcal{L}$-chain $\Phi = (\varphi_i : i < \omega)$, the limit M_Φ is an $\mathcal{L}$-ω-limit (hence $\mathcal{K}$-Mittag-Leffler). Conversely, every $\mathcal{L}$-ω-limit is of the form M_Φ for some $\mathcal{L}$-chain Φ.*

PROOF. To prove the first assertion, we show M_Φ is $\mathcal{L}$-pure on images. The image of g_i is the submodule of M_{i+1} generated by $\overline{p}_0^{i+1}, \ldots, \overline{p}_{i+1}^{i+1}$. That is, we have to verify that the pp type of this tuple of $i+2$ tuples in M_{i+1} is $\mathcal{L}$-equivalent to $\mathrm{pp}_{M_{i+2}}(\overline{p}_0^{i+2}, \ldots, \overline{p}_{i+1}^{i+2})$. Now, $(M_{i+1}, \overline{p}_0^{i+1}, \ldots, \overline{p}_{i+1}^{i+1})$ is a free realization of φ_{i+1}, hence this formula generates $\mathrm{pp}_{M_{i+1}}(\overline{p}_0^{i+1}, \ldots, \overline{p}_{i+1}^{i+1})$. Similarly, φ_{i+2} generates the type $\mathrm{pp}_{M_{i+2}}(\overline{p}_0^{i+2}, \ldots, \overline{p}_{i+2}^{i+2})$, hence $\exists \overline{x}_{i+2} \varphi_{i+2}$ generates $\mathrm{pp}_{M_{i+2}}(\overline{p}_0^{i+2}, \ldots, \overline{p}_{i+1}^{i+2})$. But in an $\mathcal{L}$-chain, the formulas $\exists \overline{x}_{i+2} \varphi_{i+2} \leq \varphi_{i+1}$ are in fact $\mathcal{L}$-equivalent, hence so are the generated types, as desired.

For the converse, consider any $\mathcal{L}$-ω-limit M_ω as in 7.4. Being finitely presented, the module M_0, has a tuple of generators, $\overline{p}_0^0$, that freely realizes a quantifier-free pp formula $\alpha_0(\overline{x}_0)$. Extend its image $\overline{p}_0^1 := g_0(\overline{p}_0^0)$ to a generating tuple $(\overline{p}_0^1, \overline{p}_1^1)$ freely realizing a quantifier-free pp formula $\alpha_1(\overline{x}_0, \overline{x}_1)$, and so on. As g_1 is $\mathcal{L}$-pure on $\operatorname{im} g_0$, the submodule of M_1 generated by $\overline{p}_0^1$, we see that its pp type in M_1 is $\mathcal{L}$-equivalent to the pp type of $\overline{p}_0^2 := g_1(\overline{p}_0^1)$ in M_2. Now the former is generated by $\exists \overline{x}_1 \alpha_1(\overline{x}_0, \overline{x}_1)$, while the latter is generated by $\exists \overline{x}_1 \exists \overline{x}_2 \alpha_2(\overline{x}_0, \overline{x}_1, \overline{x}_2)$, which are therefore $\mathcal{L}$-equivalent too. Continuing like this we see that the chain $\exists \overline{x}_1 \alpha_1 \geq \exists \overline{x}_2 \alpha_2 \geq \ldots$ forms an $\mathcal{L}$-chain, whose limit is M_ω. (One may have to drop M_0 to make the indices match, but that is irrelevant in the limit.) □

7.3. Structure of countably generated Mittag-Leffler modules.

THEOREM 7.1 (Characterization).

(1) *The following are equivalent for any given module N.*
  (i) *N is a countably generated $\mathcal{K}$-Mittag-Leffler module.*
  (ii) *N is a countably generated and some (in fact every) countable sequence of generators can be arranged as a sequence of tuples satisfying an $\mathcal{L}$-chain of pp formulas.*
  (iii) *N is a uniform $\mathcal{L}$-pure image of an $\mathcal{L}$-ω-limit.*

(2) *If $R_R \in \langle \mathcal{K} \rangle$ or $M_\Phi \in \langle \mathcal{L} \rangle$, the countably generated $\mathcal{K}$-Mittag-Leffler modules are precisely the $\mathcal{L}$-ω-limits, which, in turn, are unions of $\mathcal{L}$-pure chains of finitely presented modules.*

PROOF. (1). As $\mathcal{L}$-ω-limits are countably generated, Lemmas 5.13 and 7.7 show that (iii) implies (i). By Lemma 7.3, (i) implies (ii). For the remaining implication, assume (ii), that is, a countable sequence of tuples $\overline{a}_0, \overline{a}_1, \overline{a}_2, \ldots$ realizing an $\mathcal{L}$-chain Φ as in Remark 7.1. Construct the $\mathcal{L}$-ω-limit M_Φ according to Notation 7.9. By Lemma 7.12, M_Φ is an $\mathcal{L}$-ω-limit and we are going to exhibit a uniform $\mathcal{L}$-pure epimorphism from it to N.

As $(P_i, \overline{p}^i_0, \ldots, \overline{p}^i_{i+1})$ is a free realization of α_i, we may send it to $(N, \overline{a}_0, \ldots, \overline{a}_{i+1})$ by a map h_i. Clearly, the maps h_i and $h_j f^j_i$ coincide on the generators $\overline{p}^i_0, \ldots, \overline{p}^i_{i+1}$ of P_i. Hence $h_i = h_j f^j_i$, and the universal property of direct limit yields a map $h : M_\Phi \longrightarrow N$ such that $h_i = hf_i$. Note, $h(\overline{q}_i) = \overline{a}_i$ for all i.

That h is surjective is immediate from generator considerations. By Lemma 7.5, every tuple in N has a preimage under h with $\mathcal{L}$-equivalent pp type, whence h is uniformly $\mathcal{L}$-pure.

(2) Under the extra assumption on $\mathcal{K}$, the epimorphism h is a monomorphism by Lemma 5.11, hence an isomorphism. The same holds true if $M_\Phi \in \langle\mathcal{L}\rangle$, for then the $\mathcal{L}$-equivalent types of, say $\overline{m}$, in M_Φ and its image in N are equivalent on M_Φ, so equality of two entries in $h(\overline{m})$ would imply that of the corresponding entries in $\overline{m}$.

Further, Lemma 7.5 implies that $\mathcal{L}$-ω-limits are unions of the finitely presented $g_i(M_i)$. □

7.4. The case $\mathcal{L} = \flat$. In general, the connection between the limit and its uniformly pure image may be quite loose, as we demonstrate in a final note.

Consider $R = \mathbb{Z}$ (or any other domain) and let $\mathcal{L} = \flat$ (the torsion-free abelian groups). As was noted in [**Rot3**], any torsion group is $\sharp$-Mittag-Leffler. Equivalently, every torsion group is $\flat$-atomic, which can also be seen thus: the formula $rx \doteq 0$ is $\flat$-equivalent to $x \doteq 0$ and thus $\flat$-implies every pp formula. So, whatever torsion formula is contained in the pp type of an element, it $\flat$-generates its pp type. The same applies to finite tuples of torsion elements, since the conjunction of the annihilation formulas $\flat$-implies that the tuple must be the zero tuple and therefore $\flat$-implies the entire pp type of the tuple. For the same reason, any two tuples of the same length of torsion elements have $\flat$-equivalent pp types, which is responsible for the fact that uniform $\flat$-purity does not say much.

Let me illustrate this further by looking at two different kinds of $\flat$-limit that have a uniform $\flat$-pure image in common. For the first, consider any quantifier-free chain of the form $r_0x_0 \doteq 0 \geq r_0x_0 \doteq 0 \wedge r_1x_1 \doteq 0 \geq \ldots$. Let τ_i denote its ith member, i.e., the conjunction $\bigwedge_{j\leq i} r_jx_j \doteq 0$. As the variables are completely separate and 0 is always there as a witness, we see that this chain is a $\flat$-chain (as a matter of fact, even an R-Mod-chain). Denote the corresponding limit by T. It is not hard to see from the construction in Notation 7.9 that T is the union of the finitely generated groups $P_i = \bigoplus_{j\leq i} \mathbb{Z}/r_j\mathbb{Z}$, i.e., T is isomorphic to $\bigoplus_{j<\omega} \mathbb{Z}/r_j\mathbb{Z}$. This group is pure-projective and can be mapped onto any other group generated by a sequence of elements annihilated by those ring elements r_i, respectively. As noted before, any such group is $\flat$-atomic (i.e., $\sharp$-Mittag-Leffler) and hence, by (the proof of) Thm. 7.1, a uniform $\flat$–pure image of T.

So *every* image of T is uniformly $\flat$-pure. Setting $r_i = p^{i+1}$ in the above, we see that the Prüfer group $\mathbb{Z}_{p^\infty}$ is a uniform $\flat$–pure image of $T = \bigoplus_{i<\omega} \mathbb{Z}/p^{i+1}\mathbb{Z}$.

Of course, letting α_i be the formula $px_0 \doteq 0 \ \wedge\ \bigwedge_{j<i} x_j \doteq px_{j+1}$, we obtain a much more meaningful pp chain in $\mathbb{Z}_{p^\infty}$. As each of these formulas implies torsion in all variables, again they $\flat$-imply just any pp formula, whence these form a $\flat$-chain. This shows that the Prüfer groups are themselves $\flat$-limits.

Namely, a free realization of α_i is given by $(\mathbb{Z}/p^i\mathbb{Z}, p^{i-1}c_i, p^{i-2}c_i, \ldots, c_i)$, where c_i is a generator of that group. Obviously, the corresponding map ϱ_i from the ith free realization to the $(i+1)$st sends p^jc_i to $p^{j+1}c_{i+1}$ and the limit of this chain is (isomorphic to) $\mathbb{Z}_{p^\infty}$ itself. This isomorphism is also the uniform $\flat$-pure epimorphism in question.

Note, the chain of these α_i is obviously also a $\sharp$-chain, whence the Prüfer groups are also $\sharp$-limits.

Acknowledgments

I wish to thank the referee for helpful comments resulting in improved readability of this work.

References

[AH] L. Angeleri Hügel and D. Herbera, *Mittag-Leffler conditions on modules*, Indiana Univ. Math. J. **57** (2008), no. 5, 2459–2517, DOI 10.1512/iumj.2008.57.3325. MR2463975

[EM] P. C. Eklof and A. H. Mekler, *Almost free modules: Set-theoretic methods*, Revised edition, North-Holland Mathematical Library, vol. 65, North-Holland Publishing Co., Amsterdam, 2002. MR1914985

[Fac] A. Facchini, *Mittag-Leffler modules, reduced products, and direct products*, Rend. Sem. Mat. Univ. Padova **85** (1991), 119–132. MR1142536

[Goo] K. R. Goodearl, *Distributing tensor product over direct product*, Pacific J. Math. **43** (1972), 107–110. MR0311714

[GJ] L. Gruson and C. U. Jensen, *Modules algébriquement compacts et foncteurs* $\varprojlim^{(i)}$, C. R. Acad. Sci. Paris Sér. A-B **276** (1973), A1651–A1653. MR0320112

[HT] D. Herbera and J. Trlifaj, *Almost free modules and Mittag-Leffler conditions*, Adv. Math. **229** (2012), no. 6, 3436–3467, DOI 10.1016/j.aim.2012.02.013. MR2900444

[Her] I. Herzog, *Elementary duality of modules*, Trans. Amer. Math. Soc. **340** (1993), no. 1, 37–69, DOI 10.2307/2154545. MR1091706

[HR] I. Herzog and Ph. Rothmaler, *When cotorsion modules are pure injective*, J. Math. Log. **9** (2009), no. 1, 63–102, DOI 10.1142/S0219061309000835. MR2665782
and its Applications, vol. 42, Cambridge University Press 1993.

[H] W. Hodges, *Model theory*, Encyclopedia of Mathematics and its Applications, vol. 42, Cambridge University Press, Cambridge, 1993. MR1221741

[P1] M. Prest, *Model theory and modules*, London Mathematical Society Lecture Note Series, vol. 130, Cambridge University Press, Cambridge, 1988. MR933092

[P2] M. Prest, *Purity, spectra and localisation*, Encyclopedia of Mathematics and its Applications, vol. 121, Cambridge University Press, Cambridge, 2009. MR2530988

[PR] G. Puninski and Ph. Rothmaler, *Pure-projective modules*, J. London Math. Soc. (2) **71** (2005), no. 2, 304–320, DOI 10.1112/S0024610705006290. MR2122430

[RG] M. Raynaud and L. Gruson, *Critères de platitude et de projectivité. Techniques de "platification" d'un module* (French), Invent. Math. **13** (1971), 1–89, DOI 10.1007/BF01390094. MR0308104

[R] Ph. Rothmaler, *Mittag-Leffler modules and positive atomicity*, Habilitationsschrift, Christian-Albrechts-Universität zu Kiel 1994.

[Rot1] Ph. Rothmaler, *Mittag-Leffler modules*, Ann. Pure Appl. Logic **88** (1997), no. 2-3, 227–239, DOI 10.1016/S0168-0072(97)00024-9. Joint AILA-KGS Model Theory Meeting (Florence, 1995). MR1600915

[Rot2] Ph. Rothmaler, *Purity in model theory*, Advances in algebra and model theory (Essen, 1994; Dresden, 1995), Algebra Logic Appl., vol. 9, Gordon and Breach, Amsterdam, 1997, pp. 445–469. MR1687736

[Rot3] Ph. Rothmaler, *Torsion-free, divisible, and Mittag-Leffler modules*, Comm. Algebra **43** (2015), no. 8, 3342–3364, DOI 10.1080/00927872.2014.918990. MR3354094

[S] B. Stenström, *Rings of quotients: An introduction to methods of ring theory*, Springer-Verlag, New York-Heidelberg, 1975. MR0389953

[Zie] M. Ziegler, *Model theory of modules*, Ann. Pure Appl. Logic **26** (1984), no. 2, 149–213, DOI 10.1016/0168-0072(84)90014-9. MR739577

[Zim] W. Zimmermann, *Rein injektive direkte Summen von Moduln* (German), Comm. Algebra **5** (1977), no. 10, 1083–1117, DOI 10.1080/00927877708822211. MR0450327

[Z-HZ] B. Zimmermann-Huisgen and W. Zimmermann, *On the sparsity of representations of rings of pure global dimension zero*, Trans. Amer. Math. Soc. **320** (1990), no. 2, 695–711, DOI 10.2307/2001697. MR965304

CUNY, The Graduate Center, 365 Fifth Avenue, Room 4208, New York, New York 10016

Email address: philipp.rothmaler@bcc.cuny.edu

Contemporary Mathematics
Volume **730**, 2019
https://doi.org/10.1090/conm/730/14717

Intrinsic valuation entropy

Luigi Salce and Simone Virili

Dedicated to Mike Prest on the occasion of his 65th *birthday*

ABSTRACT. We extend the notion of intrinsic entropy for endomorphisms of Abelian groups to endomorphisms of modules over an Archimedean non-discrete valuation domain R, using the natural non-discrete length function introduced by Northcott and Reufel for such a category of modules. We prove that this notion of entropy is a length function for the category of $R[X]$-modules, it satisfies (a suitably adapted version of) the Intrinsic Algebraic Yuzvinski Formula and that it is essentially the unique invariant for $\mathrm{Mod}(R[X])$ with these properties.

1. Introduction

Several different notions of entropy have been introduced in the past few years in the algebraic setting, allowing in-depth investigation of the dynamical behavior of endomorphisms of Abelian groups (see [**DGSZ09**, **DGB16**, **DGBSV15**, **SZ09**]). Among these notions, the *intrinsic algebraic entropy*, introduced and investigated in [**DGBSV15**] and denoted by $\widetilde{\mathrm{ent}}$, is helpful for several reasons: it is meaningful for any kind of groups (not only for torsion groups); it satisfies the Addition Theorem on the whole category of Abelian groups; it is easily computable for linear transformations of finite dimensional vector spaces over the field of the rational numbers via the Intrinsic Algebraic Yuzvinski Formula (see [**Y68**, **LW88**, **DGBSV15**, **GBV12**, **SV17**]).

The first notion of algebraic entropy for endomorphisms of discrete groups, introduced in [**AKM65**, **Wei74**] and denoted here by ent, has been deeply investigated in [**DGSZ09**], and was then extended to modules over arbitrary rings via the tool of "sub-additive invariants" (see [**SZ09**]), and later on widely explored in [**SVV13**] using the notion of "discrete length functions". On the other hand, the intrinsic algebraic entropy $\widetilde{\mathrm{ent}}$ did not receive a similar attention. Instead of trying to develop a general theory of the intrinsic algebraic entropy induced by a length function on the category of modules over an arbitrary ring, we pursue

---

2010 *Mathematics Subject Classification.* Primary 13F30, 16S50; Secondary 13C60, 37A35.

Key words and phrases. Valuation domain, endomorphism, intrinsic valuation entropy, inert submodule.

The second named author is supported by the Ministerio de Economía y Competitividad of Spain via a grant 'Juan de la Cierva-Formación'. He is also supported by a research project from the Ministerio de Economía y Competitividad of Spain (MTM2014-53644-P) and from the Fundación 'Séneca' of Murcia (19880/GERM/15), both with a part of FEDER funds.

an intermediate goal, approaching this matter from a particular, but significant, point of view. The goal of this paper is to introduce and investigate the intrinsic algebraic entropy $\widetilde{\mathrm{ent}}_v$ over an archimedean non-discrete valuation domain R with valuation $v\colon Q \to \Gamma \cup \{\infty\}$, where Q denotes the field of quotients of R and Γ is a dense subgroup of the totally ordered additive group $\mathbb{R}$ of the real numbers. In this setting there is a natural length function $L_v\colon \mathrm{Mod}(R) \to \mathbb{R}_{\geq 0} \cup \{\infty\}$ induced by the valuation v, as proved by Northcott-Reufel in [**NR65**]. We will call the entropy $\widetilde{\mathrm{ent}}_v$ the *intrinsic valuation entropy.*

This setting has been already considered by Zanardo in [**Zan11**], who studied the valuation entropy ent_v induced by the length function L_v. Similarly to the Abelian group setting, this entropy is useful only for endomorphisms ϕ of torsion R-modules M. In particular, Zanardo computed $\mathrm{ent}_v(\phi)$ when M is the cyclic ϕ-trajectory generated by a torsion element (for these notions see the next section).

The particular point of view considered in this paper is significant for three reasons. Firstly, as proved by Northcott-Reufel in [**NR65**], in this setting all the non-trivial non-discrete length functions over arbitrary valuation domains are multiples of the length functions L_v mentioned above. Secondly, non-discrete valuation domains are a first nice example of non-Noetherian rings, for which some techniques used in the investigation of $\widetilde{\mathrm{ent}}$ and holding for Noetherian rings are not applicable. We need new techniques blending those used in [**DGBSV15**] for Abelian groups and those used in [**SV16**] working with non-discrete length functions. Finally, in [**SV16**] we were able to prove the Uniqueness Theorem for algebraic entropies induced by non-discrete length functions only in the particular setting of archimedean non-discrete valuation domains and of the length function L_v above; this fact indicates that this setting is the appropriate first step to be investigated in the non-Noetherian case.

Structure of the paper. In Section 2 we give the definition of the intrinsic algebraic entropy $\widetilde{\mathrm{ent}}_v$ and prove its basic properties needed in the proof of the main results appearing in the next sections. We also prove that, similarly to what happens in the Abelian group setting, for endomorphisms ϕ of torsion R-modules the equality $\widetilde{\mathrm{ent}}_v(\phi) = \mathrm{ent}_v(\phi)$ holds.

In Section 3 we prove that $\widetilde{\mathrm{ent}}_v$ is an upper-continuous invariant for the category $\mathrm{Mod}(R[X])$, looking as usual at an R-module M with an endomorphism $\phi : M \to M$ as an $R[X[$-module via the action induced by ϕ; we denote this module by M_ϕ and we set $\widetilde{\mathrm{ent}}_v(M_\phi) = \widetilde{\mathrm{ent}}_v(\phi)$. We also provide a formula characterizing $\widetilde{\mathrm{ent}}_v(M_\phi)$ as a supremum of lengths of suitable R-modules, a crucial tool in proving the Addition Theorem.

The long and articulate demonstration of this fundamental result takes all Section 4. It is divided in two parts: first we prove the sub-additivity of $\widetilde{\mathrm{ent}}_v$, and then we prove with more efforts its super-additivity. The idea of the Addition Theorem can be found already in the Yuzvinski's paper [**Y68**] for particular cases. It was proved for the algebraic entropy for torsion Abelian groups in [**DGSZ09**], for the whole category of Abelian groups for the intrinsic entropy in [**DGBSV15**], for the L-entropy for an arbitrary category of modules with a discrete length function L in [**SV16**], and elsewhere.

In Section 5 we prove the Intrinsic Algebraic Yuzvinski Formula (IAYF, for short), in a suitable formulation adapted to the present situation. We will explain where this theorem comes from, comparing its statement with the analogous formula for the intrinsic algebraic entropy $\widetilde{\mathrm{ent}}$ for Abelian groups.

In Section 6 we prove the Uniqueness Theorem, which states that the intrinsic valuation entropy is the unique length function for the category $\mathrm{Mod}(R[X])$ satisfying the IAYF and such that, composed with the tensor functor $-\otimes_R R[X]$, equals the length function L_v. In the literature, various forms of uniqueness theorems have been proved; for the algebraic entropy for discrete torsion Abelian groups in [**DGSZ09**], for the intrinsic algebraic entropy for arbitrary groups in [**DGBSV15**], for the valuation entropy in [**SV16**], etc.; they have been inspired by and have a topological counterpart in [**ST87**].

2. Definitions and preliminary facts

Let R be a ring and let $\mathcal{C}$ be a Serre subcategory of $\mathrm{Mod}(R)$. A function

$$L\colon \mathcal{C} \to \mathbb{R}_{\geq 0} \cup \{\infty\} = \mathbb{R}^*,$$

is an *invariant* if $L(0) = 0$ and $L(M) = L(M')$ whenever $M \cong M'$. Furthermore,

- L is *additive* if $L(M) = L(M') + L(M'')$ for any short exact sequence $0 \to M' \to M \to M'' \to 0$ in $\mathcal{C}$;
- L is *upper continuous* if $L(M) = \sup\{L(F) : F \leq M,\ F$ finintely generated$\}$.

An invariant which is both additive and upper continuous is said to be a *length function.* If the finite values of a given length function L form a discrete subset of $\mathbb{R}_{\geq 0}$, then L is said to be *discrete.*

2.1. The general setting. In this paper we always assume that R is a non-discrete archimedean valuation domain, that is, R is 1-dimensional and its value group $\Gamma(R)$ is a dense subgroup of the additive group of the real numbers $\mathbb{R}$, with valuation $v : Q \to \Gamma(R) \cup \{\infty\}$, where Q is the field of fractions of R. We denote by P the maximal ideal of R, which is not cyclic; every ideal in R is either cyclic or countably generated. For a comprehensive treatment of modules over valuation domains we refer to [**FS85**] and [**FS01**].

Northcott and Reufel [**NR65**] proved that, under our hypotheses on R, there is a unique (up to scalar multiplication) length function in $\mathrm{Mod}(R)$ whose values do not form a discrete subset of $\mathbb{R}^*$. This function, that we denote here by

$$L_v : \mathrm{Mod}(R) \to \mathbb{R}^*,$$

is determined by the values on cyclic modules via the following formula:

$$L_v(R/I) := \inf_{r\in I} v(r).$$

For an R-module M, the value $L_v(M)$ will be called the *valuation length* of M; if $L_v(M) < \infty$, we will sometime say that M is L_v-finite.

Obviously, $L_v(R) = \infty$ and $L_v(R/P) = 0$. More generally, any R-module which is not torsion has infinite valuation length, since it contains a submodule isomorphic to R, while it is immediate to check that $L_v(M) = 0$ if and only if M is semisimple. Furthermore, as noticed by Zanardo [**Zan11**], every finitely generated torsion R-module has finite valuation length. The following lemmas give criteria for a torsion module to have finite valuation length and they will be quite important for us in the forthcoming sections.

LEMMA 2.1. *Let F be a finitely generated R-module. Then, any torsion submodule of F has finite valuation length.*

PROOF. It is well known (see [**FS01**, V.2.9]) that the torsion part $t(F)$ of F is a summand of F, so it is finitely generated. By the above discussion, $L_v(t(F)) < \infty$ and, given a torsion submodule T of F, the inclusion $T \leq t(F)$ gives that $L_v(T) \leq L_v(t(F)) < \infty$. □

LEMMA 2.2. *Let $M = Q^n$ for some positive integer n, and let $R^n \leq K \leq N \leq Q^n$ be submodules such that N/K is a bounded R-module. Then, $L_v(N/K) < \infty$.*

PROOF. By [**FS01**, XII.1.1], N/K is weakly polyserial, that is, it has a chain of submodules $0 = H_0 < H_1 < \cdots < H_t = N/K$ such that H_i/H_{i-1} is uniserial for all $i \leq t$. As N/K is bounded, every factor H_i/H_{i-1} is such. Since uniserial modules over an archimedean valuation domain are either cyclic or countably generated (see [**FS85**, VII.1]), every factor H_i/H_{i-1} is standard uniserial, that is, isomorphic to a module of the form J/I, for $0 < I < J \leq R$ (see [**FS85**, VII.1.3]), so it has finite valuation length. Thus, $L_v(N/K) = \sum_{i=1}^t L_v(H_i/H_{i-1}) < \infty$, as desired. □

Let us also remark that there are plenty of non-finitely generated modules of finite valuation length; for example, if $\Gamma(R)$ contains $\mathbb{Z}[1/2]$, the subring of $\mathbb{Q}$ generated by $1/2$, and $M = \bigoplus_{n\geq 1} R/I_n$, with $L_v(R/I_n) = 2^{-n}$, then

$$L_v(M) = \sum_{n\geq 1} L_v(R/I_n) = \sum_{n\geq 1} 2^{-n} = 1.$$

Let us conclude this subsection recalling that there are, up to scalar multiplication, only two more non-trivial length functions of $\mathrm{Mod}(R)$ (where by "non-trivial" we mean functions that assume some finite non-zero value): the *composition length*

$$\ell\colon \mathrm{Mod}(R) \to \mathbb{R}^*,$$

characterized by the fact that $\ell(R/P) = 1$, and the *torsion-free rank*

$$\mathrm{rk}_R\colon \mathrm{Mod}(R) \to \mathbb{R}^*,$$

characterized by the fact that $\mathrm{rk}_R(R) = 1$. Notice that the set of finite values of both these functions is $\mathbb{N}$, which is clearly a discrete subset of $\mathbb{R}$.

2.2. Trajectories, anti-trajectories, and inert submodules. Let us start fixing some conventions for $R[X]$-modules. Indeed, we use the notation M_ϕ, with M an R-module and $\phi \in \mathrm{End}_R(M)$, to denote the $R[X]$-module $M_{R[X]}$, where X acts on M as ϕ. An R-homomorphism $\alpha\colon M \to N$ induces an $R[X]$-homomorphism $M_\phi \to N_\psi$ if and only if $\alpha \cdot \phi = \psi \cdot \alpha$. For more details on these notions we refer to [**SVV13**].

Given an $R[X]$-module M_ϕ, in this section we introduce and study a class of R-submodules of M called ϕ-inert. This family of submodules is meant to substitute the family of L_v-finite submodules of M in the definition of the valuation entropy ent_v. For this reason we will proceed as follows: we first recall the definition of ent_v and the role played by the L_v-finite submodules and their trajectories; in the second part of the subsection we introduce the (valuation) inert submodules, setting the bases for defining the intrinsic valuation entropy $\widetilde{\mathrm{ent}}_v$, and we use the notion of anti-trajectory to produce a new, better behaved, inert submodule from a given one.

Fix an $R[X]$-module M_ϕ. For any R-submodule K of M and any integer $n \geq 1$ we define the *partial n-th trajectory* of K as follows

$$T_n(\phi, K) = K + \phi K + \phi^2 K + \ldots + \phi^{n-1} K.$$

Similarly, the (full) *trajectory* of K is the following submodule of M:

$$T(\phi, K) = \bigcup_{n>0} T_n(\phi, K) = \sum_{n>0} \phi^n K.$$

Notice that $T(\phi, K)$ is the $R[X]$-submodule of M_ϕ generated by K. The trajectory of a cyclic submodule $xR \leq M$ is simply denoted by $T(\phi, x)$.

Following the general treatment of [**SZ09**] and [**SVV13**], the *(algebraic) valuation entropy* $\mathrm{ent}_v(M_\phi)$ was defined in [**Zan11**] as follows. Let

$$\mathrm{Fin}_v(M) = \{K \leq M : L_v(K) < \infty\}.$$

Then, given $K \in \mathrm{Fin}_v(M)$, set

$$\mathrm{ent}_v(\phi, K) = \lim_{n\to\infty} \frac{L_v(T_n(\phi, K))}{n};$$

this limit exists finite by the well-known Fekete's Lemma. In fact, one can prove as in [**SV16**, Prop. 3.2] that

$$\mathrm{ent}_v(\phi, K) = \inf_n L_v\left(\frac{T_{n+1}(\phi, K)}{T_n(\phi, K)}\right). \tag{2.1}$$

One finally defines the *valuation entropy* of ϕ as

$$\mathrm{ent}_v(M_\phi) = \sup_{K\in \mathrm{Fin}_v(M)} \mathrm{ent}_v(\phi, K).$$

We will commonly use also the notation $\mathrm{ent}_v(\phi)$ to denote $\mathrm{ent}_v(M_\phi)$.

The formula (2.1), which has its analog for the classical algebraic entropy ent in the setting of Abelian groups, suggests that $\mathrm{ent}_v(\phi, K)$ can be defined not just for L_v-finite submodules, but for any submodule K such that $L_v(T_{n+1}(\phi, K)/T_n(\phi, K))$ is finite for all $n \geq 1$. As it turns out, this happens precisely when $L_v((K+\phi K)/K)$ is finite. This motivates the following definition.

DEFINITION 2.3. In the above setting, an R-submodule K of M is *(valuation) ϕ-inert* provided $L_v((K + \phi K)/K) < \infty$.

The notion of inert subgroups defined in [**DGBSV15**] was originated by [**B93**] and gave rise to related notions in non-commutative group theory investigated in [**DR12**]. The family of all the ϕ-inert submodules of M is denoted by $\mathcal{I}_\phi(M)$. Proceeding as in [**DGBSV15**, Lem. 2.1], one can verify that, if H is ϕ-inert in M, then $L_v(T_n(\phi, H)/H) < \infty$ for all $n \geq 1$ and, consequently, $T(\phi, H)/H$ is a torsion module. This fact will play a crucial role in the proof of Proposition 4.5, a main step in proving the super-additivity of the intrinsic valuation entropy.

LEMMA 2.4. *Let $\phi : M \to M$ be an endomorphism of an R-module M, H a ϕ-invariant submodule and K a ϕ-inert submodule of M. Then $K \cap H$ is $\phi_{\restriction H}$-inert in H.*

PROOF. Consider the following isomorphisms

$$\frac{(K\cap H)+\phi(K\cap H)}{K\cap H}\cong\frac{\phi(K\cap H)}{\phi(K\cap H)\cap K\cap H}\cong\frac{K+\phi(K\cap H)}{K}.$$

The last term is a submodule of $(K+\phi K)/K$, therefore $L_v\left(\frac{(K\cap H)+\phi(K\cap H)}{K\cap H}\right)<\infty$. □

The next technical lemma will be useful in the proofs of Popositions 4.7 and 4.10.

LEMMA 2.5. *Let M_ϕ be an $R[X]$-module such that $\mathrm{rk}_R(M)=k<\infty$. Then any R-submodule $K\leq M$ such that $K/tK\cong R^k$ and $L_v(tK)<\infty$ is ϕ-inert. In particular, if M is torsion-free and F is a free submodule of maximal rank, then F is ϕ-inert.*

PROOF. Since $K/tK\cong R^k$, we get $K=tK\oplus F$, where $F\cong R^k$. Consider the following short exact sequence

$$0\to\frac{t(K+\phi K)}{tK}\to\frac{K+\phi K}{K}\to\frac{(K+\phi K)+tM}{K+tM}\to 0$$

where the module on the right-hand side is L_v-finite, since it is torsion ($K+\phi K+tM$ and $K+tM$ have the same rank) and finitely generated (it is isomorphic to a quotient of R^k). On the other hand, to prove that the module on the left-hand side is L_v-finite, it is enough to show that $L_v(t(K+\phi K))<\infty$, since by hypothesis $L_v(tK)<\infty$. But $t(K+\phi K)=tK+\phi(tK)+t(F+\phi F)$. The first two summands are L_v-finite, while the third summand, as the torsion part of a finitely generated module, is also L_v-finite by Lemma 2.1. Then the conclusion follows. □

We are now going to introduce a construction that takes as input a given inert submodule K and produces as output a bigger inert submodule $A(\phi,K)$ with better properties. This procedure is dual to a construction introduced by Willis in [**Wil15**], in studying the scale function of continuous endomorphisms of totally disconnected locally compact groups. Indeed, for an $R[X]$-module M_ϕ and an R-submodule $K\leq M$, we define, by induction on $n\geq 1$, the *(partial) n-th anti-trajectory of K* as follows:

- $A_1(\phi,K)=K$;
- $A_{n+1}(\phi,K)=K+\phi^{-1}A_n(\phi,K)$.

The *(full) anti-trajectory of K* is then defined by

$$A(\phi,K)=\bigcup_n A_n(\phi,K).$$

In the following lemma we show that there is a close relationship between trajectories and anti-trajectories of a given submodule.

LEMMA 2.6. *In the above notation, the following statements hold true:*

(1) $\phi^n(\phi^{-1}A_n(\phi,K))\cap\phi^nK=T_n(\phi,K)\cap\phi^nK$;
(2) $T_{n+1}(\phi,K)/T_n(\phi,K)\cong A_{n+1}(\phi,K)/\phi^{-1}A_n(\phi,K)$, *for all* $n\geq 1$.

PROOF. (1) Let us start proving that, for all $n\geq 1$, $\phi^n(\phi^{-1}A_n(\phi,K))\subseteq T_n(\phi,K)$. We proceed by induction on n. For $n=1$ we get $\phi(\phi^{-1}K)=K\cap\mathrm{Im}(\phi)\subseteq$

K. For $n > 1$,

$$\begin{aligned}\phi^n(\phi^{-1}A_n(\phi,K)) &= \phi^{n-1}(A_n(\phi,K)) \cap \mathrm{Im}(\phi) \subseteq \phi^{n-1}A_n(\phi,K)\\ &= \phi^{n-1}K + \phi^{n-1}(\phi^{-1}A_{n-1}(\phi,K))\\ &\subseteq \phi^{n-1}K + T_{n-1}(\phi,K) = T_n(\phi,K).\end{aligned}$$

On the other hand, a generic element $x \in T_n(\phi,K) \cap \phi^n K$ can be written as $x = \phi^n k_n = k_0 + \phi k_1 + \ldots + \phi^{n-1}k_{n-1}$, for $k_0, k_1, \ldots, k_n \in K$. Hence, the equality $\phi(\phi^{n-1}k_n - \phi^{n-2}k_{n-1} - \ldots - k_1) = k_0$ implies that

$$\phi^{n-1}k_n - \phi^{n-2}k_{n-1} - \ldots - k_1 \in \phi^{-1}K.$$

Furthermore, the fact that $\phi(\phi^{n-2}k_n - \phi^{n-3}k_{n-1} - \ldots - k_2) \in k_1 + \phi^{-1}K \subseteq K + \phi^{-1}K$, implies that

$$\phi^{n-2}k_n - \phi^{n-3}k_{n-1} - \ldots - k_2 \in \phi^{-1}(K + \phi^{-1}K)$$

Going on this way, one proves that $k_n \in \phi^{-1}A_n(\phi,K)$, and so $x = \phi^n k_n \in \phi^n K \cap \phi^n(\phi^{-1}A_n(\phi,K))$. This show that $\phi^n(\phi^{-1}A_n(\phi,K)) \cap \phi^n K \supseteq T_n(\phi,K) \cap \phi^n K$, so we are done.

(2) This is a consequence of part (1); in fact,

$$\begin{aligned}\frac{T_{n+1}(\phi,K)}{T_n(\phi,K)} &\cong \frac{\phi^n K}{T_n(\phi,K) \cap \phi^n K}\\ &= \frac{\phi^n K}{\phi^n(\phi^{-1}A_n(\phi,K)) \cap \phi^n K}\\ &\cong \frac{\phi^n(K + \phi^{-1}A_n(\phi,K))}{\phi^n(\phi^{-1}A_n(\phi,K))}\\ &\cong \frac{\phi^n(A_{n+1}(\phi,K))}{\phi^n(\phi^{-1}A_n(\phi,K))}.\end{aligned}$$

To conclude we should just prove that the following map, induced by ϕ^n:

$$f \colon \frac{A_{n+1}(\phi,K)}{\phi^{-1}A_n(\phi,K)} \to \frac{\phi^n(A_{n+1}(\phi,K))}{\phi^n(\phi^{-1}A_n(\phi,K))}$$

is an isomorphism. In fact, it is clearly surjective, while its injectivity follows from the fact that $\ker(\phi^n) \subseteq \phi^{-n}K \subseteq \phi^{-1}A_n(\phi,K)$. □

We describe in the next proposition the trajectories of $A(\phi,K)$, showing that $A(\phi,K)$ is ϕ-inert provided K has the same property.

PROPOSITION 2.7. *In the above notation, the following statements hold true:*

(1) $\phi^{-1}A(\phi,K) \subseteq A(\phi,K)$, *so* $A(\phi,K) \cap \phi A(\phi,K) = A(\phi,K) \cap \mathrm{Im}(\phi)$ *and, more generally, for every* $n \geq 1$, $\phi^{n-1}A(\phi,K) \cap \phi^n A(\phi,K) = \phi^{n-1}A(\phi,K) \cap \mathrm{Im}(\phi^n)$;
(2) $T_n(\phi,A(\phi,K)) \cap \phi^n A(\phi,K) = \phi^{n-1}(\phi A(\phi,K) \cap A(\phi,K))$, *for all* $n \geq 1$;
(3) $T_{n+1}(\phi,A(\phi,K))/T_n(\phi,A(\phi,K)) \cong A(\phi,K)/\phi^{-1}A(\phi,K)$, *for all* $n \geq 1$;
(4) *if* K *is* ϕ*-inert, then so is* $A(\phi,K)$.

PROOF. (1) By definition, $\phi^{-1}A_n(\phi, K) \subseteq A_{n+1}(\phi, K) \subseteq A(\phi, K)$, for all $n \geq 1$. Thus,

$$\phi^{-1}A(\phi, K) = \phi^{-1}\left(\bigcup_{n\geq 1} A_n(\phi, K)\right) = \bigcup_{n\geq 1} \phi^{-1}A_n(\phi, K) \subseteq A(\phi, K).$$

The inclusion $A(\phi, K) \cap \phi A(\phi, K) \subseteq A(\phi, K) \cap \mathrm{Im}(\phi)$ is obvious; the converse follows from the inclusion $\phi(\phi^{-1}A(\phi, K)) \subseteq \phi(A(\phi, K))$ and $\phi(\phi^{-1}A(\phi, K)) = A(\phi, K) \cap \mathrm{Im}(\phi)$. A similar argument proves that $\phi^{n-1}A(\phi, K) \cap \phi^n A(\phi, K) = \phi^{n-1}A(\phi, K) \cap \mathrm{Im}(\phi^n)$.

(2) The inclusion $\supseteq$ is clear. For the converse inclusion, let $x = \phi^{n-1}k_{n-1} + \ldots + \phi k_1 + k_0 = \phi^n k_n \in T_n(\phi, A(\phi, K)) \cap \phi^n A(\phi, K)$ ($k_i \in A(\phi, K)$), and let us show that $x \in \phi^{n-1}(\phi A(\phi, K) \cap A(\phi, K))$. Since $\ker(\phi^{n-1}) \subseteq \phi^{-n+1}A(\phi, K) \subseteq A(\phi, K)$, we get

$$\phi^{n-1}(\phi A(\phi, K) \cap A(\phi, K)) = \phi^n A(\phi, K) \cap \phi^{n-1}A(\phi, K).$$

Hence, it is enough to show that $x \in \phi^{n-1}A(\phi, K)$. Indeed, the equality

$$k_0 = \phi^n k_n - (\phi^{n-1}k_{n-1} + \ldots + \phi k_1)$$

shows that $k_0 \in A(\phi, K) \cap \mathrm{Im}(\phi) = A(\phi, K) \cap \phi A(\phi, K)$. Hence, there is $k_1' \in A(\phi, K)$ such that $\phi k_1' = \phi k_1 + k_0$. Hence,

$$\phi k_1' = \phi^n k_n - (\phi^{n-1}k_{n-1} + \ldots + \phi^2 k_2)$$

that is, $\phi(k_1') \in \phi A(\phi, K) \cap \mathrm{Im}(\phi^2) = \phi A(\phi, K) \cap \phi^2(A(\phi, K))$. Hence, there is $k_2' \in A(\phi, K)$, such that $\phi^2 k_2' = \phi^2 k_2 + \phi k_1' = \phi^2 k_2 + \phi k_1 + k_0$. Going on this way, we obtain an element $k_{n-1}' \in A(\phi, K)$ such that

$$\phi^{n-1}k_{n-1}' = \phi^{n-1}k_{n-1} + \ldots + \phi k_1 + k_0 = \phi^n k_n$$

concluding the proof of our claim.

(3) Consider the following sequence of isomorphisms

$$\begin{aligned}\frac{T_{n+1}(\phi, A(\phi, K))}{T_n(\phi, A(\phi, K))} &\cong \frac{\phi^n A(\phi, K)}{\phi^n A(\phi, K) \cap T_n(\phi, A(\phi, K))}\\ &= \frac{\phi^n A(\phi, K)}{\phi^{n-1}(\phi A(\phi, K) \cap A(\phi, K))}\\ &\cong \frac{\phi^{n-1}(\phi A(\phi, K) + A(\phi, K))}{\phi^{n-1}A(\phi, K)}\\ &\cong \frac{\phi A(\phi, K) + A(\phi, K)}{A(\phi, K)}\end{aligned}$$

where the last isomorphism is induced by ϕ^{n-1}, using that $\ker(\phi^{n-1}) \subseteq A(\phi, K)$. Finally, by part (1),

$$\frac{\phi A(\phi, K) + A(\phi, K)}{A(\phi, K)} \cong \frac{\phi A(\phi, K)}{A(\phi, K) \cap \phi A(\phi, K)} = \frac{\phi A(\phi, K)}{A(\phi, K) \cap \mathrm{Im}(\phi)} \cong \frac{A(\phi, K)}{\phi^{-1}A(\phi, K)}$$

where the last isomorphism is induced by ϕ.

(4) Notice that $K + \phi^{-1}A(\phi, K) = A(\phi, K)$. In fact, $K + \phi^{-1}A(\phi, K) \subseteq A(\phi, K)$ by the first part, while for all $n \in \mathbb{N}$, $A_{n+1}(\phi, K) = K + \phi^{-1}A_n(\phi, K) \subseteq K + \phi^{-1}A(\phi, K)$. Applying ϕ to the equality $K + \phi^{-1}A(\phi, K) = A(\phi, K)$, we obtain

that $\phi K + (A(\phi, K) \cap \mathrm{Im}(\phi)) = \phi A(\phi, K)$; in particular, $\phi A(\phi, K) + A(\phi, K) = \phi K + A(\phi, K)$. Therefore

$$\frac{A(\phi, K) + \phi A(\phi, K)}{A(\phi, K)} = \frac{\phi K + A(\phi, K)}{A(\phi, K)} \cong \frac{\phi K}{\phi K \cap A(\phi, K)}$$

is an epic image of

$$\frac{\phi K}{\phi K \cap K} \cong \frac{K + \phi K}{K}$$

showing that $(A(\phi, K) + \phi A(\phi, K))/A(\phi, K)$ has finite valuation length, that is, $A(\phi, K)$ is ϕ-inert. □

2.3. The intrinsic valuation entropy. Let us begin this subsection defining the main object of study for this paper.

DEFINITION 2.8. Consider an $R[X]$-module M_ϕ, and let $H \in \mathcal{I}_\phi(M)$. The intrinsic valuation entropy of ϕ with respect to H is

$$\widetilde{\mathrm{ent}}_v(\phi, H) = \lim_{n\to\infty} \frac{L_v(T_n(\phi, H)/H)}{n}.$$

The intrinsic valuation entropy of M_ϕ is then defined as

$$\widetilde{\mathrm{ent}}_v(M_\phi) = \sup_{H \in \mathcal{I}_\phi(M)} \widetilde{\mathrm{ent}}_v(\phi, H).$$

We will commonly use also the notation $\widetilde{\mathrm{ent}}_v(\phi)$ to denote $\widetilde{\mathrm{ent}}_v(M_\phi)$.

Mixing the proofs of [**DGBSV15**, Lem. 3.2] and [**SV16**, Prop. 3.2], we obtain the following

PROPOSITION 2.9. *Given an $R[X]$-module M_ϕ and $H \in \mathcal{I}_\phi(M)$,*

$$\widetilde{\mathrm{ent}}_v(\phi, H) = \inf_n L_v\left(\frac{T_{n+1}(\phi, H)}{T_n(\phi, H)}\right).$$

PROOF. In order to simplify notation set $T_n(\phi, H) = T_n$ for all $n \geq 1$. As proved in [**SZ09**], T_{n+1}/T_n is a quotient of T_n/T_{n-1} for all $n > 1$, hence

$$L_v((T_{n+1}/H)/(T_n/H)) \leq L_v((T_n/H)/(T_{n-1}/H)).$$

Let $\alpha = \inf_n L_v((T_n/H)/(T_{n-1}/H))$ and fix (arbitrarily) $\epsilon > 0$. Then there exists an index n_0 such that

$$L_v((T_n/H)/(T_{n-1}/H)) < \alpha + \epsilon$$

for all $n \geq n_0$. Using the additivity of L_v, one can prove by induction on k that

$$L_v(T_{n_0+k}/H) = L_v(T_{n_0}/H) + \sum_{0\leq i\leq k-1} L_v((T_{n_0+i+1}/H)/(T_{n_0+i}/H))$$

which implies

$$L_v(T_{n_0}/H) + k\alpha \leq L_v(T_{n_0+k}/H) \leq L_v(T_{n_0}/H) + k(\alpha + \epsilon). \tag{2.2}$$

Using the second inequality in (2.2) we obtain:

$$\widetilde{\mathrm{ent}}_v(\phi, H) = \lim_{k\to\infty} \frac{L_v(T_{n_0+k}/H)}{n_0 + k} \leq \lim_{k\to\infty} \frac{L_v(T_{n_0}/H) + k(\alpha + \epsilon)}{n_0 + k} = \alpha + \epsilon.$$

(2) Being M torsion, the finitely generated submodule $F \leq H$ has finite valuation length, hence it is ϕ-inert in M. For each $n \geq 1$, using the exact sequence

$$0 \to \frac{T_n(\phi, F) \cap H}{F} \to \frac{T_n(\phi, F)}{F} \to \frac{T_n(\phi, F) + H}{H} \to 0$$

we get:

$$\begin{aligned} &L_v\left(\frac{T_n(\phi, H)}{H}\right) - L_v\left(\frac{T_n(\phi, F)}{F}\right) \\ &= L_v\left(\frac{T_n(\phi, H)}{H}\right) - \left(L_v\left(\frac{(T_n(\phi, F) + H)}{H}\right) + L_v\left(\frac{(T_n(\phi, F) \cap H)}{F}\right)\right) \\ &\leq L_v\left(\frac{T_n(\phi, H)}{H}\right) - L_v\left(\frac{T_n(\phi, F) + H}{H}\right) < n\epsilon. \end{aligned}$$

Dividing by n and passing to the limit we get $\widetilde{\mathrm{ent}}_v(\phi, H) - \widetilde{\mathrm{ent}}_v(\phi, F) \leq \epsilon$. □

We can now prove the announced result for torsion modules.

PROPOSITION 2.14. *If $\phi : M \to M$ is an endomorphism of a torsion R-module M, then* $\mathrm{ent}_v(\phi) = \widetilde{\mathrm{ent}}_v(\phi)$.

PROOF. The inequality $\mathrm{ent}_v(\phi) \leq \widetilde{\mathrm{ent}}_v(\phi)$ is always true. In order to prove the converse inequality, it is enough to prove that, given a ϕ-inert submodule H of M, $\widetilde{\mathrm{ent}}_v(\phi, H) \leq \mathrm{ent}_v(\phi)$. Fix a real number $\epsilon > 0$ and choose, according to Lemma 2.13, a finitely generated submodule F of H (hence $L_v(F) < \infty$, being M torsion) such that $L_v((H + \phi H)/H) - L_v((H + \phi F)/H) < \epsilon$. By Lemma 2.13 (2), $\widetilde{\mathrm{ent}}_v(\phi, H) - \widetilde{\mathrm{ent}}_v(\phi, F) < \epsilon$. Furthermore, by Corollary 2.10 (2), $\widetilde{\mathrm{ent}}_v(\phi, F) = \mathrm{ent}_v(\phi, F)$. Therefore, for each $\epsilon > 0$ there exists a finitely generated submodule F of H such that

$$\widetilde{\mathrm{ent}}_v(\phi, H) \leq \mathrm{ent}_v(\phi, F) + \epsilon \leq \mathrm{ent}_v(\phi) + \epsilon.$$

As ϵ was arbitrary, we get that $\widetilde{\mathrm{ent}}_v(\phi, H) \leq \mathrm{ent}_v(\phi)$. □

3. Some tools for the computation of entropy

3.1. The intrinsic valuation entropy is upper continuous. In the following proposition, as a consequence of Lemma 2.13, we prove that $\widetilde{\mathrm{ent}}_v$ is continuous with respect to direct limits of ϕ-invariant submodules.

PROPOSITION 3.1. *Let $\phi : M \to M$ be an endomorphism of an R-module which is the direct limit of a family of ϕ-invariant submodules $\{M_i : i \in I\}$. Then*

$$\widetilde{\mathrm{ent}}_v(\phi) = \sup_i \widetilde{\mathrm{ent}}_v(\phi_i),$$

where $\phi_i = \phi \restriction_{M_i}$ for all $i \in I$.

PROOF. The inequality $\geq$ is an immediate consequence of Proposition 2.12. In order to prove the inequality $\leq$, let H be a ϕ-inert submodule of M. Then $H \cap M_i$ is ϕ_i-inert in M_i for all i, by Lemma 2.4. By Lemma 2.13 (1), there exists a finitely generated submodule F of H such that

$$L_v\left(\frac{T_n(\phi, H)}{H}\right) - L_v\left(\frac{T_n(\phi, F) + H}{H}\right) < n\epsilon$$

for each $n \geq 1$. Then F is contained in M_i for a certain i. Consequently $F \leq H \cap M_i$ and $T_n(\phi, F) \leq T_n(\phi_i, H \cap M_i)$ for all $n \geq 1$. As in the proof of [**DGBSV15**, Lem. 3.14] one can show that

$$T_n(\phi_i, H \cap M_i) = (H \cap M_i) + T_n(\phi, F),$$

consequently we have the equality

$$\frac{(H \cap M_i) + T_n(\phi, F)}{H \cap M_i} = \frac{T_n(\phi_i, H \cap M_i)}{H \cap M_i}$$

Let us consider the epimorphism:

$$\frac{(H \cap M_i) + T_n(\phi, F)}{H \cap M_i} \cong \frac{T_n(\phi, F)}{H \cap M_i \cap T_n(\phi, F)} \twoheadrightarrow \frac{T_n(\phi, F)}{H \cap T_n(\phi, F)} \cong \frac{T_n(\phi, F) + H}{H}$$

which shows that $L_v(\frac{T_n(\phi,F)+H}{H}) \leq L_v(\frac{T_n(\phi,F)}{H \cap M_i \cap T_n(\phi,F)}) = L_v(\frac{T_n(\phi_i, H \cap M_i)}{H \cap M_i})$. In conclusion, for each $n \geq 1$:

$$L_v\left(\frac{T_n(\phi, H)}{H}\right) \leq L_v\left(\frac{T_n(\phi, F) + H}{H}\right) + n\epsilon \leq L_v\left(\frac{T_n(\phi_i, H \cap M_i)}{H \cap M_i}\right) + n\epsilon.$$

Dividing by n and passing to the limit, we get:

$$\widetilde{\mathrm{ent}}_v(\phi, H) \leq \widetilde{\mathrm{ent}}_v(\phi_i, H \cap M_i) + \epsilon.$$

Being ϵ arbitrary, we deduce that $\widetilde{\mathrm{ent}}_v(\phi, H) \leq \widetilde{\mathrm{ent}}_v(\phi_i, H \cap M_i)$. From this inequality the conclusion easily follows. □

As an immediate consequence we deduce the upper continuity of $\widetilde{\mathrm{ent}}_v$.

COROLLARY 3.2. *The intrinsic valuation entropy* $\widetilde{\mathrm{ent}}_v$ *is an upper continuous invariant of* $\mathrm{Mod}(R[X])$.

PROOF. Let $\phi : M \to M$ be an endomorphism of the R-module M. Then M, viewed as an $R[X]$-module, is the direct union of the family of its finitely generated $R[X]$-submodules, which are exactly the trajectories $T(\phi, F)$ for F a finitely generated R-submodule of M. Then, Proposition 3.1 ensures that $\widetilde{\mathrm{ent}}_v(\phi) = \sup_{F \in \mathcal{F}(M)} \widetilde{\mathrm{ent}}_v(\phi \restriction_{T(\phi,F)})$. □

3.2. The Limit-Free Formula. In this subsection we prove that, given an $R[X]$-module M_ϕ, the limit computation in the definition of the intrinsic valuation entropy of ϕ can be avoided (see Proposition 3.4). A first, imprecise, claim of a limit-free formula for the entropy of endormorphisms of Abelian groups can be found in the Yuzvinski's paper [**Y68**]. This claim was than proved in a precise and general form in [**DGB12**]. Our statement here is a generalization of [**SV16**, Prop. 5.2], that proves the same formula in case M is torsion.

LEMMA 3.3. *Let* M_ϕ *be an* $R[X]$*-module. Given a* ϕ*-inert submodule* K *of* M*, the following statements hold:*

(1) $\widetilde{\mathrm{ent}}_v(\phi, A(\phi, K)) = L_v(A(\phi, K)/\phi^{-1}A(\phi, K))$;
(2) $\widetilde{\mathrm{ent}}_v(\phi, K) = \widetilde{\mathrm{ent}}_v(\phi, A(\phi, K))$.

PROOF. Part (1) follows from Proposition 2.9 and parts (3) and (4) of Proposition 2.7:

$$\widetilde{\mathrm{ent}}_v(\phi, A(\phi, K)) = \lim_{n\to\infty} L_v\left(\frac{T_{n+1}(\phi, A(\phi, K))}{T_n(\phi, A(\phi, K))}\right) = L_v\left(\frac{A(\phi, K)}{\phi^{-1}A(\phi, K)}\right).$$

(2) For all $n \geq 1$ let $S_n = K \cap \phi^{-1}A_n(\phi, K)$, and $S_\infty = \bigcup_n S_n = K \cap \phi^{-1}A(\phi, K)$. Clearly we have the isomorphisms

$$\frac{K}{S_n} \cong \frac{K + \phi^{-1}A_n(\phi, K)}{\phi^{-1}A_n(\phi, K)} = \frac{A_{n+1}(\phi, K)}{\phi^{-1}A_n(\phi, K)} \quad \text{and} \quad \frac{K}{S_\infty} \cong \frac{A(\phi, K)}{\phi^{-1}A(\phi, K)}.$$

The claim now follows from the following equalities:

$$\begin{aligned}\widetilde{\mathrm{ent}}_v(\phi, K) &= \lim_{n\to\infty} L_v(T_{n+1}(\phi, K)/T_n(\phi, K)) = \lim_{n\to\infty} L_v\left(\frac{A_{n+1}(\phi, K)}{\phi^{-1}A_n(\phi, K)}\right) \\ &= \lim_{n\to\infty} L_v(K/S_n) = L_v(K/S_\infty) = L_v(A(\phi, K)/\phi^{-1}A(\phi, K)),\end{aligned}$$

where the second equality uses part (2) of Lemma 2.6, while the fourth equality uses [**SV16**, Lem. 2.2]. □

It follows from part (1) of the above lemma that, when we consider a ϕ-inert submodule K such that $\phi^{-1}K \subseteq K$ (and thus $K = A(\phi, K)$), we do not need to compute limits to evaluate the intrinsic valuation entropy with respect to K. In fact, it is a consequence of part (2) of the lemma that these ϕ-inert submodules alone suffice to compute the intrinsic valuation entropy.

PROPOSITION 3.4 (Limit-Free Formula). *Let $M_\phi \in \mathrm{Mod}(R[X])$. Then*

$$\widetilde{\mathrm{ent}}_v(M_\phi) = \sup\{L_v(N/\phi^{-1}N) : \ N \leq M \textit{ is } \phi\textit{-inert and } \phi^{-1}N \subseteq N\}.$$

PROOF. Given a ϕ-inert submodule $N \leq M$ such that $\phi^{-1}N \subseteq N$, Lemma 3.3 shows that $\widetilde{\mathrm{ent}}_v(\phi, N) = L_v(N/\phi^{-1}N)$, so that the inequality "$\geq$" in the statement is clear. For the converse inequality, let $K \leq M$ be a ϕ-inert submodule. Then, by Proposition 2.7 and Lemma 3.3, $A(\phi, K)$ is ϕ-inert, $\phi^{-1}A(\phi, K) \subseteq A(\phi, K)$ and $\widetilde{\mathrm{ent}}_v(\phi, A(\phi, K)) = \widetilde{\mathrm{ent}}_v(\phi, K)$. □

4. The Addition Theorem

This section is devoted to the proof of the following result.

THEOREM 4.1. *The intrinsic valuation entropy*

$$\widetilde{\mathrm{ent}}_v \colon \mathrm{Mod}(R[X]) \to \mathbb{R}_{\geq 0} \cup \{\infty\}$$

is a length function.

As we have already verified in Corollary 3.2 that $\widetilde{\mathrm{ent}}_v$ is an upper continuous invariant, we just need to prove its additivity. Indeed, given an $R[X]$-module M_ϕ and an $R[X]$-submodule $N_{\phi\restriction_N}$, we have to verify that

$$\widetilde{\mathrm{ent}}_v(\phi) = \widetilde{\mathrm{ent}}_v(\phi\restriction_N) + \widetilde{\mathrm{ent}}_v(\bar{\phi}) \tag{4.1}$$

where $\bar{\phi} \colon M/N \to M/N$ is the induced map. The proof of this fact is quite involved so we divide it in several steps. In particular, in Subsection 4.1 we use the Limit Free Formula to show that $\widetilde{\mathrm{ent}}_v$ is sub-additive (that is, the inequality $\leq$ in (4.1)) and, in Subsection 4.2, we verify the super-additivity of $\widetilde{\mathrm{ent}}_v$, ending the proof of Theorem 4.1.

4.1. Sub-additivity of $\widetilde{\mathrm{ent}}_v$. Given an R-module M and an endomorphism $\phi\colon M\to M$, the following submodule

$$\ker_\infty(\phi) := \bigcup_{n\in\mathbb{N}} \ker(\phi^n)$$

is called the *hyperkernel of* ϕ. It is a ϕ-invariant submodule of M and, in fact, it is the smallest ϕ-invariant submodule such that the induced endomorphism

$$\bar\phi\colon M/\ker_\infty(\phi)\to M/\ker_\infty(\phi)$$

is injective. Consider now the ring of *Laurent polynomials* $R[X^{\pm1}]$, which can be viewed as the localization of $R[X]$ at the multiplicative set $\{X^n : n\in\mathbb{N}\}$. Consider the tensor product

$$\mathfrak{M}_\Phi = M_\phi\otimes_{R[X]} R[X^{\pm1}].$$

As an $R[X]$-module, $\mathfrak{M}_\Phi$ can be viewed as the direct limit of the following direct system:

$$M_\phi \overset{\phi}{\to} M_\phi \overset{\phi}{\to} M_\phi \overset{\phi}{\to} M_\phi \to \cdots \tag{4.2}$$

It is not difficult to show that the kernel of the canonical map $M_\phi\to\mathfrak{M}_\Phi$ is precisely $\ker_\infty(\phi)$ and that, in fact, identifying $(M/\ker_\infty(\phi))_{\bar\phi}$ as an $R[X]$-submodule of $\mathfrak{M}_\Phi$, we have the following isomorphism in $\mathrm{Mod}(R[X])$:

$$\mathfrak{M}_\Phi \cong \bigcup_{n\in\mathbb{N}} \Phi^{-n}((M/\ker_\infty(\phi))_{\bar\phi}).$$

LEMMA 4.2. *Let M_ϕ be an $R[X]$-module. The following equalities hold:*

(1) $\widetilde{\mathrm{ent}}_v(M_\phi) = \widetilde{\mathrm{ent}}_v((M/\ker_\infty(\phi))_{\bar\phi})$;

(2) $\widetilde{\mathrm{ent}}_v(M_\phi) = \widetilde{\mathrm{ent}}_v(M_\phi\otimes_{R[X]} R[X^{\pm1}])$.

PROOF. (1) By Proposition 3.4,

$$\widetilde{\mathrm{ent}}_v(M_\phi) = \sup\{L_v(N/\phi^{-1}N) :\ N\le M \text{ is } \phi\text{-inert and } \phi^{-1}N\subseteq N\}.$$

Given a ϕ-inert submodule $N\le M$ such that $\phi^{-1}N\subseteq N$, it is clear that $\ker_\infty(\phi)\le N$, so that, letting $\bar N = N/\ker_\infty(\phi)$, we get

$$L_v(N/\phi^{-1}N) = L_v(\phi\bar N/\bar\phi^{-1}\bar N) \le \widetilde{\mathrm{ent}}_v((M/\ker_\infty(\phi))_{\bar\phi}).$$

Thus, $\widetilde{\mathrm{ent}}_v(M_\phi) \le \widetilde{\mathrm{ent}}_v((M/\ker_\infty(\phi))_{\bar\phi})$. The converse inequality follows from Proposition 2.12.

(2) We have proved in Proposition 3.1 that $\widetilde{\mathrm{ent}}_v$ is continuous with respect to direct unions. Hence, by the description of $\mathfrak{M}_\Phi$ as a direct union of copies of $(M/\ker_\infty(\phi))_{\bar\phi}$ and by part (1), we get $\widetilde{\mathrm{ent}}_v(\mathfrak{M}_\Phi) = \widetilde{\mathrm{ent}}_v((M/\ker_\infty(\phi))_{\bar\phi}) = \widetilde{\mathrm{ent}}_v(M_\phi)$. □

LEMMA 4.3. *Let $0\to N_{\phi\restriction_N}\to M_\phi\to (M/N)_{\bar\phi}\to 0$ be a short exact sequence in* $\mathrm{Mod}(R[X^{\pm1}])$. *Then,*

$$\widetilde{\mathrm{ent}}_v(M_\phi) \le \widetilde{\mathrm{ent}}_v(N_{\phi\restriction_N}) + \widetilde{\mathrm{ent}}_v((M/N)_{\bar\phi})\,.$$

PROOF. Let $T\le M$ be a ϕ^{-1}-invariant ϕ-inert submodule; let $T' = T\cap N$ and $\bar T = (T+N)/N$. Clearly T' and $\bar T$ are ϕ^{-1}-invariant and $\bar\phi^{-1}$-invariant, respectively.

Using the equality $N = \phi^{-1}N$ and the fact that ϕ^{-1} commutes with intersection of submodules, we get the following isomorphisms:

$$\frac{T'}{\phi^{-1}T'} \cong \frac{T\cap(\phi^{-1}T+N)}{\phi^{-1}T} \quad , \quad \frac{T}{T\cap(\phi^{-1}T+N)} \cong \frac{\bar{T}}{\bar{\phi}^{-1}\bar{T}} \;.$$

From these isomorphisms we obtain the exact sequence

$$0 \to \frac{T'}{\phi^{-1}T'} \to \frac{T}{\phi^{-1}T} \to \frac{\bar{T}}{\bar{\phi}^{-1}\bar{T}} \to 0 \tag{4.3}$$

that, together with Proposition 3.4, shows that

$$L_v(T/\phi^{-1}T) = L_v(T'/\phi^{-1}T') + L_v(\bar{T}/\bar{\phi}^{-1}\bar{T}) \leq \widetilde{\mathrm{ent}}_v(\phi\restriction_N) + \widetilde{\mathrm{ent}}_v(\bar{\phi}).$$

Being T arbitrary, we obtain that $\widetilde{\mathrm{ent}}_v(\phi) \leq \widetilde{\mathrm{ent}}_v(\phi\restriction_N) + \widetilde{\mathrm{ent}}_v(\bar{\phi})$. □

In order to complete the proof of the sub-additivity, we must pass form $R[X^{\pm1}]$-modules considered in Lemma 4.3 to $R[X]$-modules.

PROPOSITION 4.4. *Consider a short exact sequence in* $\mathrm{Mod}(R[X])$

$$0 \to N_{\phi\restriction_N} \to M_\phi \to (M/N)_{\bar{\phi}} \to 0.$$

Then, $\widetilde{\mathrm{ent}}_v(\phi) \leq \widetilde{\mathrm{ent}}_v(\phi\restriction_N) + \widetilde{\mathrm{ent}}_v(\bar{\phi})$.

PROOF. Since $R[X^{\pm1}] = \bigcup_n X^{-n}R[X]$ is a flat $R[X]$-module, $-\otimes_{R[X]} R[X^{\pm1}]$ is an exact functor, so the following sequence is exact in $\mathrm{Mod}(R[X^{\pm1}])$:

$$0 \to N_{\phi\restriction_N} \otimes_{R[X]} R[X^{\pm1}] \to M_\phi \otimes_{R[X]} R[X^{\pm1}] \to (M/N)_{\bar{\phi}} \otimes_{R[X]} R[X^{\pm1}] \to 0.$$

By Lemma 4.2,

$$\widetilde{\mathrm{ent}}_v(N_{\phi\restriction_N}) = \widetilde{\mathrm{ent}}_v(N_{\phi\restriction_N} \otimes_{R[X]} R[X^{\pm1}]) \quad , \quad \widetilde{\mathrm{ent}}_v(M_\phi) = \widetilde{\mathrm{ent}}_v(M_\phi \otimes_{R[X]} R[X^{\pm1}]) \quad ,$$

$$\widetilde{\mathrm{ent}}_v((M/N)_{\bar{\phi}}) = \widetilde{\mathrm{ent}}_v((M/N)_{\bar{\phi}} \otimes_{R[X]} R[X^{\pm1}]) \quad ,$$

while, by Lemma 4.3,

$$\widetilde{\mathrm{ent}}_v(M_\phi \otimes_{R[X]} R[X^{\pm1}]) \leq \widetilde{\mathrm{ent}}_v(N_{\phi\restriction_N} \otimes_{R[X]} R[X^{\pm1}]) + \widetilde{\mathrm{ent}}_v((M/N)_{\bar{\phi}} \otimes_{R[X]} R[X^{\pm1}]).$$

Hence, the desired inequality follows. □

4.2. Super-additivity of $\widetilde{\mathrm{ent}}_v$. The next technical lemma deals with the intrinsic valuation entropy with respect to finitely generated ϕ-inert submodules; notice that part (2) is particularly important, as it says that, given such a submodule H of M, the quantity $\widetilde{\mathrm{ent}}_v(\phi, H)$ does not depend on H but only on $T(\phi, H)$, that is, for any other finitely generated a ϕ-inert submodule H' with $T(\phi, H) = T(\phi, H')$, one has $\widetilde{\mathrm{ent}}_v(\phi, H) = \widetilde{\mathrm{ent}}_v(\phi, H') = \widetilde{\mathrm{ent}}_v(\phi\restriction_{T(\phi,H)})$.

LEMMA 4.5. *Consider an* $R[X]$*-module* M_ϕ *and let* $H' \subseteq H \in \mathcal{I}_\phi(M)$,

1. *if* H/H' *is finitely generated, then* $\widetilde{\mathrm{ent}}_v(\phi, H') \leq \widetilde{\mathrm{ent}}_v(\phi, H)$;
2. *if* H *is finitely generated and* $M = T(\phi, H)$, *then* $\widetilde{\mathrm{ent}}_v(\phi) = \widetilde{\mathrm{ent}}_v(\phi, H)$.

PROOF. (1) Consider the following exact sequence

$$0 \to \frac{H \cap T_n(\phi, H')}{H'} \to \frac{T_n(\phi, H')}{H'} \to \frac{T_n(\phi, H)}{H}.$$

As both H and H' are ϕ-inert, all the modules appearing in the above sequence are L_v-finite. This implies that, for every $n \geq 1$, the following inequality holds

$$L_v\left(\frac{T_n(\phi,H')}{H'}\right) - L_v\left(\frac{H \cap T_n(\phi,H')}{H'}\right) \leq L_v\left(\frac{T_n(\phi,H)}{H}\right).$$

As remarked before Lemma 2.4, $T(\phi,H')/H'$ is a torsion module, hence $(H/H') \cap (T(\phi,H')/H')$ is a torsion submodule of the finitely generated module (H/H'); so, by Lemma 2.1, $L_v((H/H') \cap (T(\phi,H')/H')) < \infty$. For every $n \geq 1$ there is an inclusion $(H \cap T_n(\phi,H'))/H' \subseteq (H/H') \cap (T(\phi,H')/H')$, which implies

$$0 \leq \lim_{n\to\infty} \frac{L_v((H \cap T_n(\phi,H'))/H')}{n} \leq \lim_{n\to\infty} \frac{L_v((H/H') \cap (T(\phi,H')/H'))}{n} = 0.$$

Hence we obtain:

$$\begin{aligned}
\widetilde{\mathrm{ent}}_v(\phi,H') &= \lim_{n\to\infty} \frac{L_v(T_n(\phi,H')/H')}{n} \\
&= \lim_{n\to\infty} \frac{L_v(T_n(\phi,H')/H')}{n} - \lim_{n\to\infty} \frac{L_v((H \cap T_n(\phi,H'))/H')}{n} \\
&= \lim_{n\to\infty} \frac{L_v(T_n(\phi,H')/H') - L_v((H \cap T_n(\phi,H'))/H')}{n} \\
&\leq \lim_{n\to\infty} \frac{L_v(T_n(\phi,H)/H)}{n} = \widetilde{\mathrm{ent}}_v(\phi,H).
\end{aligned}$$

(2) Given a ϕ-inert submodule K of M, we have to prove that $\widetilde{\mathrm{ent}}_v(\phi,K) \leq \widetilde{\mathrm{ent}}_v(\phi,H)$. First of all, we can suppose that $H \leq K$, as $\widetilde{\mathrm{ent}}_v(\phi,K) \leq \widetilde{\mathrm{ent}}_v(\phi,H+K)$ by part (1) (being $H+K$ ϕ-inert and $(H+K)/K$ finitely generated). By Lemma 2.13, there exists a finitely generated submodule F of K such that, for any $n \geq 1$,

$$L_v\left(\frac{T_n(\phi,K)}{K}\right) < L_v\left(\frac{T_n(\phi,F)+K}{K}\right) + n\epsilon.$$

Since $M = T(\phi,H)$ and F is finitely generated, there exists $k \geq 1$ such that $F \leq T_k(\phi,H)$ and so

$$L_v\left(\frac{T_n(\phi,F)+K}{K}\right) + n\epsilon \leq L_v\left(\frac{T_{n+k}(\phi,H)+K}{K}\right) + n\epsilon$$

for every $n \geq 1$. Since $(K + T_{n+k}(\phi,H))/K$ is a quotient of $T_{n+k}(\phi,H)/H$, as we supposed that $H \leq K$, the inequality

$$L_v\left(\frac{T_n(\phi,K)}{K}\right) \leq L_v\left(\frac{T_{n+k}(\phi,H)}{H}\right) + n\epsilon$$

holds for every $n \geq 1$. Hence, $\widetilde{\mathrm{ent}}_v(\phi,K) \leq \widetilde{\mathrm{ent}}_v(\phi,T_k(\phi,H)) + \epsilon$ and so, by Corollary 2.10 (1), $\widetilde{\mathrm{ent}}_v(\phi,K) \leq \widetilde{\mathrm{ent}}_v(\phi,T_k(\phi,H)) + \epsilon = \widetilde{\mathrm{ent}}_v(\phi,H) + \epsilon$. We can now conclude by the arbitrariness of ϵ. □

As a consequence of Lemma 4.5 we derive one direction of the following proposition.

PROPOSITION 4.6. *Let $M_\phi = T(\phi,F)$ be a finitely generated $R[X]$-module, with F a finitely generated R-submodule of M. Then $\widetilde{\mathrm{ent}}_v(\phi) < \infty$ if and only if $\mathrm{rk}_R(M) < \infty$.*

PROOF. Assume $\widetilde{\mathrm{ent}}_v(\phi) < \infty$. The factor module $(M/t(M))_{\bar{\phi}}$ is finitely generated and $\mathrm{rk}_R(M) = \mathrm{rk}_R(M/t(M))$. By Proposition 2.11 (2), $\widetilde{\mathrm{ent}}_v(\bar{\phi}) \leq \widetilde{\mathrm{ent}}_v(\phi)$, so it is enough to prove that $\mathrm{rk}_R(M/t(M)) < \infty$. Assume, by way of contradiction, that $\mathrm{rk}_R(M/t(M)) = \infty$. Since $(M/t(M))_{\bar{\phi}}$ is a finite sum of cyclic trajectories, at least one of them, say $T(\bar{\phi}, x)$, must have infinite rank. Then necessarily $T(\bar{\phi}, x) = \oplus_{n\geq 0}\bar{\phi}^n xR$, and clearly $\widetilde{\mathrm{ent}}_v(\bar{\phi} \restriction T(\bar{\phi}, x)) = \infty$, so $\widetilde{\mathrm{ent}}_v(\bar{\phi}) = \infty$ by Proposition 2.11 (1), absurd.

Conversely, assume that $\mathrm{rk}_R(M) < \infty$. Then there exists an index k such that $rk_R(T_m(\phi, F)) = \mathrm{rk}_R(M)$ for all $m \geq k$. This implies that $T_{k+1}(\phi, F)/T_k(\phi, F)$ is torsion, so it is of finite valuation length, being finitely generated. This implies that $T_k(\phi, F)$ is ϕ-inert. From Lemma 4.5 we derive that $\widetilde{\mathrm{ent}}_v(\phi) = \widetilde{\mathrm{ent}}_v(\phi, T_k(\phi, F))$, which is obviously finite. □

Below is another application of Lemma 4.5, where we use the fact that finitely generated torsion-free R-modules are free (see [**FS01**, V.2.8]).

PROPOSITION 4.7. *Let M_ϕ be an $R[X]$-module such that, as an R-module, M is torsion-free and of finite rank. If F is a finitely generated R-submodule of maximum rank, then F is ϕ-inert and*

$$\widetilde{\mathrm{ent}}_v(\phi) = \widetilde{\mathrm{ent}}_v(\phi \restriction_{T(\phi,F)}) = \widetilde{\mathrm{ent}}_v(\phi, F).$$

PROOF. We know that F is ϕ-inert from Lemma 2.5. The divisible envelope $D(M)$ of M is the same as the divisible envelope of $T(\phi, F)$. Denote by $\widetilde{\phi}\colon D(M) \to D(M)$ the extension of ϕ to the divisible envelope. With the obvious modifications, the same argument of [**DGBSV15**, Cor. 3.15] shows that $\widetilde{\mathrm{ent}}_v(\phi) = \widetilde{\mathrm{ent}}_v(\widetilde{\phi}) = \widetilde{\mathrm{ent}}_v(\phi \restriction_{T(\phi,F)})$. It is now a consequence of Proposition 4.5 that $\widetilde{\mathrm{ent}}_v(\phi \restriction_{T(\phi,F)}) = \widetilde{\mathrm{ent}}_v(\phi, F)$. □

We are in position to prove AT for torsion-free modules of finite rank.

COROLLARY 4.8. *Let M be a torsion-free R-module of finite rank, and let $\phi\colon M \to M$ be an endomorphism. Given a ϕ-invariant submodule N of M such that M/N is torsion-free, we have that*

$$\widetilde{\mathrm{ent}}_v(\phi) = \widetilde{\mathrm{ent}}_v(\phi \restriction_N) + \widetilde{\mathrm{ent}}_v(\bar{\phi})$$

where $\bar{\phi}\colon M/N \to M/N$ is the induced map.

PROOF. Let $F \leq M$ be a finitely generated R-submodule of maximum rank. Then, $\bar{F} = (F + N)/N$ is finitely generated and torsion free, so it is free, showing that $F \cong (F \cap N) \oplus \bar{F}$ and so also $F \cap N$ is free. Notice also that $\mathrm{rk}_R(F \cap N) = \mathrm{rk}_R(N)$ and $\mathrm{rk}_R(\bar{F}) = \mathrm{rk}_R(M/N)$. By Proposition 2.12 (3) and Proposition 4.7,

$$\widetilde{\mathrm{ent}}_v(\phi) = \widetilde{\mathrm{ent}}_v(\phi, F) \geq \widetilde{\mathrm{ent}}_v(\phi, F \cap N) + \widetilde{\mathrm{ent}}_v(\bar{\phi}, \bar{F}) = \widetilde{\mathrm{ent}}_v(\phi \restriction_N) + \widetilde{\mathrm{ent}}_v(\bar{\phi}).$$

The converse inequality is proved in Proposition 4.4. □

The torsion case of AT could be deduced by one of the main results of [**SV16**], since in that case $\widetilde{\mathrm{ent}}_v = \mathrm{ent}_v$, by Proposition 2.14. But, for the sake of completeness, we prefer to give here a direct argument.

PROPOSITION 4.9. *Let M be a torsion R-module, and let $\phi\colon M \to M$ be an endomorphism. Given a ϕ-invariant submodule N of M, we have that*

$$\widetilde{\mathrm{ent}}_v(\phi) = \widetilde{\mathrm{ent}}_v(\phi\restriction_N) + \widetilde{\mathrm{ent}}_v(\bar{\phi})$$

where $\bar{\phi}\colon M/N \to M/N$ is the induced map.

PROOF. By Proposition 4.4 we already know that $\widetilde{\mathrm{ent}}_v(\phi) \leq \widetilde{\mathrm{ent}}_v(\phi\restriction_N) + \widetilde{\mathrm{ent}}_v(\bar{\phi})$, so it is enough to show the converse inequality. Let $K_1 \leq N$ and $\bar{K}_2 \leq M/N$ be finitely generated (so L_v-finite), and let $K \leq M$ be a finitely generated submodule such that $(K+N)/N = \bar{K}_2$ and $K_1 \subseteq N \cap K$. By Proposition 2.12 (3), $\widetilde{\mathrm{ent}}_v(\phi, K) \geq \widetilde{\mathrm{ent}}_v(\phi, K\cap N) + \widetilde{\mathrm{ent}}_v(\bar{\phi}, \bar{K}_2)$ and, since $\widetilde{\mathrm{ent}}_v(\phi, N\cap K) = \mathrm{ent}_v(\phi, N\cap K) \geq \mathrm{ent}_v(\phi, K_1) = \widetilde{\mathrm{ent}}_v(\phi, K_1)$, we get $\widetilde{\mathrm{ent}}_v(\phi, K) \geq \widetilde{\mathrm{ent}}_v(\phi, K_1) + \widetilde{\mathrm{ent}}_v(\bar{\phi}, \bar{K}_2)$. □

The next proposition proves AT in the particular case when the ϕ-invariant submodule is the torsion part.

PROPOSITION 4.10. *Let M_ϕ be an $R[X]$-module such that $\mathrm{rk}_R(M) = k < \infty$. Then $\widetilde{\mathrm{ent}}_v(\phi) = \widetilde{\mathrm{ent}}_v(\phi\restriction_{tM}) + \widetilde{\mathrm{ent}}_v(\bar{\phi})$, with $\bar{\phi}\colon M/tM \to M/tM$ the induced map.*

PROOF. By Proposition 4.4 we already know that $\widetilde{\mathrm{ent}}_v(\phi) \leq \widetilde{\mathrm{ent}}_v(\phi\restriction_{tM}) + \widetilde{\mathrm{ent}}_v(\bar{\phi})$, so it is enough to show the converse inequality.

We must prove that, fixed a ϕ-inert submodule K_1 of tM and a $\bar{\phi}$-inert submodule $\bar{K}_2$ of M/tM, there exists a ϕ-inert submodule K of M such that $\widetilde{\mathrm{ent}}_v(\phi, K) \geq \widetilde{\mathrm{ent}}_v(\phi, K_1) + \widetilde{\mathrm{ent}}_v(\bar{\phi}, \bar{K}_2)$. By [**SV16**, Proposition 4.2], we may assume that K_1 is finitely generated and, by Proposition 4.7, that $\bar{K}_2 \cong R^k$. Let now K be a finitely generated submodule of M such that $\bar{K}_2 = (K + tM)/tM$. Adding K_1 to K, we can assume, without loss of generality, that $K_1 \leq K$. Then K is ϕ-inert, by Lemma 2.5. By Proposition 2.12 (3), $\widetilde{\mathrm{ent}}_v(\phi, K) \geq \widetilde{\mathrm{ent}}_v(\phi, tK) + \widetilde{\mathrm{ent}}_v(\bar{\phi}, \bar{K}_2)$ and, since $\widetilde{\mathrm{ent}}_v(\phi, tK) = \mathrm{ent}_v(\phi, tK) \geq \mathrm{ent}_v(\phi, K_1) = \widetilde{\mathrm{ent}}_v(\phi, K_1)$, we get $\widetilde{\mathrm{ent}}_v(\phi, K) \geq \widetilde{\mathrm{ent}}_v(\phi, K_1) + \widetilde{\mathrm{ent}}_v(\bar{\phi}, \bar{K}_2)$. □

We have now all the ingredients needed to prove the general form of AT.

THEOREM 4.11 (Addition Theorem). *Consider the following short exact sequence of $R[X]$-modules:*

$$0 \to N_{\phi\restriction_N} \to M_\phi \to (M/N)_{\bar{\phi}} \to 0.$$

Then, $\widetilde{\mathrm{ent}}_v(\phi) = \widetilde{\mathrm{ent}}_v(\phi\restriction_N) + \widetilde{\mathrm{ent}}_v(\bar{\phi})$.

PROOF. In view of Propositon 4.4, it is enough to prove the inequality "$\geq$". If either $\widetilde{\mathrm{ent}}_v(\phi\restriction_N) = \infty$ or $\widetilde{\mathrm{ent}}_v(\bar{\phi}) = \infty$ then, by Proposition 2.12, $\widetilde{\mathrm{ent}}_v(\phi) = \infty$ and so the desired equality holds. Hence, suppose that $\widetilde{\mathrm{ent}}_v(\phi\restriction_N) < \infty$ and $\widetilde{\mathrm{ent}}_v(\bar{\phi}) < \infty$. By Corollary 3.2, these entropies are the supremum of the entropies of the restrictions to finitely generated submodules and so, given $\epsilon > 0$, there exist finitely generated R-submodules $F_1 \leq N$ and $\bar{F}_2 \leq M/N$ such that

$$\widetilde{\mathrm{ent}}_v(\phi\restriction_N) \leq \widetilde{\mathrm{ent}}_v(\phi\restriction_{T(\phi,F_1)}) + \epsilon/2 \qquad \widetilde{\mathrm{ent}}_v(\bar{\phi}) \leq \widetilde{\mathrm{ent}}_v(\bar{\phi}\restriction_{T(\bar{\phi},\bar{F}_2)}) + \epsilon/2.$$

Let F be a finitely generated R-submodule of M such that $(F+N)/N = \bar{F}_2$ and $F_1 \leq F \cap N$. Let $M' = T(\phi, F)$, $N' = N \cap T(\phi, F)$, and notice that $M'/N' \cong$

$T(\bar\phi, \bar F_2)$. By the arbitrariness of $\epsilon > 0$, it is enough to prove the first inequality below, since the other two inequalities follow by the above arguments:

$$\begin{aligned}\widetilde{\mathrm{ent}}_v(\phi\restriction_{M'}) &\geq \widetilde{\mathrm{ent}}_v(\phi\restriction_{N'}) + \widetilde{\mathrm{ent}}_v(\bar\phi\restriction_{M'/N'})\\ &\geq \widetilde{\mathrm{ent}}_v(\phi\restriction_{T(\phi,F_1)}) + \widetilde{\mathrm{ent}}_v(\bar\phi\restriction_{T(\bar\phi,\bar F_2)})\\ &\geq \widetilde{\mathrm{ent}}_v(\phi\restriction_N) + \widetilde{\mathrm{ent}}_v(\bar\phi) - \epsilon.\end{aligned}$$

In other words, we have reduced our problem to the case when $M = T(\phi, F)$ is a finitely generated $R[X]$-module. Now, if $\mathrm{rk}_R(M) = \infty$, then $\widetilde{\mathrm{ent}}_v(\phi) = \infty$ by Proposition 4.6, and so clearly $\widetilde{\mathrm{ent}}_v(\phi) \geq \widetilde{\mathrm{ent}}_v(\phi\restriction_N) + \widetilde{\mathrm{ent}}_v(\bar\phi)$. Hence, suppose $\mathrm{rk}_R(M) = k < \infty$.

Let $K = tM + N$; then K/tM is a $\bar\phi$-invariant submodule of M/tM, as well as its purification $(K/tM)^* = \{\bar x \in M/tM \mid r\bar x \in K/tM \text{ for some } r \neq 0\}$.
Let us list some short exact sequences of $R[X]$-modules on which $\widetilde{\mathrm{ent}}_v$ is additive:

(1) $0 \to tM \to M \to M/tM \to 0$, by Proposition 4.10;
(2) $0 \to tN \to N \to K/tM \to 0$, by Proposition 4.10;
(3) $0 \to t(M/N) \to M/N \to (M/N))/t(M/N) \to 0$, by Proposition 4.10;
(4) $0 \to tN \to tM \to tM/tN \to 0$, by Proposition 4.9;
(5) $0 \to (K/tM)^* \to M/tM \to (M/K)/t(M/K) \to 0$, by Corollary 4.8.

Furthermore, we also need the following three observations, in which, as in the next part of the proof, by abusing notation, we eliminate the subscript of the endomorphism for $\mathbb{Z}[X]$-modules when we apply $\widetilde{\mathrm{ent}}_v$.

(6) $\widetilde{\mathrm{ent}}_v((K/tM)^*/(K/tM)) = 0$. By Proposition 2.14 and [**SV16**, Proposition 4.2], it is enough to show that, for any finitely generated R-submodule C of $(K/tM)^*$,

$$L_v\left(\frac{T(\bar\phi, C) + (K/tM)}{K/tM}\right) < \infty.$$

For that we use Lemma 2.2. Indeed, we know that $\mathrm{rk}_R(M) = k < \infty$, so that $(K/tM)^* \leq M/tM \leq Q^k$; furthermore $\frac{T(\bar\phi,C)+(K/tM)}{K/tM} \leq \frac{(K/tM)^*}{K/tM}$ are torsion modules. Choose a finite set $\{c_1, \dots, c_t\}$ of generators for C, then there is an element $r \neq 0$ in R such that $rc_i \in K/tM$, for all i. Hence r annihilates $\frac{T(\bar\phi,C)+(K/tM)}{K/tM}$, which is then a bounded sub-module of $Q^k/(K/tM)$ and Lemma 2.2 applies.

(7) $\widetilde{\mathrm{ent}}_v((K/tM)^*) = \widetilde{\mathrm{ent}}_v(K/tM)$. This follows by point (6) and by the sub-additivity of $\widetilde{\mathrm{ent}}_v$;

(8) $\widetilde{\mathrm{ent}}_v(t(M/N)) = \widetilde{\mathrm{ent}}_v(tM/tN)$. This follows by point (6) and by the fact that $t(M/N)/(K/N)$ embeds into $(K/tM)^*/(K/tM)$; in fact, $t(M/N)/(K/N)$ embeds into $t(M/K)$, which is isomorphic to

$$t((M/tM)/(K/tM)) = (K/tM)^*/(K/tM).$$

Hence $\widetilde{\mathrm{ent}}_v((K/tM)^*/(K/tM)) = 0$ implies $\widetilde{\mathrm{ent}}_v(t(M/N)/(K/N)) = 0$ and consequently $\widetilde{\mathrm{ent}}_v(t(M/N)) = \widetilde{\mathrm{ent}}_v(K/N) = \widetilde{\mathrm{ent}}_v(tM/tN)$, since $tM/tN \cong K/N$.

Now we can conclude the proof with the following series of equalities depending on the associated overset points:

$$\begin{aligned}
\widetilde{\mathrm{ent}}_v(M) &\overset{(1)}{=} \widetilde{\mathrm{ent}}_v(tM) + \widetilde{\mathrm{ent}}_v(M/tM) \\
&\overset{(4,5)}{=} \widetilde{\mathrm{ent}}_v(tN) + \widetilde{\mathrm{ent}}_v(tM/tN) + \widetilde{\mathrm{ent}}_v((K/tM)^*) + \widetilde{\mathrm{ent}}_v((M/K)/t(M/K)) \\
&\overset{(7)}{=} \widetilde{\mathrm{ent}}_v(tN) + \widetilde{\mathrm{ent}}_v(tM/tN) + \widetilde{\mathrm{ent}}_v(K/tM) + \widetilde{\mathrm{ent}}_v((M/N)/t(M/N)) \\
&\overset{(2,3)}{=} \widetilde{\mathrm{ent}}_v(N) + \widetilde{\mathrm{ent}}_v(tM/tN) + \widetilde{\mathrm{ent}}_v(M/N) - \widetilde{\mathrm{ent}}_v(t(M/N)) \\
&\overset{(8)}{=} \widetilde{\mathrm{ent}}_v(N) + \widetilde{\mathrm{ent}}_v(M/N)
\end{aligned}$$

□

5. The Intrinsic Algebraic Yuzvinski Formula

In order to state the Intrinsic Algebraic Yuzvinski Formula (IAYF for short) in a simple way, we introduce the following terminology. If $\phi\colon Q^n \to Q^n$ is a linear transformation, where Q denotes the field of quotients of the valuation domain R, let $p_\phi(X) \in Q[X]$ be the (monic) characteristic polynomial of ϕ over Q. Since R is a valuation domain, there exists an element $s \in R$ of minimal value such that $sp_\phi(X) \in R[X]$; thus the polynomial $sp_\phi(X)$ is primitive (i.e., its content $c(sp_\phi(X))$ equals R) with leading coefficient s, and it is called the *characteristic polynomial of ϕ over R*. Our goal in this section is to prove the following version of the IAYF.

THEOREM 5.1 (IAYF). *Let $\phi : Q^n \to Q^n$ be a linear transformation. Then $\widetilde{\mathrm{ent}}_v(\phi) = v(s)$, where $s \in R$ is the leading coefficient of the characteristic polynomial of ϕ over R.*

Notice that this formula shows that $\widetilde{\mathrm{ent}}_v(\phi) = 0$ exactly when $p_\phi(X) \in R[X]$, that is, when ϕ is integral over R, while, if some $q_i \in Q \setminus R$, the intrinsic valuation entropy takes a positive value.

Let us briefly explain where the statement of Theorem 5.1 comes from. Recall that the Algebraic Yuzvinski Formula, proved in [**GBV12**], deals with another algebraic entropy for Abelian groups, denoted by h and deeply investigated in [**DGB16**]. It states that the entropy $h(\phi)$ of an endomorphism ϕ of a finite dimensional vector space over the rational field $\mathbb{Q}$ coincides with the Mahler measure of the characteristic polynomial of ϕ (we refer to [**GBV12, DGBSV15, L33, M62**] for the notions of the entropy h and of the Mahler measure of a rational polynomial). In fact, this connection between the values of h and the Mahler measure reflects a dual property of the topological entropy on solenoidal endomorphisms proved in the pioneering Yuzvinski's paper [**Y68**] (see also [**LW88**]).

The Intrinsic Algebraic Yuzvinski Formula for the intrinsic entropy $\widetilde{\mathrm{ent}}$ proved in [**DGBSV15**] (see also [**GBV15, SV17**]), states that the entropy $\widetilde{\mathrm{ent}}(\phi)$ coincides with $\log(s)$, where s is the minimal common multiple of the denominators of the rational numbers appearing in the characteristic polynomial $p_\phi(X)$ of ϕ over $\mathbb{Q}$. This s is the minimal positive integer such that $sp_\phi(X)$ is a primitive polynomial of $\mathbb{Z}[X]$. This shows the strict analogy between this result and Theorem 5.1.

The IAYF is a consequence of the following result, which is analogous to Lemma 3.3 in [**SV17**], with the due modifications.

PROPOSITION 5.2. *Let M be a torsion-free R-module and $\phi\colon M \to M$ an endomorphism. Let $x \in M$ be an element generating a ϕ-trajectory $T(\phi, x)$ of finite rank $n \geq 1$. Then,*

(1) *$T_n(\phi, x) = \bigoplus_{0\leq i\leq n-1} \phi^i xR$;*
(2) *there exists a polynomial $f(X) \in R[X]$ of degree n, with content $c(f(X)) = R$ and leading coefficient $s \in R$, such that $f(\phi)(x) = 0$;*
(3) *$R[X]/(f(X)) \cong T(\phi, x)$;*
(4) *$T_{k+1}(\phi, x)/T_k(\phi, x) \cong R/sR$ for every $k \geq n$;*
(5) *if s is a unit, then $T(\phi, x) = T_n(\phi, x)$ is free of rank n, otherwise the quotient $T(\phi, x)/T_n(\phi, x)$ is a uniserial divisible module isomorphic to Q/R.*

PROOF. (1) Let $m \geq 1$ be the minimal positive integer such that $t\phi^m x \in T_m(\phi, x)$ for some $0 \neq t \in R$. Then $T_m(\phi, x) = \bigoplus_{1\leq i\leq m-1} \phi^i xR$ and a simple computation shows that, for each $k \geq 1$, $t^k \phi^{m+k-1} x \in T_m(\phi, x)$; consequently $T(\phi, x)/T_m(\phi, x)$ is a torsion module, so $T_m(\phi, x)$ has rank m. This shows that $m = n$.

(2) Choose an element $t \in R$ in such a way that $t\phi^n x \in T_n(\phi, x)$, and let $t\phi^n x = \sum_{0\leq i\leq n-1} r_i \phi^i x$. Let r_j be such that $v(r_j) \leq v(r_i)$ for every $i \leq n-1$. Setting $s = tr_j^{-1}$, the polynomial $f(X) = s\phi^n X - \sum_{0\leq i\leq n-1} r_j^{-1} r_i \phi^i X$ clearly satisfies $c(f(X)) = R$ and $f(\phi)(x) = 0$.

(3) Consider the ideal of $R[X]$:

$$K_x = \{g(X) \mid g(\phi) = 0\}.$$

The map $R[X] \to T(\phi, x)$ sending $g(X)$ into $g(\phi)$ is surjective and has kernel K_x, hence $R[X]/K_x \cong T(\phi, x)$. Clearly the ideal $(f(X))$ is contained in K_x, so, in order to conclude, it is enough to prove that, if $g(\phi) = 0$, then $g(X)$ is a multiple of $f(X)$. If $g(\phi) = 0$, then obviously $g(X)$ has degree $k \geq n$; without loss of generality we can assume $g(X)$ primitive, that is, $c(g(X)) = R$. The division algorithm gives:

$$s^{k-n+1} g(X) = f(X)q(X) + r(X)$$

where $r(X)$ has degree $< n$. This implies that $r(X) = 0$, since $r(\phi) = 0$. Therefore $s^{k-n+1} g(X) = f(X)q(X)$, thus s^{k-n+1} divides $c(f(X)q(X)) = c(q(X))$ (see property (b) of the content in [**FS01**, pg. 7]). Dividing the above equality by s^{k-n+1} we derive that $g(X)$ is a multiple of $f(X)$, as desired.

(4) The proof goes by induction on $k \geq n$, the starting case $k = n$, which follows from (2). Let $k > n$ and consider the cyclic factor module $T_{k+1}(\phi, x)/T_k(\phi, x) \cong R/I_k$, where I_k is a non-zero ideal of R. By [**SZ09**, Lem. 1.9],

$$L_v(T_{k+1}(\phi, x)/T_k(\phi, x)) \leq L_v(T_{n+1}(\phi, x)/T_n(\phi, x)) = v(s),$$

therefore $I_k \geq sR$. Let us assume, by way of contradiction, that $I_k > sR$. Then there exists $t \in I_k \setminus sR$ such that a polynomial $g(X)$ of degree k and leading coefficient t satisfies $g(\phi)(x) = 0$. By point (3), $g(X)$ is a multiple of $f(X)$, hence $t \in sR$, a contradiction.

(5) The first claim is obvious. If $v(s) > 0$, then (4) implies that $T(\phi, x)/T_n(\phi, x)$ is isomorphic to the direct limit of the direct system $\{R/s^n R\}_{n\geq 1}$, where the connecting maps $R/s^n R \to R/s^{n+1} R$ are induced by the multiplication by s, and this direct limit is isomorphic to Q/R, being R archimedean. $\square$

PROOF OF THEOREM 5.1. The Q-vector space $V = Q^n$, endowed with the structure of $Q[X]$-module induced by ϕ, is finitely generated and torsion. Since $Q[X]$ is a PID, V_ϕ decomposes into the direct sum of cyclic $Q[X]$-modules:

$$V_\phi = V_1 \oplus \cdots \oplus V_r.$$

If ϕ_i denotes the restriction of ϕ to the submodule V_i, then for the characteristic polynomials over Q we have the factorization:

$$p_\phi(X) = \prod_{1 \le i \le r} p_{\phi_i}(X).$$

Then the Addition Theorem ensures that it is enough to prove the result for the cyclic summands, that is, we can assume that V_ϕ is a cyclic $Q[X]$-module. Then there exists an element $x \in V$ such that:

$$V = xQ \oplus \phi xQ \oplus \cdots \oplus \phi^{n-1} xQ.$$

Let $F = \bigoplus_{0 \le i \le n-1} \phi^i xR$ be the free partial n-th trajectory of x of full rank in V. Then, by Proposition 4.7, F is ϕ-inert in V and $\widetilde{\mathrm{ent}}_v(\phi) = \widetilde{\mathrm{ent}}_v(\phi, F)$. Now Proposition 2.9 ensures that

$$\widetilde{\mathrm{ent}}_v(\phi, F) = \inf_k L_v\left(\frac{T_{k+1}(\phi, F)}{T_k(\phi, F)}\right).$$

But for each $k \ge 1$ we have that $T_k(\phi, F) = T_{n+k-1}(\phi, x)$, therefore Proposition 5.2 (4) implies that $T_{k+1}(\phi, F)/T_k(\phi, F) \cong R/sR$ for every $k \ge 1$. We deduce that $\widetilde{\mathrm{ent}}_v(\phi) = L_v(R/sR) = v(s)$, that gives the proof, since the polynomial $f(X)$ of Proposition 5.2 (2) coincides, in case V is a cyclcic ϕ-trajectory, with the characteristic polynomial of ϕ over R. □

6. The Uniqueness Theorem

The next result it is used at the beginning of the proof of [**SV16**, Thm. 7.3] without an explicit proof, and it is a consequence of [**Zan11**, Prop. 2.6] in the particular case in which M is finitely generated and torsion; so, for the sake of completeness, we give here a detailed proof.

PROPOSITION 6.1. *Let M be a R-module of finite valuation length and let*

$$\beta\colon \bigoplus_{n \ge 1} M_n \to \bigoplus_{n \ge 1} M_n$$

be the right Bernoulli shift, where $M_n = M$ for all n, i.e. $\left(\bigoplus_{n\ge1} M_n\right)_\beta \cong M \otimes_R R[X]$. Then, $\widetilde{\mathrm{ent}}_v(\beta) = L_v(M)$.

PROOF. By Proposition 2.14, it is enough to check that $\mathrm{ent}_v(\beta) = L_v(M)$, since M must be a torsion module, so also $\bigoplus_{n\ge1} M_n$ is torsion. From the equality

$$\mathrm{ent}_v(\beta, M_1) = \inf_n L_v\left(\frac{T_{n+1}(\beta, M_1)}{T_n(\beta, M_1)}\right)$$

and since $\frac{T_{n+1}(\beta, M_1)}{T_n(\beta, M_1)} \cong M$ for any n, we derive that $L_v(M) = \mathrm{ent}_v(\beta, M_1) \le \mathrm{ent}_v(\beta)$. On the other hand, by [**SV16**, Prop. 4.2], $\mathrm{ent}_v(\beta) = \sup_F \mathrm{ent}_v(\beta, F)$, with F finitely generated R-submodule of $\bigoplus_{n\ge1} M_n$. Given such an $F \le \bigoplus_{n\ge1} M_n$, then $F \leq T_j(\beta, M_1) = T_j$ for a certain $j \geq 1$ so, by Corollary 2.10, we can conclude that

$$\mathrm{ent}_v(\beta, F) \leq \mathrm{ent}_v(\beta, T_j) = \mathrm{ent}_v(\beta, M_1) = L_v(M). \qquad \square$$

The next theorem is based on [**SV16**, Thm. 7.3], that is, the Uniqueness Theorem for the valuation entropy ent_v on the category of torsion modules.

THEOREM 6.2 (Uniqueness Theorem). *Let R be the valuation domain of an archimedean non-discrete valuation $v : Q \to \mathbb{R} \cup \{\infty\}$ on its field of quotient Q, and let L_v be the induced non-discrete length function on* $\mathrm{Mod}(R)$. *The intrinsic valuation entropy $\widetilde{\mathrm{ent}}_v$ is the unique length function $L_X \colon \mathrm{Mod}(R[X]) \to \mathbb{R}^*$ such that:*

(1) *$L_X(M \otimes_R R[X])) = L_v(M)$ for all modules M (of finite length);*
(2) *$L_X(M_\phi) = v(s)$, for any linear transformation $\phi\colon M \to M$ of a finite dimensional Q-vector space, where s is the leading coefficient of the characteristic polynomial of ϕ over R.*

PROOF. Since the $R[X]$-module $M \otimes_R R[X]$ is isomorphic to $\bigoplus_{n\geq 1} M_n$ endowed with the right Bernoulli shift β, where $M_n = M$ for all n, Proposition 6.1 shows that $\widetilde{\mathrm{ent}}_v$ satisfies condition (1) when M has finite valuation length, and it is immediate to derive that (1) follows also when $L_v(M) = \infty$. Condition (2) is satisfied by $\widetilde{\mathrm{ent}}_v$ by Theorem 5.1.

Concerning uniqueness, since a length function is determined by the values it takes on cyclic modules, it is enough to prove that $L_X(T(\phi, x)_\phi) = \widetilde{\mathrm{ent}}_v(T(\phi, x)_\phi)$ for any $x \in M$, where $\phi\colon M \to M$ is an endomorphism of an R-module M. In view of the Addition Theorem, we can assume that M is either torsion or torsion-free. In the first case [**SV16**, Thm. 7.3] shows that $L_X(T(\phi, x)_\phi) = \mathrm{ent}_v(T(\phi, x)_\phi)$, so the conclusion follows by Proposition 2.14. In the latter case, either $T(\phi, x)_\phi \cong R[X]$, in case $T(\phi, x)$ is an R-module of infinite rank, or $T(\phi, x)_\phi \cong R[X]/(f(X))$, by Proposition 5.2, where $f(X)$ is the characteristic polynomial of $\phi\restriction_{T(\phi,x)}$ over R.

If $T(\phi, x)_\phi \cong R[X]$, then [**NR65**, Thm. 2] ensures that $L_X(T(\phi, x)_\phi) = \infty$, otherwise L_X would be equivalent to $\mathrm{rk}_{R[X]}$, which is impossible. Also $\widetilde{\mathrm{ent}}_v(T(\phi, x)_\phi) = \infty$, as a consequence of Proposition 2.9.

If $T(\phi, x)_\phi \cong R[X]/(f(X))$, $L_X(T(\phi, x)_\phi) = v(s)$ by hypothesis, where s is the leading coefficient of $f(X)$, and also $\widetilde{\mathrm{ent}}_v((T(\phi, x)_\phi)) = v(s)$, by the IAYF. So we are done. $\square$

References

[AKM65] R. L. Adler, A. G. Konheim, and M. H. McAndrew, *Topological entropy*, Trans. Amer. Math. Soc. **114** (1965), 309–319, DOI 10.2307/1994177. MR0175106

[B93] V. V. Belyaev, *Inert subgroups in infinite simple groups* (Russian, with English and Russian summaries), Sibirsk. Mat. Zh. **34** (1993), no. 4, 17–23, i, vii, DOI 10.1007/BF00975160; English transl., Siberian Math. J. **34** (1993), no. 4, 606–611. MR1248784

[DR12] U. Dardano and S. Rinauro, *Inertial automorphisms of an abelian group*, Rend. Semin. Mat. Univ. Padova **127** (2012), 213–233, DOI 10.4171/rsmup/127-11. MR2978006

[DGBSV15] D. Dikranjan, A. Giordano Bruno, L. Salce, and S. Virili, *Intrinsic algebraic entropy*, J. Pure Appl. Algebra **219** (2015), no. 7, 2933–2961, DOI 10.1016/j.jpaa.2014.09.033. MR3313513

[DGB13] D. Dikranjan and A. Giordano Bruno, *Discrete dynamical systems in group theory*, Note Mat. **33** (2013), no. 1, 1–48. MR3071309

[DGB12] D. Dikranjan and A. Giordano Bruno, *Limit free computation of entropy*, Rend. Istit. Mat. Univ. Trieste **44** (2012), 297–312. MR3019564

[DGB16] D. Dikranjan and A. Giordano Bruno, *Entropy on abelian groups*, Adv. Math. **298** (2016), 612–653, DOI 10.1016/j.aim.2016.04.020. MR3505752

[DGSZ09] D. Dikranjan, B. Goldsmith, L. Salce, and P. Zanardo, *Algebraic entropy for abelian groups*, Trans. Amer. Math. Soc. **361** (2009), no. 7, 3401–3434, DOI 10.1090/S0002-9947-09-04843-0. MR2491886

[FS85] L. Fuchs and L. Salce, *Modules over valuation domains*, Lecture Notes in Pure and Applied Mathematics, vol. 97, Marcel Dekker, Inc., New York, 1985. MR786121

[FS01] L. Fuchs and L. Salce, *Modules over non-Noetherian domains*, Mathematical Surveys and Monographs, vol. 84, American Mathematical Society, Providence, RI, 2001. MR1794715

[GBV12] A. Giordano Bruno and S. Virili, *Algebraic Yuzvinski formula*, J. Algebra **423** (2015), 114–147, DOI 10.1016/j.jalgebra.2014.10.007. MR3283712

[GBV15] A. Giordano Bruno and S. Virili, *About the algebraic Yuzvinski formula*, Topol. Algebra Appl. **3** (2015), no. 1, 86–103, DOI 10.1515/taa-2015-0008. MR3430946

[L33] D. H. Lehmer, *Factorization of certain cyclotomic functions*, Ann. of Math. (2) **34** (1933), no. 3, 461–479, DOI 10.2307/1968172. MR1503118

[LW88] D. A. Lind and T. Ward, *Automorphisms of solenoids and p-adic entropy*, Ergodic Theory Dynam. Systems **8** (1988), no. 3, 411–419, DOI 10.1017/S0143385700004545. MR961739

[M62] K. Mahler, *On some inequalities for polynomials in several variables*, J. London Math. Soc. **37** (1962), 341–344, DOI 10.1112/jlms/s1-37.1.341. MR0138593

[NR65] D. G. Northcott and M. Reufel, *A generalization of the concept of length*, Quart. J. Math. Oxford Ser. (2) **16** (1965), 297–321, DOI 10.1093/qmath/16.4.297. MR0195905

[P79] J. Peters, *Entropy on discrete abelian groups*, Adv. in Math. **33** (1979), no. 1, 1–13, DOI 10.1016/S0001-8708(79)80007-9. MR540634

[SV16] L. Salce and S. Virili, *The addition theorem for algebraic entropies induced by non-discrete length functions*, Forum Math. **28** (2016), no. 6, 1143–1157, DOI 10.1515/forum-2015-0118. MR3567862

[SV17] L. Salce and S. Virili, *Two new proofs concerning the intrinsic algebraic entropy*, Comm. Algebra **46** (2018), no. 9, 3939–3949, DOI 10.1080/00927872.2018.1430805. MR3820607

[SVV13] L. Salce, P. Vámos, and S. Virili, *Length functions, multiplicities and algebraic entropy*, Forum Math. **25** (2013), no. 2, 255–282, DOI 10.1515/form.2011.117. MR3031785

[SZ09] L. Salce and P. Zanardo, *A general notion of algebraic entropy and the rank-entropy*, Forum Math. **21** (2009), no. 4, 579–599, DOI 10.1515/FORUM.2009.029. MR2541474

[ST87] L. N. Stojanov, *Uniqueness of topological entropy for endomorphisms on compact groups* (English, with Italian summary), Boll. Un. Mat. Ital. B (7) **1** (1987), no. 3, 829–847. MR916296

[Wei74] M. D. Weiss, *Algebraic and other entropies of group endomorphisms*, Math. Systems Theory **8** (1974/75), no. 3, 243–248, DOI 10.1007/BF01762672. MR0385834

[Wil15] G. A. Willis, *The scale and tidy subgroups for endomorphisms of totally disconnected locally compact groups*, Math. Ann. **361** (2015), no. 1-2, 403–442, DOI 10.1007/s00208-014-1074-y. MR3302624

[Y68] S. A. Juzvinskiĭ, *Metric properties of the endomorphisms of compact groups* (Russian), Izv. Akad. Nauk SSSR Ser. Mat. **29** (1965), 1295–1328. MR0194588

[Zan11] P. Zanardo, *Multiplicative invariants and length functions over valuation domains*, J. Commut. Algebra **3** (2011), no. 4, 561–587, DOI 10.1216/JCA-2011-3-4-561. MR2863387

Dipartimento di Matematica, Università di Padova, Via Trieste 63 - 35121 Padova, Italy
Email address: salce@math.unipd.it

Dept. de Matemáticas, Universidad de Murcia, 30100 Espinardo, Murcia, Spain
Email address: virili.simone@gmail.com

Contemporary Mathematics
Volume **730**, 2019
https://doi.org/10.1090/conm/730/14718

Decidability and modules over Bézout domains

Carlo Toffalori

This paper is dedicated to Mike Prest and to the memory of Gena Puninski.

Abstract. I survey recent results on the decidability of first order theories of modules over several noteworthy Bézout domains.

1. Introduction

First of all, I would like to thank Mike for four joint papers with him and Annalisa Marcja [**MPT1**], [**MPT2**], [**MPT3**], [**MPT4**], for his constant generous help and encouragement, for uncountably many suggestions, and more generally as a polestar of the model theory of modules [**P1**], [**P2**]. But let me also dedicate a tribute to the memory of Gena (I do not remember how many joint papers...).

This note is devoted to decidability of first order theories of modules over a given ring R. This topic may look out of fashion. How effective a decision algorithm can be, provided it is exists? Are there "really" decidable first order theories with equality? It is well known that an answer to this question overlaps computational complexity and the P versus NP problem. In fact, if any first order theory with equality admits a decision procedure running in at most polynomial time, then P = NP.

In the case of Abelian groups, and more generally of modules over Dedekind domains, decidability results date back to the seminal works of W. Szmielew [**Sz**] and then P. Eklof and E. Fischer [**EF**]. However their procedures are far from being effective, see the introduction of [**Lo**]. In fact, M.J. Fischer and M. Rabin proved [**FR**] that the time complexity of the first order theory of the addition of real numbers, i.e. of divisible torsionfree Abelian groups, is at least exponential (see on that also the final part of [**S**]). Thus, even if Lo [**Lo**] provided a double exponential upper bound for the space complexity of the first order theory of all Abelian groups, procedures handling decidability of theories of modules seem far from being feasible.

Moreover, the decision question for modules over a ring R requires strong assumptions on R, for instance countability, otherwise the theory of R-modules is undecidable, or the decision problem makes no sense (see [**P1**], Chapter 17). In

2010 *Mathematics Subject Classification.* Primary 03C60; Secondary 03C98, 03B25, 13F05.
Key words and phrases. Bézout domain, Prüfer domain, valuation domain, Ziegler topology, decidability, prime radical relation, double prime radical relation.
The author was supported in part by Italian PRIN 2012 and GNSAGA INdAM.

detail R should be *effectively given*, which means, when R is for simplicity a commutative domain with unity, that its elements can be listed, possibly with repetitions, as $a_0 = 0$, $a_1 = 1$, $a_2, \ldots, a_n, \ldots$ $(n \in \mathbb{N})$ so that suitable algorithms effectively perform the following, when m, n range over non negative integers:

(1) Deciding whether $a_m = a_n$ or not.
(2) Producing $a_m + a_n$ and $a_m \cdot a_n$ (or rather indices of these elements in the list).
(3) Establishing whether a_m divides a_n.

Then other familiar procedures can be effectively carried out, such as determining units, calculating additive and (when possible) multiplicative inverses, calculating in the right frameworks greatest common divisor gcd and least common multiple lcm.

On the other hand, for a given domain R, and beyond any countability restriction, decidability of R-modules deeply overlaps fundamental model theoretic and algebraic questions on R-modules, basic steps towards a full comprehension of them, such as

- the analysis of pp-formulae over R,
- the description of the Ziegler spectrum of R (both the points and the topology),
- through them, the classification of R-modules up to elementary equivalence,

which highly fits with the spirit of [**Sz**] and [**EF**]. Prest's first Bible of model theory of modules [**P1**] devotes a whole chapter to decidability and undecidability, and explains their connection with pp-formulae and Ziegler topology. We will recall its main facts later, over the domains we are going to consider. Also Mike's more recent fundamental book on module theory [**P2**] extensively updates on (un)decidability results and related references.

The conclusion is that decidability of modules is still alive and, as far as I am concerned, *"close to my heart"*. So let me deal with decidability.

In this note I mainly focus on a research program developed in the latest years with Lorna Gregory, Sonia L'Innocente, Francoise Point and Gena Puninski on the decidability of modules on Bézout domains. The next sections recall some necessary basic facts on model theory of modules and then introduce and comment on this program. But I'll devote a short intermezzo to decidability of representations of the Lie algebra $\mathrm{sl}(2,k)$ where k is an algebraically closed field of characteristic 0.

In the remainder of the paper, I tacitly assume some familiarity with model theory of modules, as presented in the initial chapters of [**P1**] (and partly in [**P2**]).

I would like to thank Lorna Gregory and Sonia L'Innocente for their help and their comments, and the referee for her/his valuable suggestions.

2. Basic facts on modules

2.1. Model theory. Let me fix my notation: for a given ring R, L_R is the first order language of (say right) R-modules and T_R is the first order theory of R-modules in L_R.

For $a \in R$, $a \mid x$ denotes the *divisibility formula* $\exists y(ya = x)$ of L_R, defining in an R-module M the submodule Ma. Similarly $xa = 0$ is the *annihilator formula* defining in M the kernel of the multiplication by a.

Both exemplify *pp-formulae* in L_R. In fact a *pp-formula* $\varphi(x)$ of L_R in one free variable x singles out in any R-module M the elements making a fixed finite linear system of equations with coefficients in R in the indeterminates x and possibily more $z_1, \ldots, z_m$ solvable in the module.

Then each pp-formula $\varphi(x)$ defines in a given R-module M a subgroup $\varphi(M)$, called a *pp-subgroup*. Sometimes, and when R is commutative, this subgroup is even a submodule.

As said, I will mainly deal below with divisibility and annihilator formulas, including possible variants, like $\exists y(ya = x \wedge yb = 0)$ or $c \mid xd$ with $a, b, c, d \in R$. I will also consider finite sums and conjunctions of pp-formulae, such as $a \mid x + xb = 0 \doteq \exists x_1 \exists x_2(a \mid x_1 \wedge x_2 b = 0 \wedge x = x_1 + x_2)$, or $c \mid x \wedge xd = 0$, which is clearly equivalent to a pp-formula. The reader will see later that I will mainly be concerned with these simple formulas because of the rings I am working over.

If one factorizes the set of pp-formulae of L_R in one free variable identifying logically equivalent formulae, then one gets a quotient set $L(R)$ which is a lattice with respect to the partial order determined by implication. In this lattice the meet is given by conjunction of formulae, and the join by their sum.

An *invariant statement* is a sentence of L_R saying, for $\varphi(x), \psi(x)$ pp-formulae and l a positive integer, that the index of the subgroup determined by $\varphi(x) \wedge \psi(x)$ in the subgroup determined by $\varphi(x)$ is exactly l, or at least l. Let us denote by $\mathrm{Inv}(\varphi, \psi) = l$, $\mathrm{Inv}(\varphi, \psi) \geq l$ these statements, respectively. According to a celebrated result by Baur-Monk:

THEOREM 2.1. *Every sentence of L_R is equivalent modulo T_R to a Boolean combination of invariant statements, and even to a disjunction of conjunctions of statements* $\mathrm{Inv}(\varphi, \psi) \geq l$, $\mathrm{Inv}(\varphi, \psi) = l$ *with $\varphi(x)$, $\psi(x)$ pp-formulae, l positive integer.*

The Ziegler spectrum, Zg_R, of a ring R is a quasi-compact topological space whose points are (isomorphism classes of) indecomposable pure injective R-modules, and a basis of the topology is given by the compact open sets

$$(\varphi(x)/\psi(x)) := \{N \in \mathrm{Zg}(R) : \varphi(N) \supset \psi(N) \cap \varphi(N)\},$$

where $\varphi(x)$, $\psi(x)$ range over the pp-formulae of L_R in the free variable x.

Pure injective R-modules are those R-modules N satisfying the following weak saturation property: every partial pp-type in one variable over N, which is finitely satisfiable over N, is actually realised in N. Recall that a pp-type is a set of pp-formulae.

Every R-module admits up to isomorphism a unique minimal pure injective elementary extension, which is called its *pure injective hull.* Moreover every pure injective module decomposes, essentially in a unique way, as the pure injective hull of a direct sum of indecomposable pure injective modules, possibly plus a *superdecomposable* pure injective summand, that is, a pure injective module without indecomposable summands.

When R is commutative, every indecomposable pure injective R-module N localizes, in the sense that the set $\mathfrak{p}(N)$ of elements of R which act as nonisomorphisms on N is a prime ideal of R and N has a natural structure of a module (still indecomposable and pure injective) over the localization of R at $\mathfrak{p}(N)$.

The existence of superdecomposable pure injective R-modules is closely related to a possible *dimension* function from the lattice $L(R)$ to ordinals or ∞, which is called *width*. See [**P1**], Chapter 10, or [**P2**], Chapter 7 for a detailed definition; basically, width progressively eliminates diamonds in $L(R)$. It turns out that, if there is a superdecomposable pure injective module over R, then the width of $L(R)$ remains undefined, or equivalently is meant ∞. The converse is also true when $L(R)$ is countable, so, in particular, when R is countable, even if in the uncountable case the equivalence between the two conditions is still an open question.

Another important dimension function on $L(R)$ is its *m-dimension*, that is, the Krull-Gabriel dimension of R, also treated in [**P1**], Chapter 10, and in [**P2**], Chapter 7. This time, the m-dimension – again an ordinal number or ∞ – is determined by iterative factoring $L(R)$, and what is obtained from it, by congruence relations that, this time, collapse intervals of finite length. In the next section we will provide an equivalent characterization of it in topological terms, valid in our setting, that is, over Bézout domains.

2.2. Decidability. I follow here [**P1**, 17.3]. Assume R is effectively given, as explained in the previous section 1. Then T_R, that is, the set of sentences of L_R true in **all** R-modules, can be effectively listed, just as the set of logical consequences of the axioms of L_R-modules. Hence, in order to deduce that T_R is decidable, it suffices to produce a similar list of sentences of L_R true in **some** R-module. The previous considerations on model theory of L_R-modules, both on pp-formulae and on indecomposable pure injective modules, easily lead to restrict the attention to

- sentences that are finite conjunctions of invariament statements $\text{Inv}(\varphi, \psi) \geq l$, $\text{Inv}(\varphi, \psi) = l$.
- modules that are finite direct sums of indecomposable pure injectives.

A general machinery deciding the existence of a module M of this kind realizing a given sentence as above consists of:

- a recursive enumeration of all points N in Zg_R,
- a recursive enumeration of all the open sets (φ/ψ) of a basis for the topology of Zg_R,
- an algorithm that, for every choice of N, φ, ψ and of a positive integer l, effectively answers whether $|\varphi(N) : \varphi(N) \cap \psi(N)| = l$ or not.

Warning: in order to get decidability one does not necessarily need, for instance, a recursive presentation of Zg_R, but this is often helpful, as it accompanies and supports the decidability analysis.

When the only occurring indexes of pp-subgroups for a ring R are 1 or ∞, hence ∞ when > 1 (we will see soon rings allowing this simplification), things get easier. In fact we are given

(1) finitely many pairs of pp-formulae $(\varphi(x), \psi(x))$ for which we require $\text{Inv}(\varphi, \psi) > 1$,
(2) finitely many pairs of pp-formulae $(\varphi'(x), \psi'(x))$ for which we expect $\text{Inv}(\varphi', \psi') = 1$,

and we look for a finite direct sum M of indecomposable pure injective R-modules making the conjunction of all these formulae true. Assume there exists such an indecomposable pure injective module for every statement in (1), so satisfying that particular sentence and all the ones in (2). Build the direct sum of the modules resulting in this way, and get a module respecting all the previous requirements. Hence the general question reduces itself to check, for every choice of φ and ψ in (1), the existence of a module lying in (φ/ψ) but out of all the open sets (φ'/ψ'), equivalently to exclude, for every φ and ψ, that the open set (φ/ψ) is contained in the finite union of the neighborhoods (φ'/ψ') - a topological question.

Let me also say that undecidability results of modules in general rely on the interpretation of the word problem for groups (see [**P1**, 17.2]). A typical ring with an undecidable theory of modules is that $k\langle X, Y\rangle$ of polynomials in two non commuting indeterminates X and Y over a field k.

3. Intermezzo

Let me spend here some words on an intriguing decidability project on modules, different from the main one in this paper. It regards representations of the Lie algebra $\mathrm{sl}(2,k)$ of 2×2 traceless matrices over a field of characteristic 0. They can be viewed as modules over the universal enveloping algebra of $\mathrm{sl}(2,k)$, $U(\mathrm{sl}(2,k))$.

The general theory of these representations over any field k of characteristic 0 is undecidable, as it interprets $k\langle X, Y\rangle$-modules (as proved by Prest and Puninski in [**PrP**, Corollary 6.3]).

On the other hand Herzog developed in [**Her**] the model theory of the so called *pseudo-finite dimensional* representations of $\mathrm{sl}(2,k)$ where k is an algebraically closed field. These are the possibly infinite dimensional (over k) models of the theory of finite dimensional representations. The latter are completely understood in the setting of $\mathrm{sl}(2,k)$, so of $U(\mathrm{sl}(2,k))$. Among other things, Herzog equipped them with an explicit and nice axiomatization [**Her**, Theorem 63].

L'Innocente and Macintyre [**LM**] started from this analysis with the aim to approach the decidability problem of pseudo-finite representations. Their paper [**LM**] contains interesting positive partial results, based on some widely accepted conjectures of Diophantine geometry.

A beautiful theorem in this direction is that obtained by Herzog and L'Innocente in [**HL**]. For every positive integer n, let $L(n)\subseteq k[X,Y]$ be the representation of $\mathrm{sl}(2,k)$ of homogeneous polynomials of total degree n. Thus [**HL**] proves that, if $\psi(x)\to\varphi(x)$ is an implication of pp-formulae of $L_{U(\mathrm{sl}(2,k))}$ true in every module, then the function mapping every n into the dimension of $\varphi(L(n))/\psi(L(n))$ over k is primitive recursive. In particular, when M is a finitely generated representation of $U(\mathrm{sl}(2,k))$, then the function mapping any n into the dimension over k of $\mathrm{Hom}_{U(\mathrm{sl}(2,k))}(M, L(n))$ is primitive recursive. As a consequence the set of positive integers n for which $\varphi(L(n))$ properly includes $\psi(L(n))$, so associated with a basic open subset of the Ziegler spectrum of $U(\mathrm{sl}(2,k))$, is computable, and therefore Diophantine. This answers positively a question set in [**LM**].

4. Bézout domains

I focus now on Bézout domains, more generally on Prüfer domains.

A commutative domain R with identity is *Bézout* if every 2-generated ideal, and consequently every finitely generated ideal, is principal. A Bézout domain is *coherent*: the intersection of 2 principal ideals is also principal. Then one can determine, for every $a,b\in R$ a greatest common divisor $\gcd(a,b)$ and a least common multiple $\mathrm{lcm}(a,b)$, both defined up to invertible factors, and satisfying the Bézout identities: for some suitable $u,v,g,h\in R$,

$$au+bv=\gcd(a,b),\quad a=g\gcd(a,b),\quad b=h\gcd(a,b).$$

Bézout domains include several noteworthy examples, such as

- principal ideal domains,
- the ring of algebraic integers (not a principal ideal domain, but a directed union of Dedekind domains),
- the ring of entire (complex or real) functions in 1 variable,
- $\mathbb{Z}+X\mathbb{Q}[X]$, and more generally the rings coming from the so-called $D+M$-construction, namely $D+XQ[X]$ where D is a principal ideal domain that is not a field and Q is its field of fractions (one gets in this way a Bézout domain, which is neither Noetherian nor a unique factorization domain),
- (commutative) valuation domains.

Prüfer domains are a larger setting. In fact a domain is Prüfer if all its localizations at maximal ideals, and consequently at non zero prime ideals, are commutative valuation domains.

Notably over a Prüfer domain (in particular over a Bézout domain) R the Cantor-Bendixson rank of the Ziegler spectrum equals the m-dimension of $L(R)$: this follows from [**P2**, Corollary 5.3.29 and Corollary 5.3.60]. Recall that the Cantor-Bendixson rank of a topological space is again either an ordinal or ∞. At any (successor) step isolated points are forgotten, so producing the so called *derivative* of the preceding space. Limit steps are handled forming intersections.

The model theory of modules over Bézout domains is considered in [**PT1**]. That paper is largely devoted to nonexistence of width and existence of superdecomposable pure injective modules. But at the beginning it recalls that

- over a Prüfer domain R every pp-formula is equivalent to a finite sum of formulas $\exists y(ya = x \wedge yb = 0)$, or also to a finite conjunction of formulas $c \mid xd$;
- over a Bézout domain R every pp-formula is equivalent to a finite sum of formulas $c \mid x \wedge xd = 0$, or also to a finite conjunction of formulas $a \mid x + xb = 0$.

The former statement dates back to old results by Warfield, while the latter uses gcd. As a consequence, over a Bézout domain R a basis for Zg_R is given by the open sets $(c \mid x \wedge xd = 0 \,/\, a \mid x + xb = 0)$, with $a, b, c, d \in R$.

In the next subsection I recall and comment on some decidability results regarding some of the Bézout domains in the previous list. The case of Dedekind domains is classical. It was considered by Eklof and Fischer, and is illustrated in detail in [**P1**, Section 2.$\mathbb{Z}$]. Actually [**P1**, Corollary 2.$\mathbb{Z}$.11] provides a description of the Ziegler spectrum of a Dedekind domain R. Let $\mathfrak{p}$ range over maximal ideals of R. Then the indecomposable pure injective R-modules are

- $R/\mathfrak{p}^n$ with n a positive integer,
- adic modules, that is, $\mathfrak{p}$-adic completions of localizations of R at $\mathfrak{p}$,
- Prüfer modules $Q/R_{\mathfrak{p}}$,
- and the generic module Q (the field of fractions of R and of each localization).

It turns out that the Cantor-Bendixson rank of Zg_R is 2, with Q being the unique point of maximal rank. Decidability easily follows for an effectively given R [**P1**, Theorem 17.13].

On the other hand there exist countable Dedekind domains, and indeed countable principal ideal domains, with an undecidable theory of modules. Hence these domains cannot be effectively given. For instance, let K be any undecidable subset of the set of all primes. Let $\mathcal{F}_K$ be the family of finite fields $\mathbb{F}_p$ with $p \in K$. By [**He1**, Theorem A] there is a countable principal ideal domain R_K whose collection of residue fields is just $\mathcal{F}_K$. Following [**PoPr**, Chapter 3], construct for every prime p the sentence of L_{R_K} saying *"there exists an R_K-module of size p"*. Clearly for every prime p this sentence is in T_{R_K} if and only is $p \in K$. Consequently, if T_{R_K} is decidable, then K is – a contradiction. It follows that R_K cannot be effectively given. So undecidability depends on the lack of an effective presentation.

4.1. Valuation domains. In 2015 Lorna Gregory [**G**] extended over valuation domains previous partial results by Gena Puninski, Vera Puninskaya and myself of 2007 [**PPT**] and, answering in the most general form a conjecture in that paper, proved the following beautiful result.

THEOREM 4.1. *The theory T_V of modules over a(n effectively given) valuation domain V is decidable if and only if there is an algorithm which decides the prime radical relation in V, namely, for every $a, b \in V$, decides whether $a \in \mathrm{rad}(b)$ (equivalently whether the prime ideals of V containing b also include a).*

Note that the direction from left to right is easy to prove: for every effectively given commutative ring R with unity, if T_R is decidable, then there is an algorithm deciding $a \in \mathrm{rad}(bR)$ for $a, b \in R$. This is because, for every a and b,

$$a \in \mathrm{rad}(bR)$$

if and only if T_R contains the following sentence of L_R

$$\exists u(u \neq 0 \wedge ub = 0) \rightarrow \exists v(v \neq 0 \wedge va = 0).$$

Hence a decision algorithm for T_R implies an algorithm for the radical relation.

The proof of the previous claim is simple. Let $a^n \in bR$ for some positive integer n. Take $u \neq 0$ such that $ub = 0$, then $ua^n = 0$. Let $m \leq n$ be the minimal positive integer such that $ua^m = 0$. Put $v = ua^{m-1}$, clearly $v \neq 0$ but $va = 0$.

Conversely we can assume without loss of generality that b is not a unit of R. Let $\mathfrak{p}$ be a prime ideal of R including b. Then $1 + \mathfrak{p}$ is annihilated by b in $R/\mathfrak{p}$. Then there is some element of $R/\mathfrak{p}$ annihilated by a, in other words some $v \in R$, $v \notin \mathfrak{p}$ such that $va \in \mathfrak{p}$. As $\mathfrak{p}$ is prime, $a \in \mathfrak{p}$. Thus $a \in \mathrm{rad}(bR)$.

Of course the opposite direction of the Gregory Theorem is by far the most difficult. We will provide more details on that when discussing the general case, over arbitrary Bézout domains. Actually that approach will focus on inclusions of basic open sets in the Ziegler topology

$$(\star) \qquad (\varphi \,/\, \psi) \subseteq \cup_{i=1}^{n} (\varphi_i \,/\, \psi_i)$$

corresponding to the case when reciprocal indexes of pp-subgroups are 1 or ∞. Some highly non trivial work has to be added over valuation domains in the general case, when finite $\neq 1$ invariants arise.

4.2. The $D + M$-construction. Let D be a principal ideal domain that is not a field, and let Q denote its field of fraction. Let $R = D + XQ[X]$ be the Bézout domain that D and Q generate via the $D + M$-construction. Thus R is consisting of the polynomials of $Q[X]$ having their constant term in D.

The decidability of the theory of R-modules is considered in [**PT2**]. It is based on a complete analysis of the Ziegler spectrum Zg_R.

The first step towards this description is the following classification of nonzero prime ideals of R:

- pR where p ranges over the prime elements of D,
- $f(X)R$ where $f(X)$ ranges over the irreducible polynomials of $Q[X]$ with constant term 1,
- finally $XQ[X]$.

The only non principal ideal is the last one, which is also the intersection of the various ideals pR.

Next recall that every indecomposable pure injective R-module can be viewed as a module over a localization of R at some prime ideal. This implies that Zg_R can be viewed as the union of two subspaces,

- the union of the closed spaces $\mathrm{Zg}_{R_{f(X)R}}$ where $f(X)$ ranges over the irreducible polynomials of $Q[X]$ having 1 as a constant term, and
- the union of the closed spaces $\mathrm{Zg}_{R_{pR}}$ where p ranges over the prime elements of D.

Tha latter subspaces contain $\mathrm{Zg}_{R_{XQ[X]}}$, which is their intersection. Moreover the intersection of two different subspaces $\mathrm{Zg}_{R_{f(X)R}}$, or of a subspace $\mathrm{Zg}_{R_{f(X)R}}$ and a subspace $\mathrm{Zg}_{R_{pR}}$, is just the point $Q(X)$ - the field of fractions of the polynomial ring $Q[X]$. Of course in all these subspaces one has to consider isolated, adic and Prüfer points in the relative topology.

On this ground one gets the main theorem about the spectrum.

THEOREM 4.2. [**PT2**] *The Cantor-Bendixson rank of* Zg_R*, that is, the m-dimension of* $L(R)$*, is 4, with* $Q(X)$ *being the unique point of maximal rank.*

Now assume R effectively given and turn to decidability. Theorem 4.2 ensures that Zg_R is countable and provides two explicit lists, of points and basic open sets of the topology. The expected effective list of sentences true in some finite direct sum of indecomposable pure injective R-modules is easier to find passing to localizations at prime ideals of R, so to valuation domains. In order to control their behaviour, some further effectiveness conditions are necessary on R, requiring algorithms

(1)′ listing all the prime elements of D,
(2)′ listing all the irreducible polynomials of $Q[X]$ (with constant term 1),
(3)′ calculating for every prime p of D the size of the residue field D/pD.

We say that D is *strongly effectively given* when it satisfies these further assumptions. For instance the ring $\mathbb{Z}$ of integers is strongly effectively given, by a procedure going back to Kronecker for checking irreducibility of rational polynomials.

THEOREM 4.3. [**PT2**] *Let* D *be a strongly effectively given principal ideal domain and let* $R = D + XQ[X]$ *be the corresponding Bézout domain. Then* T_R *is decidable.*

In particular the theory of modules over $\mathbb{Z} + X\mathbb{Q}[X]$ is decidable.

4.3. Entire complex valued functions. The case of entire functions with complex values is also worth mentioning, despite its uncountable power, because a nice description of the Ziegler spectrum of this ring is provided in [**LPPT**].

Recall that a function $f : \mathbb{C} \to \mathbb{C}$ is said to be *entire*, if it is given by an everywhere convergent power series $\sum_{n=0}^{\infty} a_n z^n$ with complex coefficients a_n, i.e. $\lim_{n\to\infty} \sqrt[n]{|a_n|} = 0$.

For instance, the exponential function $e^z = \sum_{n=0}^{\infty} \frac{z^n}{n!}$ is entire, as is the sine function $\sin z = \sum_{n=0}^{\infty}(-1)^n \frac{z^{2n+1}}{(2n+1)!}$. More examples and explanations can be found in any complex analysis textbook, such as [**A**]. For instance, each entire function is differentiable, and its derivative is of the same kind.

Entire functions are preserved under pointwise addition and multiplication, which equips them with a structure of a ring, and indeed of a Bézout domain $\mathbb{E}$. Moreover the residue fields of this domain are infinite. As a consequence the indexes of pp-subgroups are either 1 or ∞. This is also because $\mathbb{E}$ is a $\mathbb{C}$-algebra over the constant functions. Let us confirm that the cardinality of $\mathbb{E}$ is the continuum $2^{\aleph_0}$.

The first result on $\mathrm{Zg}_{\mathbb{E}}$ regards its isolated points.

THEOREM 4.4. [**LPPT**] *The isolated points of* $\mathrm{Zg}_{\mathbb{E}}$ *are just the finite length points* $\mathbb{E}/(z-t)^k\mathbb{E}$, $t \in \mathbb{C}$, k *a positive integer, and are dense in* $\mathrm{Zg}_{\mathbb{E}}$.

A full classification of closed points can also be obtained. They are:

(1) The modules $\mathbb{E}/(z-t)^k\mathbb{E}$, $t \in \mathbb{C}$, $k \geq 2$;
(2) The modules $\mathbb{E}/M^k$, still with $k \geq 2$, where M ranges over the maximal ideals of $\mathbb{E}$ that are called *free* (meaning that the intersection of the zero sets of their functions is empty);
(3) finally the generic modules $Q(\mathbb{E}/P)$ – the fields of fractions of $\mathbb{E}/P$ where P runs over prime ideals of $\mathbb{E}$; in particular, when $P = 0$, one obtains in this way the field of meromorphic functions.

The next step is to consider the first Cantor-Bendixson derivative of $\mathrm{Zg}_{\mathbb{E}}$, which is obtained by dropping from $\mathrm{Zg}_{\mathbb{E}}$ the isolated points. It turns out that the remaining modules are exactly those on $\mathbb{E}_S$, where S is the multiplicatively closed set consisting of nonzero polynomials over $\mathbb{C}$. Moreover:

THEOREM 4.5. [**LPPT**] *The first Cantor-Bendixson derivative of* $\mathrm{Zg}_{\mathbb{E}}$ *is perfect, that is, has no isolated point.*

Incidentally: the pure injective envelope of $\mathbb{E}_S$ is a superdecomposable module, over $\mathbb{E}_S$ itself and hence over $\mathbb{E}$.

Another example of superdecomposable pure injective module over $\mathbb{E}$ was already given in [**PT1**, Proposition 6.2 and Example 6.3]. It is based on the following sufficient criterion: *let R be a Bézout domain with a non trivial idempotent ideal I (so $I \neq 0, R$ and $I = I^2$), then R possesses superdecomposable pure injective modules* (see [**Pu**, Theorem 12.12]).

It is easy to build in $R = \mathbb{E}$ an ideal I as required. For every sequence $c = \{c_n : n \text{ a positive integer }\}$ of natural numbers c_n such that $c_n \leq n$ for all n and $c_n \to \infty$, consider the complex entire function f_c such that $f_c(z) = \prod_{n=1}^{\infty}(1+z/n^3)^{c_n}$ for all z. Let I be the ideal generated by these entire functions. Clearly I is not trivial. On the other hand every function f_c can be viewed as a multiple of $f_{c'}^2$, where c'_n is the integer part of $c_n/2$ for all n. Hence I is idempotent.

4.4. Algebraic integers. Another crucial example of a Bézout domain, this time countable, is the ring $\mathbb{A}$ of algebraic integers. The decision problem of $\mathbb{A}$-modules is discussed in [**LPT**]. The final statement is short and clear.

THEOREM 4.6. [**LPT**] *The theory of modules over the ring $\mathbb{A}$ of algebraic integers is decidable.*

On the other hand, the proof applies to the larger framework of the effectively given Bézout domains R that, like $\mathbb{A}$, satisfy the following conditions:

- Every non zero prime ideal $\mathfrak{p}$ is maximal (namely, R has Krull dimension 1),
- The residue field $R/\mathfrak{p}$ is infinite,
- The maximal ideal of the localization $R_{\mathfrak{p}}$ is not finitely generated.

Hence the positive result for algebraic integers opens the road to possible generalizations. This time the decidability proof avoids any analysis of $\mathrm{Zg}_{\mathbb{A}}$. Previous results by van den Dries and Macintyre ([**D**], [**DM**]) and Prestel and Schmid ([**PrS**]) on $\mathbb{A}$ inspire the approach, and localization at prime ideals is also helpful. The *prime radical relation*, $a \in \mathrm{rad}(bR)$, can be first order represented over Bézout domains such that each nonzero prime ideal is maximal, by the statement $\forall v(\gcd(a, v) = 1 \to \gcd(b, v) = 1)$ (in the language of the ring R). In fact I claim that $a \in \mathrm{rad}(bR)$ if and only if this sentence is in the first order theory of R:

In fact, assume that this sentence is true in R, but $a \notin \mathrm{rad}(bR)$, that is, $a \notin \mathfrak{p}$ for some prime ideal $\mathfrak{p}$ containing b. As $\mathfrak{p}$ is maximal, there are $c \in \mathfrak{p}$ and $r \in R$ such that $ar + c = 1$. Then a, c are coprime, while b, c are not. The converse is proved by similar arguments.

Once again, decidability matches with existence of superdecomposable pure injective objects over $\mathbb{A}$. This can be deduced, as used for $\mathbb{E}$, from the existence of non trivial idempotent ideals. In fact $\mathbb{A}$ contains such ideals, for instance, the ideal generated by all roots $\sqrt[2^n]{2}$ when n runs over positive integers.

5. The main theorem

My aim is now the enlarge the decidability analysis over general Bézout (or even Prüfer) domains with infinite residue fields. Here is the main result, by Gregory, L'Innocente, Puninski and me.

THEOREM 5.1. [**GLPT**] *Let R be a(n effectively given) Bézout domain with an infinite residue field for every maximal ideal. Then T_R is decidable if and only if there is an algorithm that answers a double prime radical relation, that is, decides, given $a, b, c, d \in R$, whether, for all prime ideals $\mathfrak{p}, \mathfrak{q}$ of R with $\mathfrak{p} + \mathfrak{q} \neq R$, $b \in \mathfrak{p}$ implies $a \in \mathfrak{p}$ or $d \in \mathfrak{q}$ implies $c \in \mathfrak{q}$.*

Generalizations to Prüfer domains are also obtained.

The hypothesis on infinite residue fields involves a large class of noteworthy algebraic examples. We already met among them algebraic integers and complex valued entire functions, but further interesting domains can be added, see [**GLPT**].

Its benefit is that, as already underlined, in order to prove decidability of modules it suffices to check effectively the inclusions of basic open sets of the Ziegler topology

$$(\star) \qquad (\varphi/\psi) \subseteq \cup_{i=1}^{n}(\varphi_i/\psi_i).$$

where φ, ψ, φ_i, ψ_i $(i = 1, \dots, n)$ range over pp-formulae in one free variable.

Here is how [**GLPT**] proceeds in this direction, moving from [**G**].

Step 1. Reduce the open sets in $(\star)$ into a simple form, very elementary as in $(x = x \,/\, xd = 0)$, $(xb = 0 \,/\, x = 0)$ and $(x = x \,/\, c \mid x)$, or a bit more complicated as $(xb = 0 \,/\, c \mid x)$. This is a quite long and intricate reduction procedure, using the following tools:

- Over Bézout domains, gcd;
- In the Prüfer setting, a result by Tuganbaev [**T**, Lemma 1.3], saying that, if R is a Prüfer domain, then for all $a, b \in R$ there exist $\alpha, r, s \in R$ such that $b\alpha = as$ and $a(\alpha - 1) = br$;
- Over a Prüfer domain R, any indecomposable pure injective R-module N is *pp-uniserial*, i.e. its lattice of pp-definable submodules is a chain.

Notably, all the required reductions can be performed effectively when R is effectively given. This is also true in the remainder of this section.

Step 2. Localize at prime ideals $\mathfrak{r}$ of a Prüfer domain R and follow Gregory's approach over valuation domains. Here the following order relation play a key role, for any relevant $\mathfrak{r}$: for $a, b \in R \setminus \{0\}$, $a \leq_{\mathfrak{r}} b$ if and only $bR_{\mathfrak{r}} \subseteq aR_{\mathfrak{r}}$.

One can express this relation in equivalent ways by using Tuganbaev's result or, over a Bézout domain, gcd. Here are these alternative characterizations, expressed in terms of membership conditions to $\mathfrak{r}$.

- (Over any Prüfer domain) Let a, b be non zero elements of R, $\alpha, r, s \in R$, $b\alpha = as$ and $a(\alpha - 1) = br$. Then $a \leq_{\mathfrak{r}} b$ if and only if either $\alpha \notin \mathfrak{r}$ or $r \notin \mathfrak{r}$.
- (Over a Bézout domain R) Let $a, b \in R \setminus \{0\}$. Then $a \leq_{\mathfrak{r}} b$ if and only if $a/\gcd(a, b) \notin \mathfrak{r}$.

Step 3. With every indecomposable pure injective module N over a Prüfer domain R two ideals of R are associated,

- $\mathrm{Ass}\ N := \{r \in R : \text{there exists } m \in N \setminus \{0\} \text{ with } mr = 0\}$ (so $b \in \mathrm{Ass}\ N$ if and only if the annihilator condition $xb = 0$ is realized by some non zero element of N), and
- $\mathrm{Div}\ N := R \setminus \{r \in R : r \mid m \text{ for all } m \in N\}$ (so $a \in \mathrm{Div}\ N$ if and only if there is some element of N that does not realize the divisibility condition $a \mid x$).

Actually both Ass N and Div N are proper prime ideals of R, and so is their union $\mathrm{Ass}\ N \cup \mathrm{Div}\ N$, as the set of elements of R that do not act as automorphisms on N. Over R as before, it turns out that either $\mathrm{Ass}\ N \subseteq \mathrm{Div}\ N$ or $\mathrm{Div}\ N \subseteq \mathrm{Ass}\ N$, so the intersection of these ideals is either Ass N or Div N respectively.

For every choice of prime ideals $\mathfrak{p}$, $\mathfrak{q}$ of R, look at the indecomposable pure injective R-modules having $\mathfrak{p}$, $\mathfrak{q}$ as Ass N, Div N respectively. Consider the prime ideal $\mathfrak{p} \cap \mathfrak{q}$.

Step 4. Now translate $(\star)$, as resulting after Step 1, in the local setting through order relations of the elements $a, b, c, d, \ldots$ with respect to the relevant $\leq_{\mathfrak{p}\cap\mathfrak{q}}$, or to equivalent membership conditions to the corresponding $\mathfrak{p} \cap \mathfrak{q}$.

Next let us recall the *double prime radical relation* DPR, which handles the membership conditions of the elements of R to the involved prime ideals: for every Bézout domain R (actually for every commutative ring R) and $a, b, c, d \in R$,

$$(a, b, c, d) \in \mathrm{DPR}(R)$$

if and only if, for all prime ideals $\mathfrak{p}, \mathfrak{q}$, if $\mathfrak{p} + \mathfrak{q} \neq R$ then $a \in \mathfrak{p}$ or $b \notin \mathfrak{p}$ or $c \in \mathfrak{q}$ or $d \notin \mathfrak{q}$.

Here are some equivalent characterizations.

- (*Algebraic, local*) Let R be any commutative domain. For $a, b, c, d \in R$, $(a, b, c, d) \notin \mathrm{DPR}(R)$ if and only if there is some maximal ideal $\mathfrak{m}$ of R such that $a \notin \mathrm{rad}(bR_{\mathfrak{m}})$ and $c \notin \mathrm{rad}(dR_{\mathfrak{m}})$.
- (*Algebraic, global*) Let R be a Prüfer domain. For $a, b, c, d \in R$, $(a, b, c, d) \notin \mathrm{DPR}(R)$ if and only if $(\mathrm{rad}(b) : a) + (\mathrm{rad}(d) : c)$ is a proper ideal of R.
- (*Topological, the most important for us*) Let R be a Prüfer domain. For $a, b, c, d \in R$, $(a, b, c, d) \in \mathrm{DPR}(R)$ if and only if $(xb = 0 \,/\, d \mid x) \subseteq (xa = 0 \,/\, x = 0) \cup (x = x \,/\, c \mid x)$.

Now let us sketch the proof of the main theorem.

($\Rightarrow$) follows from the topological characterization of DPR.

($\Leftarrow$) The previous analysis already emphasizes the role of pairs of prime ideals $\mathfrak{p}$ and $\mathfrak{q}$ of R, and in this way suggests the notion of *Boolean combination of conditions on a pair of prime ideals*, meaning a Boolean combination Δ of conditions of the form $a \in P$, $b \notin P$, $c \in Q$ and $d \notin Q$ where $a, b, c, d \in R$ and P, Q are variables for prime ideals. We say that a pair of prime ideals $(\mathfrak{p}, \mathfrak{q})$ of R satisfies Δ if when we replace all instances of P by $\mathfrak{p}$ and all instances of Q by $\mathfrak{q}$ the statement of Δ becomes true in R (so the various a, b, c, d are meant as constants of the language, taken from R).

Now let R be any effectively given Bézout domain, and Δ be a Boolean combination of conditions on a pair of prime ideals. I claim that, if $\mathrm{DPR}(R) \subseteq R^4$ is recursive, then there is an algorithm which answers whether, for all prime ideals $\mathfrak{p}, \mathfrak{q}$, $\mathfrak{p} + \mathfrak{q} \neq R$ implies that $(\mathfrak{p}, \mathfrak{q})$ satisfies Δ.

In fact one can assume that Δ is a conjunction of disjunctions, and each disjunct consists of conditions of the form

$$a_l \in P,\ b_j \notin P,\ c_h \in Q,\ d_k \notin Q,$$

where the various l, j, h, k range over suitable finite sets of indexes.

Clearly one can check whether for all primes $\mathfrak{p}, \mathfrak{q}$, $\mathfrak{p} + \mathfrak{q} \neq R$ implies that $(\mathfrak{p}, \mathfrak{q})$ satisfies Δ (that is, all the disjuncts in the conjunction of Δ) if one can decide whether, for all prime ideals $\mathfrak{p}, \mathfrak{q}$, $\mathfrak{p} + \mathfrak{q} \neq R$ implies that one of the following holds: for some l, j, h, k,

$$a_l \in \mathfrak{p},\ b_j \notin \mathfrak{p},\ c_h \in \mathfrak{q} \text{ or } d_k \notin \mathfrak{q}.$$

But this is equivalent to saying that for all prime ideals $\mathfrak{p}, \mathfrak{q}$, $\mathfrak{p} + \mathfrak{q} \neq R$ implies that

$$\text{either } \prod_l a_l \in \mathfrak{p}, \text{ or } \gcd(b_j)_j \notin \mathfrak{p}, \text{ or } \prod_h c_h \in \mathfrak{q}, \text{ or } \gcd(d_k)_k \notin \mathfrak{q},$$

that is, the 4-uple $(\prod_l a_l, \gcd(b_j)_j, \prod_h c_h, \gcd(d_k)_k$ is in $\mathrm{DPR}(R)$.

This concludes the sketch of the proof.

Over valuation domains V (the setting of [**G**]) one can restrict the analysis to the case when the ideals $\mathfrak{p}$ and $\mathfrak{q}$ are radical, say $\mathrm{rad}(tV)$, $\mathrm{rad}(sV)$ respectively. Consequently the double prime radical relation can be reduced to the single prime radical relation. For instance, a proper inclusion like $\mathfrak{p} \subset \mathfrak{q}$ holds if and only if $s \notin \mathrm{rad}(tV)$.

The case of algebraic integers is now a consequence of the main theorem.

COROLLARY 5.2. *Let R be an effectively given Bézout domain of Krull dimension 1 all of whose residue fields are infinite. Then the theory of R-modules is decidable.*

In fact, under the Krull dimension 1 hypothesis, the double prime radical relation can be easily reduced to the classical "one-dimensional" prime radical relation, and the latter can be effectively decided.

6. From Bézout to Prüfer

Let me enlarge now the scenery to Prüfer domains. I introduce a larger family of prime radical relations over any Prüfer domain R: for every positive integer n we introduce a $(2n+2)$-ary relation DPR_n in R by putting, for $a, c, b_i, d_i \in R$ $(1 \leq i \leq n)$,

$$(a, c, b_1, \ldots, b_n, d_1, \ldots, d_n) \in \mathrm{DPR}_n(R)$$

if and only if for all prime ideals $\mathfrak{p}$ and $\mathfrak{q}$ of R with $\mathfrak{p} + \mathfrak{q} \neq R$, either $a \in \mathfrak{p}$ or $c \in \mathfrak{q}$ or some b_i is not in $\mathfrak{p}$ or some d_i is not in $\mathfrak{q}$. Hence DPR, as defined in the last section, is just DPR_1.

The previous approach over Bézout domains now implies, without gcd:

THEOREM 6.1. *Let R be an effectively given Prüfer domain with an infinite residue field for every maximal ideal. If there are algorithms deciding in R the membership to* DPR_n *for every positive integer n uniformly in n, then the theory T_R of all R-modules is decidable.*

Two questions arise in a natural way from this theorem:

- First, is the condition on the DPR_n not only sufficient but also necessary to guarantee that T_R is decidable?
- Secondly, can one bound the n's to check, just as in the Bézout case where $n = 1$ is enough?

Let me recall an old result by Heitmann [**He2**], showing that if a Prüfer domain has Krull dimension d then every finitely generated ideal can be generated by $d+1$ elements. As a consequence the following can be proved.

THEOREM 6.2. *Let R be a Prüfer domain all of whose residue fields are infinite and N be a positive integer such that any finitely generated ideal of R can be generated by N elements (in particular, this is the case when R has Krull dimension $N-1$). If there are algorithms deciding membership of* $\mathrm{DPR}_N(R)$ *then T_R is decidable.*

Here is a partial converse to this result.

THEOREM 6.3. *Let R be a Prüfer domain of Krull dimension* 1 *all of whose residue fields are infinite. Then the following are equivalent:*

(1) *T_R is decidable.*
(2) $\mathrm{DPR}_2(R)$ *is recursive.*
(3) *There is an algorithm which given $a, b_1, b_2 \in R$ answers whether $a \in$* $\mathrm{rad}(b_1R + b_2R)$.

But work on Prüfer domains and Bézout domains possibly with finite residue fields is still in progress; see [**GLT**]. This also raises the following intriguing question: do there exist Prüfer domains, or even Bézout domains, whose theory of modules is undecidable because it interprets the word problem of groups? I do not know any such ring. In other words, does decidability/undecidability in this setting only depend on the choice and on the possibility of an effective presentation? Note that the case of principal ideal domains was already discussed in Chapter 4, just before 4.1.

References

[A] L. V. Ahlfors, *Complex analysis*, 3rd ed., International Series in Pure and Applied Mathematics, McGraw-Hill Book Co., New York, 1978. MR510197

[D] L. van den Dries, *Elimination theory for the ring of algebraic integers*, J. Reine Angew. Math. **388** (1988), 189–205, DOI 10.1515/crll.1988.388.189. MR944190

[DM] L. van den Dries and A. Macintyre, *The logic of Rumely's local-global principle*, J. Reine Angew. Math. **407** (1990), 33–56. MR1048527

[EF] P. C. Eklof and E. R. Fischer, *The elementary theory of abelian groups*, Ann. Math. Logic **4** (1972), 115–171, DOI 10.1016/0003-4843(72)90013-7. MR0540003

[FR] M. J. Fischer and M. O. Rabin, *Super-exponential complexity of Presburger arithmetic*, Complexity of computation (Proc. SIAM-AMS Sympos., New York, 1973), Amer. Math. Soc., Providence, R.I., 1974, pp. 27–41. SIAM-AMS Proc., Vol. VII. MR0366646

[FS] L. Fuchs and L. Salce, *Modules over non-Noetherian domains*, Mathematical Surveys and Monographs, vol. 84, American Mathematical Society, Providence, RI, 2001. MR1794715

[G] L. Gregory, *Decidability for theories of modules over valuation domains*, J. Symb. Log. **80** (2015), no. 2, 684–711, DOI 10.1017/jsl.2014.1. MR3377362

[GLPT] L. Gregory, S. L'Innocente, G. Puninski, C. Toffalori, *Decidability of the theory of modules over Prüfer domains with infinite residue fields*, J. Symb. Logic **83** (2018), no. 4, 1391–1412, DOI 10.1017/jsl.2018.58.

[GLT] Gregory, S. L'Innocente, C. Toffalori, *Decidability of the theory of modules over Prüfer domains with dense value groups*, preprint.

[He1] R. C. Heitmann, *PID's with specified residue fields*, Duke Math. J. **41** (1974), 565–582. MR0369351

[He2] R. C. Heitmann, *Generating ideals in Prüfer domains*, Pacific J. Math. **62** (1976), no. 1, 117–126. MR0407000

[Her] I. Herzog, *Pseudo-finite dimensional representations of* $\mathrm{sl}(2,k)$, Selecta Math. (N.S.) **7** (2001), no. 2, 241–290, DOI 10.1007/PL00001402. MR1860015

[HL] I. Herzog and S. L'Innocente, *Diophantine sets of representations*, Adv. Math. **255** (2014), 338–351, DOI 10.1016/j.aim.2014.01.005. MR3167485

[LM] S. L'Innocente and A. MacIntyre, *Towards decidability of the theory of pseudo-finite dimensional representations of* $\mathrm{sl}_2(k)$. *I*, Andrzej Mostowski and Foundational Studies, IOS, Amsterdam, 2008, pp. 235–260. MR2422688

[LPPT] S. L'Innocente, F. Point, G. Puninski, C. Toffalori, *The Ziegler spectrum of the ring of entire complex valued functions*, J. Symb. Logic **84** (2019), no. 1, 160–177, DOI 10.1017/jsl.2018.2.

[LPT] S. L'Innocente, C. Toffalori, and G. Puninski, *On the decidability of the theory of modules over the ring of algebraic integers*, Ann. Pure Appl. Logic **168** (2017), no. 8, 1507–1516, DOI 10.1016/j.apal.2017.02.003. MR3650351

[Lo] L. Lo, *On the computational complexity of the theory of abelian groups*, Ann. Pure Appl. Logic **37** (1988), no. 3, 205–248, DOI 10.1016/0168-0072(88)90020-6. MR934368

[MPT1] A. Marcja, M. Prest, and C. Toffalori, *Classification theory for abelian groups with an endomorphism*, Arch. Math. Logic **31** (1991), no. 2, 95–104, DOI 10.1007/BF01387762. MR1139394

[MPT2] A. Marcja, M. Prest, and C. Toffalori, *The stability classification for abelian-by-finite groups and modules over a group ring*, J. London Math. Soc. (2) **47** (1993), no. 2, 212–226, DOI 10.1112/jlms/s2-47.2.212. MR1207944

[MPT3] A. Marcja, M. Prest, and C. Toffalori, *On the undecidability of some classes of abelian-by-finite groups*, Ann. Pure Appl. Logic **62** (1993), no. 2, 167–173, DOI 10.1016/0168-0072(93)90172-A. Stability in model theory, III (Trento, 1991). MR1226305

[MPT4] A. Marcja, M. Prest, and C. Toffalori, *The torsionfree part of the Ziegler spectrum of RG when R is a Dedekind domain and G is a finite group*, J. Symbolic Logic **67** (2002), no. 3, 1126–1140, DOI 10.2178/jsl/1190150153. MR1926602

[PoPr] F. Point and M. Prest, *Decidability for theories of modules*, J. London Math. Soc. (2) **38** (1988), no. 2, 193–206, DOI 10.1112/jlms/s2-38.2.193. MR966292

[P1] M. Prest, *Model theory and modules*, London Mathematical Society Lecture Note Series, vol. 130, Cambridge University Press, Cambridge, 1988. MR933092

[P2] M. Prest, *Purity, spectra and localisation*, Encyclopedia of Mathematics and its Applications, vol. 121, Cambridge University Press, Cambridge, 2009. MR2530988

[PrP] M. Prest and G. Puninski, *Some model theory over hereditary Noetherian domains*, J. Algebra **211** (1999), no. 1, 268–297, DOI 10.1006/jabr.1998.7607. MR1656581

[PrS] A. Prestel and J. Schmid, *Existentially closed domains with radical relations: an axiomatization of the ring of algebraic integers*, J. Reine Angew. Math. **407** (1990), 178–201, DOI 10.1515/crll.1990.407.178. MR1048534

[Pu] G. Puninski, *Serial rings*, Kluwer Academic Publishers, Dordrecht, 2001. MR1855271

[PPT] G. Puninski, V. Puninskaya, and C. Toffalori, *Decidability of the theory of modules over commutative valuation domains*, Ann. Pure Appl. Logic **145** (2007), no. 3, 258–275, DOI 10.1016/j.apal.2006.09.002. MR2286415

[PT1] G. Puninski and C. Toffalori, *Some model theory of modules over Bézout domains. The width*, J. Pure Appl. Algebra **219** (2015), no. 4, 807–829, DOI 10.1016/j.jpaa.2014.04.031. MR3282111

[PT2] G. Puninski and C. Toffalori, *Decidability of modules over a Bézout domain $D + XQ[X]$ with D a principal ideal domain and Q its field of fractions*, J. Symb. Log. **79** (2014), no. 1, 296–305, DOI 10.1017/jsl.2014.4. MR3226026

[S] V. Strassen, *Algebraic complexity theory*, Handbook of theoretical computer science, Vol. A, Elsevier, Amsterdam, 1990, pp. 633–672. MR1127177

[Sz] W. Szmielew, *Elementary properties of Abelian groups*, Fund. Math. **41** (1955), 203–271, DOI 10.4064/fm-41-2-203-271. MR0072131

[T] A. A. Tuganbaev, *Distributive rings, uniserial rings of fractions, and endo-Bezout modules*, J. Math. Sci. (N. Y.) **114** (2003), no. 2, 1185–1203, DOI 10.1023/A:1021977603746. Algebra, 22. MR1967956

School of Science and Technology, University of Camerino, Camerino, Italy

Current address: Division of Mathematics, School of Science and Technology, University of Camerino, Camerino, Italy

Email address: `carlo.toffalori@unicam.it`

Selected Published Titles in This Series

730 **Alberto Facchini, Lorna Gregory, Sonia L'Innocente, and Marcus Tressl, Editors,** Model Theory of Modules, Algebras and Categories, 2019

726 **Eugene Plotkin, Editor,** Groups, Algebras and Identities, 2019

725 **Shijun Zheng, Marius Beceanu, Jerry Bona, Geng Chen, Tuoc Van Phan, and Avy Soffer, Editors,** Nonlinear Dispersive Waves and Fluids, 2019

724 **Lubjana Beshaj and Tony Shaska, Editors,** Algebraic Curves and Their Applications, 2019

723 **Donatella Danielli, Arshak Petrosyan, and Camelia A. Pop, Editors,** New Developments in the Analysis of Nonlocal Operators, 2019

722 **Yves Aubry, Everett W. Howe, and Christophe Ritzenthaler, Editors,** Arithmetic Geometry: Computation and Applications, 2019

721 **Petr Vojtěchovský, Murray R. Bremner, J. Scott Carter, Anthony B. Evans, John Huerta, Michael K. Kinyon, G. Eric Moorhouse, and Jonathan D. H. Smith, Editors,** Nonassociative Mathematics and its Applications, 2019

720 **Alexandre Girouard, Editor,** Spectral Theory and Applications, 2018

719 **Florian Sobieczky, Editor,** Unimodularity in Randomly Generated Graphs, 2018

718 **David Ayala, Daniel S. Freed, and Ryan E. Grady, Editors,** Topology and Quantum Theory in Interaction, 2018

717 **Federico Bonetto, David Borthwick, Evans Harrell, and Michael Loss, Editors,** Mathematical Problems in Quantum Physics, 2018

716 **Alex Martsinkovsky, Kiyoshi Igusa, and Gordana Todorov, Editors,** Surveys in Representation Theory of Algebras, 2018

715 **Sergio R. López-Permouth, Jae Keol Park, S. Tariq Rizvi, and Cosmin S. Roman, Editors,** Advances in Rings and Modules, 2018

714 **Jens Gerlach Christensen, Susanna Dann, and Matthew Dawson, Editors,** Representation Theory and Harmonic Analysis on Symmetric Spaces, 2018

713 **Naihuan Jing and Kailash C. Misra, Editors,** Representations of Lie Algebras, Quantum Groups and Related Topics, 2018

712 **Nero Budur, Tommaso de Fernex, Roi Docampo, and Kevin Tucker, Editors,** Local and Global Methods in Algebraic Geometry, 2018

711 **Thomas Creutzig and Andrew R. Linshaw, Editors,** Vertex Algebras and Geometry, 2018

710 **Raphaël Danchin, Reinhard Farwig, Jiří Neustupa, and Patrick Penel, Editors,** Mathematical Analysis in Fluid Mechanics, 2018

709 **Fernando Galaz-García, Juan Carlos Pardo Millán, and Pedro Solórzano, Editors,** Contributions of Mexican Mathematicians Abroad in Pure and Applied Mathematics, 2018

708 **Christian Ausoni, Kathryn Hess, Brenda Johnson, Ieke Moerdijk, and Jérôme Scherer, Editors,** An Alpine Bouquet of Algebraic Topology, 2018

707 **Nitya Kitchloo, Mona Merling, Jack Morava, Emily Riehl, and W. Stephen Wilson, Editors,** New Directions in Homotopy Theory, 2018

706 **Yeonhyang Kim, Sivaram K. Narayan, Gabriel Picioroaga, and Eric S. Weber, Editors,** Frames and Harmonic Analysis, 2018

705 **Graham J. Leuschke, Frauke Bleher, Ralf Schiffler, and Dan Zacharia, Editors,** Representations of Algebras, 2018

704 **Alain Escassut, Cristina Perez-Garcia, and Khodr Shamseddine, Editors,** Advances in Ultrametric Analysis, 2018

For a complete list of titles in this series, visit the
AMS Bookstore at **www.ams.org/bookstore/conmseries/**.